Ecological Studies, Vol. 159

Analysis and Synthesis

Edited by

I.T. Baldwin, Jena, Germany
M.M. Caldwell, Logan, USA
G. Heldmaier, Marburg, Germany
O.L. Lange, Würzburg, Germany
H.A. Mooney, Stanford, USA
E.-D. Schulze, Jena, Germany
U. Sommer, Kiel, Germany

Ecological Studies

Volumes published since 1996 are listed at the end of this book.

Springer
New York
Berlin
Heidelberg
Hong Kong
London
Milan
Paris
Tokyo

Ariel E. Lugo   Julio C. Figueroa Colón
Mildred Alayón
Editors

Big-Leaf Mahogany
Genetics, Ecology, and Management

With 82 Illustrations

 Springer

Ariel E. Lugo
International Institute of
    Tropical Forestry
USDA Forest Service
Río Piedras, PR 00928-5000
USA
alugo@fs.fed.us

Julio C. Figueroa Colón
Puerto Rico Conservation Foundation
Urbanización Puerto Nuevo
San Juan, PR 00920-4131
USA
jfigueroa@fs.fed.us

Mildred Alayón
International Institute of
    Tropical Forestry
USDA Forest Service
Río Piedras, PR 00928-5000
USA
malayon@fs.fed.us

Cover illustration: Large big-leaf mahogany tree in Chiapas, Mexico. The tree is approximately 70 meters tall. Photo credit: Julio C. Figueroa Colón.

Library of Congress Cataloging-in-Publication Data
Big-leaf mahogany: genetics, ecology, and management/editors, Ariel E. Lugo, Julio C.
Figueroa Colón, Mildred Alayón.
         p.; cm.—(Ecological studies; v. 159)
    Includes bibliographical references (p.).
    ISBN 0-387-98837-8 (hc: alk. paper)
        1. Honduras mahogany. I. Lugo, Ariel E. II. Figueroa Colón, Julio C.
    III. Alayón, Mildred. IV. Series
    SD397 .H67 B54 2002
    634.9′7377—dc21   2002019727

ISSN 0070-8356
ISBN 0-387-98837-8   Printed on acid-free paper.

Printed in the United States of America.

9 8 7 6 5 4 3 2 1   SPIN 10723668

www.springer-ny.com

Springer-Verlag   New York Berlin Heidelberg
A member of BertelsmannSpringer Science+Business Media GmbH

Preface: Knowledge—and Conserving Mahogany

Since the 1930s, the USDA (U.S. Department of Agriculture) Forest Service's International Institute of Tropical Forestry (the Institute) has studied mahogany and its management. In the 1960s, F.B. Lamb, the author of the classic book on mahogany (1966), was an Institute collaborator. Before *gene flow* and *genetic erosion* became popular terms, my predecessor Frank Wadsworth established a gene bank at the Luquillo Experimental Forest. This project required two expeditions to Central America in successive years (1958 and 1959) to collect seed from 20 populations of big-leaf and Pacific coast mahogany. These provenances plus small-leaf mahogany were planted throughout Puerto Rico in a replicated, large-scale transplant study.

This volume presents the Institute's long-term experience with mahogany plantations. We continue to sponsor research on big-leaf mahogany, and this volume contains results of those studies plus other chapters about aspects of the biology of all species of mahogany and their potential for management.

Lamb (1966) described about 1 million km² of land, comprising the range of all mahogany species, and discussed the diversity of conditions under which big-leaf mahogany grows. Its range includes dry to wet climates, lowland to montane elevations, many soil types, and from floodplain to upland habitats. Big-leaf mahogany forms stands that range from dense clumps of large ancient trees to stands of small-diameter trees scattered

among taller vegetation. This complexity of distribution and life history
means that generalizing about the ecology of big-leaf mahogany on the
basis of a few short-term studies is difficult.

Over the past three decades, we have learned a lot about the ecology
and management of mahogany. Knowledge about the ecological life history
of mahogany is not much better than the poor state of research on ecological life histories of most tropical trees (Lugo and Zimmerman, in press).
Long-term quantitative information on many parameters as discussed
by Pelton (1953) and McCormick (1995) is required for ecological understanding of tree populations, but very little information is available for
tropical trees (Clark and Clark 1992; Lugo and Zimmerman, in press).
This volume contains 23 chapters that expand the base of information on
big-leaf mahogany, but much more research is needed across the range of
the species.

A major challenge facing scientists interested in big-leaf mahogany is
how to organize available information in space and time. This task is analogous to that which scientists faced in addressing the role of tropical forests
in the global carbon cycle. The difference is that the carbon cycle group has
had more than 20 years to debate its questions, but the mahogany debate
is just beginning. Let us hope we can learn from the experience of the global
carbon cycle scientists.

Woodwell et al. (1978) extrapolated local and short-term data on tropical deforestation and biomass to global and long-term scales to create a bio-
geochemical scenario. This study suggested that the global carbon budget
was out of balance as the result of the emission of some $8\,Pg\,C\,yr^{-1}$ ($1\,Pg =
10^{15}\,g\,C$) to the atmosphere from tropical deforestation. After two decades
of scientific debate, data have become available at the proper scales of time
and space and with proper attention to the complexity of tropical forests.
Today, the science community generally agrees on the global carbon budget,
and the estimate of the carbon source from tropical forests has been
lowered to less than $2\,Pg\,C\,yr^{-1}$; the discussion is now moving into the carbon
sink function of tropical forests (Brown 1996).

The current debate on the ecology of big-leaf mahogany is at the point
where the global carbon issue was in 1978. For example, MacLellan
(1996) presents a gloomy future for big-leaf mahogany and focuses solutions on the need to list the species in Appendix II of the Convention
for International Trade of Endangered Species (Rodan and Campbell
1996; Snook 1996). MacLellan also includes statements about—but no supporting evidence for—the endangerment of big-leaf mahogany, the occurrence of genetic erosion in both small- and big-leaf mahogany, and the
failure of plantations or forest management as alternatives to the conservation of mahogany (Newton et al. 1996; Rodan and Campbell 1996; Snook
1996).

This volume provides new quantitative information on the ecology and
success of mahogany plantations (Lugo and Fu, Chapter 15; Wadsworth

et al., Chapter 17), the light adaptation and regeneration of big-leaf mahogany seedlings and saplings (Fetcher et al., Chapter 6; Medina et al., Chapters 7 and 8; Wadsworth et al., Chapter 17; Wang and Scatena, Chapter 12), and examples of both failures of enrichment plantings (Negreros-Castillo and Mize, Chapter 14) and successes of mahogany regeneration under natural conditions (Grogan et al., Chapter 10; Gullison et al., Chapter 11; Snook, Chapter 9). This volume also presents new information about genetic variation at the clump, provenance, and multiprovenance scales for all species of mahogany, as well as long-term data on the relative importance of genetic and environmental factors in the growth and resistance to the shoot borer of big-leaf mahogany (Loveless and Gullison, Chapter 2; Ward and Lugo, Chapter 3; Watt et al., Chapter 22).

The conservation of big-leaf mahogany requires enlightened policies, strong management institutions, technology, research, human interventions (such as land management), artificial propagation of the species, protection and management of natural forest stands, and a host of bilateral and multilateral agreements among nations. Many nations and professions are united in this task, which means that scientists are only one set of players in a complex political debate of international dimensions.

Given the many actions required to conserve mahogany, the question is, What is the role of science? I believe that scientists are not responsible for policy development, nor should they be advocates for particular policies. I call attention to a debate in the journal *Conservation Biology* (**10**:904–920) about the relative roles of advocacy and science in conservation. The focus on values by participants of the debate led Matsuda (1997) to conclude that conservation science was advocacy science, but science avoids advocacy through its focus on the scientific method. The preservation of biological resources has three dimensions: advocacy, conservation, and knowledge (Tracy and Brussard 1996). Science is the best method available to society for developing the knowledge base. Knowledge derived through the scientific method and subjected to independent peer review is what is needed to support enlightened resource management and policy formulation (Meffe et al. 1998).

This volume is dedicated to knowledge in support of the better understanding, management, and conservation of mahogany. It is another measure of the Institute's continuing role in developing the scientific knowledge that underpins sustainable tropical forest conservation.

Acknowledgments. We acknowledge the support given to us by the editorial staff under the direction of Cynthia Miner of the Pacific Northwest Research Staff. In particular, we thank Martha Brookes (since retired). We also thank Frank Vanni for assistance in redrafting the figures. Sandra Brown, Peter G. Murphy, and T.C. Whitmore reviewed all manuscripts.

Ariel E. Lugo

Literature Cited

Brown, S. 1996. Present and potential future role of forests in the global climate change debate. *Unasylva* **185**(47):3–10.

Clark, D.A., and Clark, D.B. 1992. Life history diversity of canopy and emergent trees in a neotropical rain forest. *Ecological Monographs* **62**:315–344.

Lamb, F.B. 1966. *Mahogany of Tropical America. Its Ecology and Management.* University of Michigan Press, Ann Arbor, MI.

Lugo, A.E., and Zimmerman, J.K. In press. Ecological life histories of tropical trees with emphasis on disturbance effects. Tropical Tree Seed Manual. U.S. Department of Agriculture, Forest Service, Washington, DC.

MacLellan, A., ed. 1996. Is there a future for mahogany? *Botanical Journal of the Linnean Society* **122**:1–87.

Matsuda, B.M. 1997. Conservation biology, values and advocacy. *Conservation Biology* **11**:1449–1450.

McCormick, J.F. 1995. A review of the population dynamics of selected tree species in the Luquillo Experimental Forest, Puerto Rico. In *Tropical Forest: Management and Ecology*, eds. A.E. Lugo and C. Lowe, pp. 224–257. Springer-Verlag, New York.

Meffe, G.K., Boersma, P.D., Murphy, D.D., Noon, B.R., Pulliam, H.R., Soulé, M.E., and Walker, D.M. 1998. Independent scientific review in natural resource management. *Conservation Biology* **12**:268–270.

Newton, A.C., Cornelius, J.P., Baker, P., Gilles, A.C.M., Hernández, M., Ramnarine, S., Mesén, J.F., and Watt, A.D. 1996. Mahogany as a genetic resource. *Botanical Journal of the Linnean Society* **122**:61–73.

Pelton, J.F. 1953. Ecological life cycle of seed plants. *Ecology* **34**:619–628.

Rodan, B.D., and Campbell, F.T. 1996. CITES and the sustainable management of *Swietenia macrophylla* King. *Botanical Journal of the Linnean Society* **122**:83–87.

Snook, L.K. 1996. Catastrophic disturbance, logging and the ecology of mahogany (*Swietenia macrophylla* King): grounds for listing a major tropical timber species in CITES. *Botanical Journal of the Linnean Society* **122**:35–46.

Tracy, R.C., and Brussard, P.F. 1996. The importance of science in conservation biology. *Conservation Biology* **10**:918–919.

Woodwell, G.M., Whittaker, R.H., Reiners, W.A., Likens, G.E., Delwiche, C.C., and Botkin, D.B. 1978. The biota and the world carbon budget. *Science* **199**:141–146.

Contents

Contributors

Salvador E. Alemañy

Department of Biology and
Environmental Sciences,
Interamerican University, San Juan,
PR 00919-1293, USA

M. Andrew

Forestry Department, Castries,
Saint Lucia

Jonathan P. Cornelius

Center for Research and Higher
Education in Tropical Agriculture,
7170 Turrialba, Costa Rica

Elvira Cuevas

Centro de Ecología, Instituto
Venezolano de Investigaciones
Científicas, Caracas 1020A, Venezuela

Francisco de Castro

CSA Architects & Engineers,
Hato Rey, PR 00918, USA

Richard C. Ennion

16 Horley Road, St. Werburghs,
Bristol BS5 9TJ, UK

Ned Fetcher   Department of Biology, University of
                                     Scranton, Scranton, PA 18510-4625,
                                     USA

Julio C. Figueroa Colón   Puerto Rico Conservation
                                     Foundation, Urbanización Puerto
                                     Nuevo, San Juan, PR 00920-4131,
                                     USA

Robert B. Floyd   Commonwealth Scientific and
                                     Industrial Research Organization,
                                     Entomology, Canberra, ACT 2601,
                                     Australia

John K. Francis   International Institute of Tropical
                                     Forestry, USDA Forest Service,
                                     Río Piedras, PR 00928-5000, USA

Shenglei Fu   International Institute of Tropical
                                     Forestry, USDA Forest Service,
                                     Río Piedras, PR 00928-5000, USA;
                                     Current address: Department of
                                     Environmental Studies, University of
                                     California, Santa Cruz, CA 95064,
                                     USA

Jurandir Galvão   Projeto Mogno, Caita Postal 94,
                                     Redenção, Pará, Brazil

Amanda Gillies   Institute of Environmental and
                                     Evolutionary Biology, University of
                                     St. Andrews, St. Andrews, Fife KY16
                                     9TH, UK

Edgardo González González   Department of Natural Sciences and
                                     the Environment, Commonwealth of
                                     Puerto Rico, San Juan, PR 00906,
                                     USA

Manon Griffiths   Queensland Forestry Research
                                     Institute, Indooroopilly, QLD 4068,
                                     Australia

James Grogan

Yale School of Forestry and
Environmental Studies, Sage Hall,
New Haven, CT 06511, USA;
and Instituto da Homem e Meio,
Ambiente da Amazônia, Caita Postal
5101, Belém, Pará, Brazil

Raymond E. Gullison

Centre for Biodiversity Research,
University of British Columbia,
Vancouver, British Columbia 6T IZ4,
Canada

Caroline Hauxwell

Farming Systems Institute,
Indooroopilly, QLD 4068, Australia

Marvin Hernández

Center for Research and Higher
Education in Tropical Agriculture,
7170 Turrialba, Costa Rica

Marianne Horak

Commonwealth Scientific and
Industrial Research Organization,
Entomology, Canberra, ACT 2601,
Australia

Agustín Lobo

Instituto de Ciencias de la Tierra
(CSI), Marti Franques S/N 08028,
Barcelona, Spain

Marilyn D. Loveless

Department of Biology, The College
of Wooster, Wooster, OH 44691, USA

Ariel E. Lugo

International Institute of Tropical
Forestry, USDA Forest Service,
Río Piedras, PR 00928-5000, USA

Javier Lugo Pérez

Department of Biology,
University of Puerto Rico,
Río Piedras, PR 00928-5000, USA

John E. Mayhew

Institute of Ecology and Resource
Management, The University of
Edinburgh, Darwin Building,
Edinburgh EH9 3JU, UK;
Current address: ECCM Ltd.,
Edinburgh EH15 3PZ, UK

Ernesto Medina   Centro de Ecología, Instituto
                                 Venezolano de Investigaciones
                                 Científicas, Caracas 1020A, Venezuela;
                                 and International Institute of
                                 Tropical Forestry, USDA Forest
                                 Service, P.O. Box 25000, Río Piedras,
                                 PR 00928-5000, USA

Carl W. Mize   Forestry Department, Iowa State
                                 University, Ames, IA 50011, USA

Adisel Montaña   CSA Architects & Engineers,
                                 Mercantile Plaza, Hato Rey, PR
                                 00918, USA

Carlos Navarro   Center for Research and Higher
                                 Education in Tropical Agriculture,
                                 7170 Turrialba, Costa Rica

Patricia Negreros-Castillo   Forestry Department, Iowa State
                                 University, Ames, IA 50011, USA

Adrian C. Newton   Institute of Ecology and Resource
                                 Management, The University of
                                 Edinburgh, Edinburgh EH9 3JU, UK;
                                 Current address: Forest Programme,
                                 UNEP-World Conservation,
                                 Cambridge CB3 0DL, UK

Nathaniel Popper   556 Kirkland Mail Center,
                                 Cambridge, MA 02138, USA

James H. Sandom   Woodmark Soil Association, Bristol
                                 BS1 6BY, UK

Don P.A. Sands   Commonwealth Scientific and
                                 Industrial Research Organization,
                                 Entomology, Idooroopilly, QLD
                                 4068, Australia

Frederick N. Scatena   International Institute of Tropical
                                 Forestry, USDA Forest Service,
                                 Río Piedras, PR 00928-5000, USA

Luciana Simões   Agricultural University, 9101 6700 HB
                                        Wageningen, The Netherlands

Laura K. Snook   Sustainable Forest Management,
                                        Programme Center for International
                                        Forestry Research, Jakarta 10065,
                                        Indonesia

Martin R. Speight   Department of Zoology, University of
                                        Oxford, Oxford OX1 3PS, UK

S. Thayaparan   Forest Department, Colombo,
                                        Sri Lanka

Adalberto Veríssimo   Instituto de Homem e Meio,
                                        Ambiente da Amazônia, 1015 Belém,
                                        Pará, Brazil

Corine Vriesendorp   Department of Forestry, Michigan
                                        State University, East Lansing, MI
                                        48824, USA

Frank H. Wadsworth   International Institute of Tropical
                                        Forestry, USDA Forest Service,
                                        Río Piedras, PR 00928-5000, USA

Hsiang-Hua Wang   International Institute of Tropical
                                        Forestry, USDA Forest Service,
                                        Río Piedras, PR 00928-5000, USA;
                                        Current address: Division of Forest
                                        Biology, Taiwan Forestry Research
                                        Institute, Taipei 100, Taiwan

Sheila E. Ward   International Institute of Tropical
                                        Forestry, USDA Forest Service,
                                        Río Piedras, PR 00928-5000, USA

Allan D. Watt   Centre for Ecology and Hydrology,
                                        Hill of Brathens, Banchory,
                                        Aberdeenshire AB31 4BW, UK

Shiyun Wen   Department of Biology, University of
                                        Puerto Rico, San Juan, PR 00931-
                                        3360, USA

Contributors

Timothy C. Whitmore   Geography Department, Cambridge
                          University, Cambridge CB2 3EN, UK

Julia Wilson   Centre for Ecology and Hydrology,
                          Bush Estate, Penicuik, Midlothian
                          EH26 OQB, UK

F. Ross Wylie   Queensland Forestry Research
                          Institute, Indooroopilly, QLD 4068,
                          Australia

1. Mahogany: Tree of the Future

Timothy C. Whitmore[1]

Abstract. In mahogany, many of the problems confronting tropical rain forests worldwide are seen in microcosm, yet in some important ways the genus is unique. It is unusual among well-known tropical timbers in having poorly known ecological requirements. Most tropical timbers have low unit value and are interchangeable. Mahogany has thrice their value and is a high-quality cabinet wood with unique properties. It is unusual among cabinet woods in being a strong light demander and fast grower that does well in plantations. Thus, mahogany holds great promise for the future. Attempts to control its trade suffer familiar problems of misinformation and missing information. The natural range is vast, and, although the species is not on the verge of extinction, genetic erosion has been alleged. Sustainable exploitation is hampered by inadequate knowledge of its regeneration ecology and the all too common practical problems of rule enforcement.

---

[1] Deceased

Keywords: *Swietenia macrophylla*, Cabinet wood, Genetic erosion, Autecology, Fast grower, CITES, Timber trade

The true mahoganies are a small genus (*Swietenia*) in the Meliaceae, with three species growing in the evergreen and semievergreen tropical rain forests of America. Big-leaf mahogany (*Swietenia macrophylla*) has the widest natural range and is the only species entering commerce today.

The history of international trade in mahogany timber started in the late eighteenth century with the exploitation of small-leaf (*S. mahagoni*) mahogany, from the Caribbean islands, for European markets. It was used as a substitute for oak in shipbuilding and for walnut in furniture making when these woods were becoming difficult to obtain (Keay 1996). The trade established the high value and high reputation of mahogany timber that continues today. Small-leaf mahogany was so heavily exploited that it has apparently become extinct in commercial sizes in some places. The trade has switched to big-leaf mahogany, to which the rest of this chapter refers.

Big-leaf mahogany is unique among rain forest timbers in having the high value of a specialty cabinet wood, yet being available in the large volumes characteristic of some normal commodity rain forest woods. Its value per cubic meter as sawn timber, up to about U.S.$1000, is double or more than that of sawn timber of commodity woods such as iroko (*Milicia* sp.), keruing (*Dipterocarpus* sp.), and meranti (*Shorea* sp.) (Adams 1996). Ecologically, it is a strongly light-demanding climax species (*sensu* Swaine and Whitmore 1988) and, as such, has the characteristics needed to be a fine subject for plantations. It is among a very small group of tropical, fast-growing, light-demanding trees with high-value timber. Other examples are *Entandrophragma* spp. (sapele, utile) of West Africa, which have similar environmental requirements and attractive cabinet-grade timber, albeit less valuable than mahogany. Big-leaf mahogany is already widely grown across the tropics, both on a research scale and as extensive plantations. Plantation-grown timber is similar to timber from natural forests, although the logs are smaller (D. Thompson, Thompson Mahogany Company, personal communication), but the species has high potential for further planting. The family Meliaceae, including big-leaf mahogany, has another unique characteristic, however, the crippling susceptibility of young open-grown trees to the pantropical shoot borer genus *Hypsipyla*, which makes success in the early years of cultivation difficult although not impossible.

Big-leaf mahogany has an immense natural range, from México southward into western South America and then over a broad crescent area of $1.5 \times 10^6 \, km^2$ across southern Amazonia from Bolivia through the Brazilian states of Acre, Rondonia, and southern Amazonas to southern Pará (Lamb

1966; Styles 1981). The whole range is in semievergreen rain forest (*sensu* Whitmore 1984), except for a narrow zone of evergreen rain forest close to the Andes in Perú, Ecuador, and Colombia. It grows mostly at a low average density of one huge tree per hectare or fewer, with no smaller trees and no seedlings or saplings.

Three recent ecological studies of big-leaf mahogany are reported in this volume. In México (Snook, Chapter 9, this volume) and in terra firma rain forest in Bolivia (Gullison et al., Chapter 11, this volume), big trees are thought to have established on abandoned fields; they are not found in the semievergreen rain forest, which here has a dense ground layer, except where it has been destroyed by fire. In the Bolivian study, seedlings were seen in a few places under closed canopy, but they only lived for about a year. Trees were found, in addition to terra firma forest, in places where the forest had been destroyed by flooding and silt deposition. In both Bolivia and México, the adult trees were mostly 80 cm or greater in diameter.

In the third study area, southeastern Pará in Brazil (Grogan et al., Chapter 10, this volume), smaller adults were also found. Here, the species is confined to the vicinity of streams. Seedling ecology seems similar to Bolivia—establishing under closed forest and dying early—but release of groups was observed where a canopy opening had formed. Adults often grew in clumps in which all trees were of about the same size and which therefore had probably established as a group. Thus, in México and Bolivia, big-leaf mahogany seems to have come in after some kind of massive canopy destruction. By contrast, in southeastern Pará, at the fringe of the Amazonian rain forest in a climate with a 5-month dry season, knowledge so far indicates confinement to wetter sites, and ability to grow from seedling to tree in canopy gaps, although apparently not those created by logging.

Given the natural range of big-leaf mahogany over such a huge area, in climates with different strengths of dry season as well as in perhumid western Amazonia, scope for extensive provenance trials would seem to exist. Care is needed in selecting a provenance appropriate to the site in future plantations. The provenance trials established in Puerto Rico have only sampled México and Central America (omitting Belize).

Mahogany acutely demonstrates problems that follow from contemporary human effects on tropical rain forests. The high value of the timber pays for the long roads needed to reach the scattered trees. These roads open the forest to clearance for agriculture. Logging of big-leaf mahogany does not of itself destroy the forest but does catalyze its destruction. In Brazil, about one-fifth of the big-leaf mahogany forests are Indian reservations. The government forest agency, Instituto Brasileiro do Meio Ambiente (IBAMA), has regulations to control forest exploitation, but the rules do not apply in indigenous reserves. Some indigenous groups have been willing sellers of their mahogany; others have been duped and cheated (Watson 1996).

Tropical rain forests (together with the Siberian boreal forest) are the frontier today of extensive forest destruction by human activity, the latest and the last of the world's forest biomes to succumb. The history of forest destruction dates from the Mediterranean forests a millennium ago, to China about 5000 B.C., through western Europe from about 0 A.D., to temperate America and Australasia in recent centuries (Whitmore 1998). Campaigning environmental pressure groups have focused attention on rain forest destruction since the late 1970s. They initially advocated bans on the use of tropical timber but have now evolved to pleading for use solely of timber from sustainably managed forests. That designation raises massive questions, summarized in the phrase "sustainable for what?" and very well epitomized by big-leaf mahogany. Timber extraction of a few trees or less per hectare does not create enough canopy opening to allow regrowth of a strongly light-demanding species such as big-leaf mahogany. It leaves the matrix of the canopy essentially intact, producing timber at about 1 to 2m^3 per year, so it would allow a sustainable harvest of 30m^3 or more every 2 to 30 years. Organizations are evolving to certify sustainable forest exploitation. Prominent among them is the Forest Stewardship Council, set up by a group of concerned campaigning groups in 1992. An important philosophical nettle these organizations have yet to grasp is whether they wish to vouch for sustainable timber production or whether it has to be sustainable production of the same species, which therefore means maintaining the same species mix in the forest. If the latter, then the degree of forest churning needed to regenerate strongly light-demanding species is highly likely to alter other values built into the concept of sustainability, many of them summed up in the term biodiversity. Big-leaf mahogany is perhaps the most important tropical timber to which this dilemma applies, although a similar problem faces exploitation of okoume (*Aucoumea klaineana*) from central Africa.

Finally, commercial sizes of Pacific coast mahogany and small-leaf mahogany have been depleted by human activity and are now very rare. Trade in their timber is monitored because they have been listed in Appendix II of the Convention on International Trade in Endangered Species (CITES). Big-leaf mahogany as a species is not endangered; it has a gigantic natural range, much of which is still inaccessible to loggers. Parts of its range are accessible, however, and there it has been exploited; because of its curious regeneration ecology, it may have already disappeared from some sites. Should big-leaf mahogany be added to Appendix II? Timber traders think not; although such listing only monitors trade, it contains overtones of "endangerment," which they fear campaigners might overplay. The trade convention was not designed to cover trade in timber and could be devalued or discredited if such a major item as big-leaf mahogany were included; a committee is currently considering the options, however.

I conclude that—because of its beautiful, high-value timber—big-leaf mahogany has great potential as a future source of tropical timber but that key problems remain to be solved in its silviculture and the resolution of social and political problems of its exploitation from natural forests. For plantation growth, too, the future is rosy, but resolution of the shoot borer problem would help.

Acknowledgments. I thank the USDA Forest Service for enabling me to participate in the Puerto Rico workshop and G. Pleydell for advice.

Literature Cited

Adams, M. 1996. Timber trends. *Tropical Forest Update* **6**(2):19.

Keay, R.W.J. 1996. The future for the genus *Swietenia* in its native forests. *Botanical Journal of the Linnean Society* **122**:3–7.

Lamb, F.B. 1966. *Mahogany of Tropical America: Its Ecology and Management.* University of Michigan Press, Ann Arbor.

Styles, B.T. 1981. Swietenioideae. In *Meliaceae*, eds. T.D. Pennington, B.T. Styles, and D.A.H. Taylor. *Flora Neotropica Monograph, Vol. 28.* New York Botanical Gardens, Bronx.

Swaine, M.D., and Whitmore, T.C. 1988. On the definition of ecological species groups in tropical rain forests. *Vegetatio* **75**:81–86.

Watson, F. 1996. A view from the forest floor: the impact of logging on indigenous peoples in Brazil. *Botanical Journal of the Linnean Society* **122**:75–82.

Whitmore, T.C. 1984. *Tropical Rain Forests of the Far East*, second edition. Clarendon Press, Oxford.

Whitmore, T.C. 1998. *Introduction to Tropical Rain Forests*, second edition. Clarendon Press, Oxford.

1. The Tree and Its Genetics

2. Genetic Variation in Natural Mahogany Populations in Bolivia

Marilyn D. Loveless and Raymond E. Gullison

Abstract. Allozyme analysis was used to measure genetic variation, population differentiation, and the mating system in natural populations of big-leaf mahogany from Bolivia. We sampled seeds from individual maternal trees in four mahogany populations separated by as much as 100 km. Data were analyzed based on estimated maternal genotypes and on randomly drawn progeny arrays from known mothers. The analysis indicated that mahogany in this part of its range was characterized by genetic variation similar to that found in other common tropical tree species. The average gene diversity in populations ranged from $H_e = 0.200$ to $H_e = 0.205$. The G_{ST} values among these four populations ranged from 0.034 to 0.063, indicating rather low population differentiation on a scale of 40 to 100 km. We also used progeny arrays to estimate the mahogany mating system in one population. Mahogany seeds were completely outcrossed; the multilocus outcrossing estimate was $t_m = 1.038$ (SE = 0.024). Mahogany thus conforms to the model emerging from other genetic studies of tropical trees: a long-lived woody species, with a highly outcrossed breeding system that generates the potential for significant pollen movement, and high variation, mostly within regionally distributed populations. We discuss the implications of these data for conserving and managing populations of mahogany and other tropical trees.

Keywords: *Swietenia macrophylla*, Genetic variation, Genetic structure, Mating system, Genetics and management

Introduction

The genetic makeup, or genetic structure, of a species is a complex product of historical processes, geography, population divergence, gene flow, and natural selection operating on that species over millennia. A species' genetic makeup thus encapsulates its evolutionary past and foreshadows its adaptive future. The genetic variability housed within and among populations gives rise to the ecological, physiological, and morphological qualities that distinguish one species from another (Newton et al. 1996). As a result, the genetic composition of a species is an essential element in conserving biodiversity (Frankel and Soulé 1981; Wilson 1988; National Research Council 1991). Loss of genetic diversity, particularly through human actions such as overharvesting and fragmentation or loss of habitat, has been aptly described as genetic erosion. Although documenting precisely what genetic information is being lost is often difficult, as population size decreases, as local populations disappear, and as historical ecological processes of dispersal and gene movement are altered, the genetic variability that characterizes virtually every species is diminished (Frankel and Soulé 1981; Frankel 1983; Barrett and Kohn 1991; Ellstrand and Elam 1993).

Conservation efforts and management plans to conserve genetic diversity thus need to be based on a clear understanding of the genetic makeup of a species (Bawa 1976; Bawa and Krugman 1991; Hamrick et al. 1991). For species to be conserved or managed, knowledge is essential about how much populations vary, about the scale on which that variation is expressed, and about the ecological processes that have produced and continue to generate those patterns of variation. Evidence suggests that most plant species harbor a great deal of variability accumulated over historical time, vastly exceeding the variation being produced by contemporary mutation rates (Brown and Schoen 1992). Loss of genetic diversity, through habitat loss, population extinction, or changes in the ecological landscape in which that species has evolved, has potential consequences for the long-term success of a species. Such genetic attrition may mean a narrowing of the spectrum of valuable traits associated with a species, thus influencing its economic value. Attrition may result in changes in the way variation is partitioned in a local population, through changes in breeding structure and amounts of inbreeding. It may affect the ability of a population to adjust, over evolutionary time, to changes in climate or to modifications of its ecological environment.

None of these effects can be predicted with certainty. As biologists or as resource managers, we probably do not even identify the genetic basis for most characters we notice in the field or follow the actual phenotypes produced by particular matings. But basic population genetics theory, as well as empirical studies of remaining populations of rare and endangered species, clearly show that such genetic diminution is likely to result in inbreeding, genetically based reproductive failures, and other negative phenotypic effects (Frankel and Soulé 1981; Hamrick and Godt 1996). When such results can be foreseen, using good information on the genetic makeup of a species is critical for guiding management and conservation decisions, and protecting, where possible, the genetic variation present in natural populations (National Research Council 1991; Bawa and Krugman 1991; Kanowski and Boshier 1997).

The importance of a sound genetic basis for effective conservation and management is well understood by foresters (Ledig 1986, 1988; Bawa and Ashton 1991), and economically important temperate trees, particularly conifers, are relatively well characterized genetically (National Research Council 1991). In contrast, genetic information on tropical trees is just beginning to accumulate (Loveless 1992; Murawski 1995; Kanoswki and Boshier 1997). Because of the high species diversity of tropical forests and the large number of tree species exploited for timber worldwide, little information on genetic structure is available for most tropical timber species. The available information we have comes disproportionately from the Neotropics, especially from Central America, and rarely have studies been done on species of importance to timber extraction. Recent investigations of *Carapa guianensis* (Hall et al. 1994) and *Cordia alliodora* (Chase et al. 1995; Boshier et al. 1995a,b) are notable exceptions to this last generalization. Because many tropical species are used for timber at least occasionally, however, any information on the genetic architecture and breeding systems of tropical trees is of potential interest in conserving and managing genetic resources.

Mahogany species (big-leaf, small-leaf, and Pacific coast) have long been a focus of the timber trade in the Neotropics, and they represent the most valuable timber species currently harvested in Central and South America (Rodan et al. 1992; Veríssimo et al. 1995). They are very poorly characterized genetically and have been the subject of only a few, small-scale investigations (Newton et al. 1993, 1996). When the mahoganies were tested in provenance trials, variation was found among populations for growth rates, susceptibility to shoot borer attack, and other quantitative traits (Newton et al. 1996). Anecdotal accounts from timber workers also suggest variation in wood quality, color, and similar traits among populations in different parts of its South American range. The vast geographic range of big-leaf mahogany in the Neotropics (Lamb 1966) and the diversity of local habitats it occupies also suggest that it is likely to show considerable variation within and among populations (Loveless and Hamrick 1984; Hamrick and Godt 1990; Newton et al. 1996). Its rapid extraction under current harvesting practices could affect its genetic structure, however, making mahogany the subject of a heated debate about its conservation status (Gullison et al. 1996; Snook 1996). Styles and Khosla (1976) suggested that "high-grading" (selectively cutting the best individuals in the population) had already resulted in genetic erosion in the genera *Cedrela* and *Swietenia* as long as 20 years ago.

Because of the economic importance of mahogany, the International Board for Plant Genetic Resources has designated big-leaf mahogany a high-priority species for genetic conservation. The Food and Agriculture Organization has given highest priority to the *in situ* conservation of mahogany, and the International Timber Trade Organization has identified the need to conserve genetic variability among mahogany populations (Rodan et al. 1992; Newton et al. 1996). Because little is known about the amount or distribution of genetic variation in this species throughout its range, however, this information cannot be used to make informed management decisions about in situ or ex situ genetic conservation, strategies for selection or breeding, or similar management issues (Newton et al. 1993; Kanowski and Boshier 1997).

In this study, we examined allozyme variability within and among several big-leaf mahogany populations in the Chimanes Forest, in the Bení Province of Bolivia. We present an analysis of genetic variability in four populations in the Chimanes region, measures of genetic differentiation among these populations, and an estimate of the outcrossing rate that characterizes one natural, unharvested population of this species. We compare our data for mahogany to similar measures on allozyme diversity and mating systems in other tropical tree species, and then discuss the possible implications of these preliminary findings for managing and genetically conserving this species in South and Central America.

Materials and Methods

Measuring Genetic Variables

The basic genetic organization of a species can be quantified by a variety of measures (Nei 1987; Hamrick 1989; Hamrick and Godt 1996). The amount of variability present for some measurable genetic trait can be compared among populations and among species. The way in which the observed variability is distributed within and among populations at different spatial scales can also be assessed. This genetic structure is the result of ecological processes that bring the gametes together during mating, disperse the progeny on the landscape, and determine which genotypes survive into future generations (Loveless and Hamrick 1984; Hamrick and Loveless 1987). Although quantifying all these elements of ecological genetics is difficult, several studies have shown clearly that the mating system, or the degree to which a species is outcrossed, is one of the principal determinants of genetic structure (Loveless and Hamrick 1984; Hamrick and Godt 1990; Loveless 1992). Thus, estimates of outcrossing rates provide insight into how mating occurs and the historical processes of gene movement that have contributed to current patterns of genetic structure.

Measures of the amount of genetic variation are typically based on measures of allelic variation at some sampled subset of loci. For allozyme loci, which are codominant, allele frequencies can be assessed directly, and the standard measure of genetic variation is thus H_e, expected heterozygosity under Hardy–Weinberg equilibrium. This measure is generally called gene diversity. It is affected by the number of alleles at each locus, as well as by the relative frequencies of those alleles. It is not affected by the particular genotypes present in the population (H_o), and thus it presents a sort of idealized measure of genetic diversity that can be compared among populations, regardless of their recent mating history. Gene diversity can be applied to quantify the variability in an individual, within a population, or within the species (Hamrick and Godt 1990).

Two other descriptive variables are often also used to describe amounts of variation. P, the percentage of loci surveyed that are polymorphic, is an easily appreciated measure of the amount of genetic variation in the population. The effective number of alleles, A_e, is a measure of allelic diversity weighted by the relative frequency of the different alleles. It is calculated as $A_e = 1/(1 - H_e)$ for each locus, and then averaged over all loci. It measures, in a sense, the richness of variation at different genetic loci. Together, these measures (H_e, P, and A_e) represent different but interrelated statistics for quantifying genetic diversity.

Population differentiation is measured by using statistics of gene diversity developed by Nei (1973). The total gene diversity for species can be partitioned into gene diversity within and between populations in a method analogous to analysis of variance. Then G_{ST}, the statistic of gene diversity, measures the fraction of the total gene diversity that is between populations. This variable ranges from 0.0, where all populations are identical (none of the variation is between populations, and all is contained within any one population) to 1.0, where populations are completely divergent and have no alleles in common. Thus, G_{ST} measures the degree to which populations are genetically different from each other.

Mating systems are measured by examining progeny arrays from individual maternal trees and estimating the proportion of those progeny arrays that resulted from outcrossing. Outcrossing measurements can be based on single-locus or multilocus estimates. In general, multilocus analyses permit more discrimination among progeny and generally give outcrossing rates that are higher than an average of single-locus estimates (Shaw et al. 1981).

Study Sites

Seed samples were collected in the Chimanes Timber Production Area, in the Province of Bení, Bolivia, and in the Bení Biosphere Reserve (the Reserve; Estación Biosfera Bení), adjacent to the Chimanes Forest (Fig. 2.1). The Chimanes Production forest was established in 1988 by the Bolivian government; it consists of a patchwork of timber concessions (principally for extracting mahogany timber) and indigenous territories. Our two sites in the Chimanes Forest were in the timber concessions of Bolivian Mahogany S.A. (Gullison et al. 1996). In an attempt to obtain genetic material from more distant populations, we made two collections at locations within the Reserve. Those trees were found with assistance from local Chimanes guides, but because of the timing of fruit fall, only a few trees were collected at each of the Reserve localities.

Sample sizes (numbers of maternal trees and total number of seedlings analyzed from each site) are given in Table 2.1. Seed from which data are reported here were mostly collected in 1994, from capsules produced by flowers in 1993. Collections were obtained by climbing, by collecting the wind-dispersed seeds beneath isolated maternal trees, or—for the population at Jaimanche Dos—by harvesting capsules from the canopies of trees cut for timber during the preceding week.

Collection and Laboratory Protocols

Seed collections were made from adult trees by field assistants who climbed them and retrieved intact capsules whenever possible. Usually, several (two to four) capsules were collected from each tree. Where climbing was impossible, seeds or parts of complete capsules were collected from the ground beneath the tree; care was taken not to mix seed shadows from nearby adults. Seeds were air-dried in the field, packaged in paper envelopes, and shipped via airfreight to Ohio (United States) where they were germinated in a greenhouse in individual pots. Seedlings were maintained under greenhouse conditions until mature leaves were available and until tissue could be sampled and prepared for electrophoresis.

Leaf tissue was collected fresh daily for preparation and crushing. Leaves were crushed under liquid nitrogen in a mortar and pestle, by using a crushing buffer (buffer 2) modified from Wendell and Weeden (1989) by Roger Lauschman (Department of Biology, Oberlin College, Oberlin, Ohio, personal communication). Leaf extract was adsorbed onto filter paper wicks,

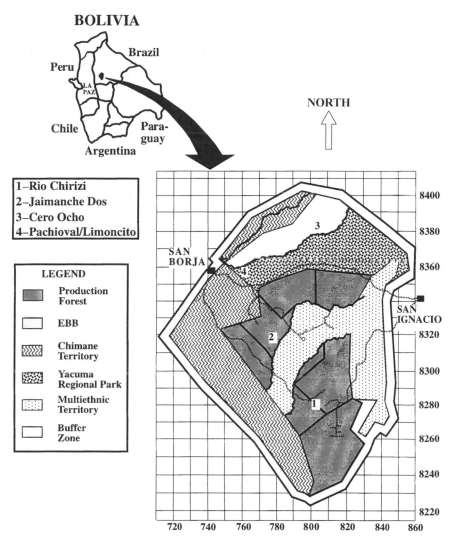

Figure 2.1. The Chimanes Timber Production Area, 66°00′ to 67°00′W and 14°30′ to 16°00′S, showing various land-use areas and our four sampling sites. The grid is given in UTM (universal transverse mercator) units, where the numbers are kilometers. The Bení Biological Reserve (Estación Biológica Bení) is *labeled*.

and wicks were stored in covered 96-well enzyme-linked immunosorbent assay (ELISA) plates at −80°C until used for electrophoresis.

A total of 12 banding systems, representing 9 enzyme systems, were resolved by using five different electrophoretic buffers. The enzyme systems, and the number of loci scored in each system, were FE, fluorescent

Table 2.1. Sample Sizes for Population Samples of Big-Leaf Mahogany from Bení Province, Bolivia, Used in Calculating Genetic Statistics (Table 2.2)

Site[a]	Adults Sampled (number)	Progeny Sampled (up to 9 per adult) (number)
Río Chirizi	93	685
Jaimanche Dos	13	110
Cero Ocho	5	24
Pachioval/Limoncito	6	32

[a] The first two sites were in the Chimanes Timber Production Forest; the second two were in or adjacent to the Estación Biológica Bení. See map in Figure 2.1.

esterase (1); ADH, alcohol dehydrogenase (1); LAP, leucine amino peptidase (1); MDH, malate dehydrogenase (3); ME, malic enzyme (1); PGI, phosphoglucose isomerase (2); GOT, glutamate oxaloacetate transaminase (1); TPI, triose phosphate isomerase (1); and β-GAL, β-galactosidase (1). Eleven loci (all except LAP) were polymorphic in this study. The genetic basis of these polymorphisms was confirmed by examining progeny arrays from individual parents and by identifying segregation patterns consistent with Mendelian expectations.

Standard genetic statistics were calculated by using a computer program written by M.D. Loveless and subsequently modified by A. Schnabel (Department of Biology, University of Indiana at South Bend, personal communication). Mating system parameters were calculated by using a multilocus program (MLT) by Ritland (1990). Variance estimates for mating system parameters were calculated using 100 bootstraps.

Two of our populations contained samples from only a few adult trees because populations were inherently small in these sites, and only a few of those trees had fruits or seeds available at the time of collection. Population allele frequencies based on the genotypes of the maternal trees do not accurately represent allele frequencies in the natural (largely unsampled) population, however. To increase the sampling in these populations, we calculated statistics of genetic variation for our data in two ways. In the first method, we used allele frequencies from the mother trees for the calculations. In the second method, we randomly chose up to eight (or occasionally nine) progeny (where available) from each mother, and used those allele frequencies to measure population characteristics. Although this strategy clearly overrepresents the maternal genotypes of the sampled trees in the analysis, it also permits sampling of genotypes in the population that were producing pollen, even if they were not sampled for seeds. Because mahogany is essentially completely outcrossed (see results, following), half the alleles in these progeny arrays were from trees other than the mother. In well-sampled populations (Chirizi and Jaimanche Dos), these alleles are likely to have come from near neighbors, which were also sampled for seeds. In the less-sampled populations, however, they allowed us to survey a wider (although imperfectly defined) population sample, thus increasing the probability of detecting alleles, especially those with low frequencies. Because one of our goals was to assess total variation and to measure population similarities, this method of data analysis was judged to be of potential value. Genetic measures from these two methods are clearly distinguished in the tables.

Results

Of the 12 enzyme loci sampled in these four populations, 11 showed detectable polymorphism. The alternate allele at *βgal* was very rare in our samples, however. This finding may be an artifact of the small sample sizes from two of our study populations. With more balanced sampling, rare alleles would perhaps be represented at higher frequencies, at least in these two populations. The single variant at this locus was found in the most completely sampled population (Chirizi).

Basic genetic parameters are calculated as a composite for the species (based on our four populations) and then as an average of population samples (Table 2.2). In addition, maternal genotypes and gene frequencies derived from progeny arrays are calculated separately. For all measures, statistics for species are slightly higher than those averaged over populations. Allozyme polymorphism (*P*) was higher when it was calculated based on progeny arrays, indicating that the seeds (seedlings) were receiving at least some pollen alleles from fathers not represented in the samples of maternal trees at these sites. This finding suggests that pollen may move long distances in mahogany populations.

The other two measures of genetic variation, H_e and A_e, were lower when calculated by using multiple progeny from each mother, probably because of the overrepresentation of individual maternal genotypes under these conditions. All the maternal alleles in any progeny array had the same

Table 2.2. Measures of Genetic Variation and Population Differentiation in Big-Leaf Mahogany Populations from Bení Province, Bolivia

	Species[a]			Population[a]			
	P	A_e	H_e	*P*	A_e	H_e	G_{ST}
Maternal genotypes	66.7	1.37	0.231	55.0	1.35	0.205	0.034
Seeds per adult (8 or 9)	83.3	1.34	0.212	66.7	1.34	0.200	0.063

A total of 12 isozyme loci were surveyed in four natural populations spread over a distance of 100 km. The first line represents genetic estimates based on maternal genotypes, which were inferred from progeny arrays. The second line is based on genotypes from up to 8 (occasionally 9) seedlings from each maternal tree sampled in the population. This second estimation method could sample pollen genotypes from trees outside the stand or from trees from which fruits were not collected.

P, percentage of polymorphic loci; A_e, effective number of alleles; H_e, gene diversity; G_{ST}, population differentiation.

[a] Measures are given for the species as a whole (combining data from all populations) and as the mean population value.

individual origin, but the paternal alleles could have come from as many as eight genetically different individuals. As a result, the effective allele frequencies and heterozygosities were not inflated, but rather diminished slightly by this method; however, they were quite close in value to estimates derived from maternal genotypes alone.

The four populations surveyed in this study differed little from one another. Based on our data, only between 3% and 6% of the total variation was unique to individual populations in this region (see Table 2.2). Sampling progeny arrays gave a higher estimate for G_{ST} than calculations based on maternal genotype because progeny arrays included alleles from a paternal, as well as the maternal, gene pool. This finding suggests that, if we were able to genotype more individuals in our two poorly sampled populations, total genetic differentiation might be larger among these populations. Because of the small sample sizes from the Reserve populations, calculations of G_{ST} mainly measured differentiation between Chirizi and Jaimanche Dos. These populations, although separated by about 40 km, were closer to each other than they were to the trees in the Reserve. Obtaining larger samples from the Reserve populations to assess the effects of geographic distance on population differentiation would clearly be of interest, as would trying to measure the scale over which progressive differentiation might occur.

Despite the fact that total G_{ST} values were low among these four populations, differences were found in allele frequencies that indicate that gene flow among these populations probably does not occur regularly. Variations in allele frequencies among the four populations at three variable loci are shown in Table 2.3. Although these four populations all share the same common allele at these loci, frequencies are somewhat different, indicating that populations of mahogany across this landscape, while similar, are not genetically identical.

We used progeny arrays from trees at Jaimanche Dos to assess the mating system of big-leaf mahogany in natural populations. The multilocus estimate for outcrossing rate was $t_m = 1.038$ (SE = 0.024), and the mean single locus rate was $t_s = 1.125$ (SE = 0.056). These rates indicated that, under these conditions of population size and adult density, mahogany was completely outcrossed. Pollen was moving among individuals, and thus pollen flow was likely to be important in preventing genetic differentiation among nearby populations. Such high outcrossing rates were consistent with the low levels of G_{ST} and population differentiation given in Table 2.2.

Discussion

Of the four populations we analyzed in this study, two were very small and do not represent a satisfactory sample from which to draw long-term management conclusions. The other two samples do provide a good mea-

Table 2.3. Variation in Allele Frequencies at Three Loci Among Populations of Big-Leaf Mahogany in Bolivia

Site	Allele[a]	Fe-2	Tpi-2	Me
Chirizi	1	0.685	0.252	0.025
	2	0.315	0.748	0.167
	3	—	—	0.790
	4	—	—	0.018
Jaimanche Dos	1	0.739	0.154	0.009
	2	0.261	0.841	0.414
	3	—	0.005	0.559
	4	—	—	0.018
Cero Ocho	1	0.841	0.296	0.095
	2	0.159	0.704	0.286
	3	—	—	0.595
	4	—	—	0.024
Pachioval-Limoncito	1	0.550	0.234	0.000
	2	0.450	0.766	0.383
	3	—	—	0.617

[a] The most common allele at each locus is the same in all four populations, but actual allele frequencies varied considerably among locations.

sure of genetic variation and population differentiation in this part of the range of mahogany in South America. To understand the regional dynamics of mahogany genetics in Bolivia, however, complete samples from several additional populations at different spatial scales would be needed.

Our data were consistent with results from other studies on tropical trees and from predictions based on the life history of this species. Loveless and Hamrick (1984) and Hamrick and Godt (1990) reviewed the correlations between various ecological traits and patterns of genetic variation in plant species. Traits correlated with high genetic variation included a wide geographic range; a long, woody life history; and an outcrossed, animal-pollinated breeding system. Based on these predictions, mahogany would be likely to show relatively high genetic variation and intermediate or even low population differentiation. Our data for mahogany were compared with summary statistics from the literature for several different groups of plant species (Table 2.4). Gene diversity (H_e) in mahogany was about the same as in common tropical tree species sampled in Panama by Loveless and Hamrick (1984), and higher than summary statistics for woody plant species generally or for tropical trees surveyed by Loveless (1992). Although the density of mahogany in our study sites was low [0.1–0.2 mahogany trees ha^{-1} greater than 80 cm diameter at breast height (dbh); Gullison et al. 1996], big-leaf mahogany was genetically much more variable in our samples than populations of rare tree species with similar low densities measured on

Table 2.4. Summary Data on Allozyme Variation in Different Categories of Plants

	N	P	H_e	G_{ST}	Reference
Long-lived woody perennials	115	50	0.149	—	Hamrick and Godt 1990
Native woody tropical species	81	39	0.109	0.109	Loveless 1992
Common trees on Barro Colorado, Panamá	16	60.9	0.211	0.055	Hamrick and Loveless 1989
Rare trees on Barro Colorado	16	42.1	0.142	—	Hamrick and Murawski 1991
Big-leaf mahogany	1	83.3	0.212	0.063	Our study

Values are based on species calculations.
N, number of species surveyed; P, percentage of polymorphism; H_e, gene diversity; G_{ST}, population differentiation.

Barro Colorado Island (Hamrick and Murawski 1991). Clearly, local density is a factor in gene movement, but historical factors of population size and distribution may also affect genetic variation. Mahogany populations in this part of the species range contain substantial genetic variation.

The lack of substantial population differentiation (low G_{ST} values) suggests that, at least in this region in Bolivia, mahogany populations have not been strongly genetically isolated from one another. This result could be the consequence of gene flow, either by pollen or by seed dispersal, over moderate to long distances. The river channels in this very flat terrain, and the seasonal flooding, could move mahogany seeds over long distances across the landscape. Also, very small amounts of migration are sufficient to inhibit population differentiation.

Lack of population divergence could also reflect the long-term population history in this region. If establishment of new populations is uncommon, depending on flooding or similar regional disturbances (Gullison et al. 1996; Gullison et al., Chapter 11, this volume; Snook 1996), and if, once established, mahogany stands may survive for hundreds of years, then populations over a region the size of our sampling area might share some common origin within only a few generations. If populations are long lived, and have a genetic "memory," strong genetic differentiation among populations as a result of selection or drift may not yet have been generated. We do not clearly understand the dynamics of how mahogany populations establish and persist in this part of the Amazon basin. One major factor that could have influenced regional genetic structure was the presence, in the last few thousand years, of farmers who practiced raised-bed cropping, and whose artifacts and earthworks can be found under today's mahogany forest. Although we have no reason to think that mahogany was moved or planted by local peoples, the patchwork of disturbance effects, flooding, stand persistence, movement of propagules, and other factors has probably had important and unquantifiable effects on the genetic architecture of

Table 2.5. Data from Recent Literature Showing Measures of Genetic Variation and Population Differentiation in a Variety of Tropical Tree Species

Species	P	H_e	G_{ST}	N	Reference
Alseis blackiana	89.3	0.340	0.048	6	Loveless and Hamrick 1987
Astrocaryum *mexicanum*	31.8	0.153	0.040	4	Eguiarte et al. 1992
Brosimum alicastrum	61.3	0.225	0.055	6	Hamrick and Loveless 1989
Dipteryx panamensis	66.3	0.205	—	2	Hamrick and Loveless 1989
Hevea brasiliensis	87.5	0.307	0.003	2	de Paiva et al. 1994
Ocotea tenera	44.0	0.225	0.128	6	Gibson and Wheelwright 1996
Quararibea asterolepis	64.4	0.256	0.022	6	Loveless and Hamrick 1987
Shorea megistophylla	93.0	0.348	—	2	Murawski et al. 1990
Stemonoporus *oblongifolius*	91.7	0.282	0.162	4	Murawski and Bawa 1994
Tachigali versicolor	29.6	0.073	0.070	6	Loveless et al. 1998

Values given are the mean measures at the population scale.
Populations were less than 10 km distant from one another; these data thus represent local variation.
P, percentage polymorphism; H_e, gene diversity; G_{ST}, population differentiation; N, number of populations measured.

these populations. Whatever the causes, the result is that our four sample sites show only weak genetic differentiation.

In its lack of interpopulation differentiation, however, mahogany is not unique. Low population differentiation has been recorded for several other species of tropical trees, both at the local scale (less than 10 km distant; Table 2.5) and at the more regional scale (Table 2.6). In fact, of the species of tropical trees surveyed to date, none has shown strong local divergence among populations, based on isozyme markers. This finding seems to suggest that gene mobility in populations of lowland tropical trees is substantial, despite their low densities, their diverse flowering phenologies, and their dependence on animal pollination (Bawa and Krugman 1991). *Carapa*

Table 2.6. Data from Recent Literature Showing Measures of Genetic Variation and Population Differentiation in a Variety of Tropical Tree Species

Species	P	H_e	G_{ST}	N	Reference
Cordia alliodora	44.2	0.127	0.117	11	Chase et al. 1995
Carapa guianensis	35.0	0.120	0.046	9	Hall et al. 1994
Pterocarpus macrocarpus	82.3	0.246	0.121	11	Liengsiri et al. 1995
Big-leaf mahogany	66.7	0.200	0.063	5	This study

Values given are the mean measures at the population scale.
Populations of these species were more than 10 km distant from one another; these data thus represent variation on a more regional Scale.
P, percentage polymorphism; H_e, gene diversity; G_{ST}, population differenation; N, number of populations measured.

guianensis (Hall et al. 1994) represents an especially interesting comparison to our data for mahogany; *C. guianensis*, also in the Meliaceae, is exploited as a valuable timber tree. Its floral morphology is similar to that of mahogany, although its pollination system is unknown. This species is apparently much more common than mahogany, at least in some sites (Hall et al. 1994), but over distances of up to 70 km between populations, it shows a lack of population differentiation (G_{ST} = 0.046) similar to the values we found for mahogany. At the same time, it has much larger and heavier seeds than the wind dispersal propagules of mahogany and so might be expected to show less dispersed mobility. For *C. guianensis* and perhaps for many tropical trees, populations are not strongly differentiated regionally. Speculating about how much of this regional genetic similarity may be historical is of interest; perhaps it dates from patterns of forest expansion and colonization in the 10,000 or so years since the start of the Holocene.

Concluding, however, that populations of tropical trees are all genetically similar, and that we need not take into account possible regional genetic differentiation in managing for genetic conservation, would be erroneous. Species such as these have huge geographic ranges, often stretching through Central America and Amazonia. No tropical tree species has had its isozyme variation surveyed throughout even a reasonable portion of its range. As a result, we have no spatial context in which to assess the scale of local population differentiation shown in Tables 2.5 and 2.6. Such large, landscape-scale information is crucial in designing sampling strategies to archive genetic variability and in proposing reserves and management plans to protect regional genetic variation. Some data suggest that, as we expect, more widely dispersed populations do show increased amounts of population differentiation. In *C. guianensis*, genetic divergence increased with increasing geographic distance among populations (Hall et al. 1994). Loveless and Hamrick (unpublished data) also found an increase in G_{ST} values for some Panamanian species as the spatial separation among populations increased. This finding suggests that population differentiation and genetic divergence may exist between different parts of a species range, which would be expected to be more strongly defined where species grow along an altitudinal gradient or where local populations might be separated by savanna or swamp forest. For species in the central Amazon basin, river channels and similar geographic barriers could also promote more marked population isolation and differentiation.

Our lack of information about the scale of geographic genetic differentiation is thus a serious impediment in making effective conservation and management recommendations for genetic conservation in tropical trees. Such data are crucial for recommendations of where centers of particular gene diversity might be found and how protected areas should be distributed to capture the largest fraction of the genetic variation in a species. We have a crucial need for genetic information on a wider geographic scale to assist in making sound management recommendations for preserving regional and species-wide genetic variation in tropical tree species such as mahogany.

The low population differentiation we found in mahogany is consistent with our data showing that big-leaf mahogany is typically outcrossed in natural populations. Outcrossing implies regular pollen movement among individuals, and such pollen mobility and gene flow would prevent genetic differentiation among local populations. Mahogany flowers are monoecious, a morphology typically thought to enhance the likelihood of outcrossing. They are extremely fragrant, and appear to be morphologically suitable for pollination by small insects. Pollination studies in natural populations of mahogany are not yet available, but big-leaf mahogany is apparently also self-compatible, at least under experimental conditions (Lee 1967). The trees produce tens of thousands of flowers in an interval of a few weeks (Grogan and Loveless, personal observation) and thus have considerable opportunity to self-pollinate. The fact that, under these conditions, virtually all the seeds are outcrossed suggests that selfing is not the favored mode of pollination. Selection may be strong against selfed ovules or fruits with selfed seeds. No studies to date have directly examined the possibility that mahogany might show inbreeding depression, although such genetic effects would be predicted in habitually outcrossing species. The outcrossed mating system of mahogany has important management consequences because isolated trees (such as those left as seed trees after timber extraction) or populations whose densities have been severely reduced might be expected to experience reproductive failures or might produce seeds with low germinability or other evidences of inbreeding depression.

Outcrossing is also the prevalent mating system in most other genetically examined species of tropical trees (Loveless 1992; Murawski 1995) (Table 2.7). Only a few tropical species show mixed selfing and outcrossing, and no species examined so far is completely selfing, although apomixis has occasionally been implied (Bawa and Ashton 1991). In addition to information on outcrossing rates, strong direct evidence indicates that pollen can move long distances in tropical tree populations. Hamrick and Murawski (1990) used genetic paternity analysis to show that pollen of *Platypodium elegans* moved, on average, between 368 and 419 m during a 3-year study. Loveless et al. (1998) used a rare marker allele to show that 21% of the pollen received by a group of *Tachigali versicolor* trees came from at least 500 m away, and Chase et al. (1996) found average pollen movement distances of 142 m in *Pithecellobium elegans* by using microsatellite markers. Such regular pollen movement would clearly inhibit local population divergence. If populations in a region are genetically similar, as our data suggest, then genetic erosion at the regional scale may be counteracted so long as some substantial, ecologically intact forest remains undisturbed in the region. At the same time, regular long-distance pollen movement seems to represent the ecological and genetic system in which lowland tropical trees have evolved, and the gene mixing from such pollen flow is likely to be a

Table 2.7. Data Compiled from Recent Literature Showing Tropical Tree Species with Outcrossed Breeding Systems

Species	Outcrossing Rate (±s.e.)	Reference
Astrocaryum mexicanum	1.007 ± 0.053	Eguiarte et al. 1992
Beilschmedia pendula	0.918	Hamrick and Murawski 1990
Bertholletia excelsa	0.85 ± 0.03	O'Malley et al. 1988
Brosimum alicastrum	0.876	Hamrick and Murawski 1990
Calophyllum longifolium	1.030 ± 0.085	Stacy et al. 1996
Carapa guianensis	0.967 ± 0.022	Hall et al. 1994
Cordia alliodora	0.966 ± 0.027	Boshier et al. 1995a,b
Pithecellobium pedicilare	0.951 ± 0.02	O'Malley and Bawa 1987
Platypodium elegans	0.921	Hamrick and Murawski 1990
Quararibea asterolepis	1.01 ± 0.01	Murawski et al. 1990
Sorocea affinis	1.089 ± 0.045	Murawski and Hamrick 1991
Shorea megistophylla	0.866 ± 0.058	Murawski et al. 1994
Spondias mombin	0.989 ± 0.163	Stacy et al. 1996
Stemonoporus oblongifolia	0.844 ± 0.021	Murawski and Bawa 1994
Big-leaf mahogany	1.038 ± 0.024	This study
Tachigali versicolor	0.998 ± 0.054	Loveless et al. 1998
Trichilia tuberculata	1.08	Hamrick and Murawski 1990
Turpinia occidentalis	1.006 ± 0.090	Stacy et al. 1996

An outcrossing rate of at least 0.85 suggests that virtually all seedlings result from cross-pollination. Very few tropical tree species have mixed mating systems (outcrossing rates $0.0 < t_m < 0.85$), and no completely self-pollinated tropical trees have been reported.

critical element in population genetic structure. Effective population sizes in trees with such widespread pollen movement are likely to be quite large, especially if adults are at low density. Hamrick and Murawski (1990) estimated that, for *Platypodium elegans*, with pollen commonly moving to distances of 1 km, the natural breeding unit (deme) would be about 25 to 50 ha surrounding any particular focal tree, and *Platypodium elegans* is a common species in their study area. For species with much lower adult densities, such as mahogany in parts of its range, the breeding range of a single tree could be as much as 1000 ha or more (Hamrick and Murawski 1990). An area much larger than that would be required to maintain evolutionary dynamics and genetic diversity for a population of tens or hundreds of adult individuals.

Our data for mahogany thus suggest that, although genetic variation is rather widely distributed through the Bení region, an evolutionarily viable population with intact ecological processes and normal gene movement would require a large protected area for adequate pollination and genotypic regeneration. The loss of a single population in a region is not likely to seriously erode the genetic base of the species. Population boundaries for tropical trees are still poorly understood, however. The natural breeding unit, within which pollen movement is common, is much smaller than the genetically effective population size, that size which resists loss of allelic heterozygosity by genetic drift. Variation among years in the size of the flowering population or the phenological overlap of flowering between individuals could substantially increase the effective population size (N_e). These data are poorly quantified for virtually all tropical trees (Bawa and Krugman 1991). Almost certainly, changes in adult densities from selective harvest in such a reserve would alter the genetic structure of the remaining population and undermine, or at least alter, the gene conservation that such a reserve would hope to serve. Regional habitat fragmentation will carve up large forest tracts into smaller, more isolated habitats, thus potentially impeding gene movement. Even under a scenario of selective timber extraction or local land-use change, large uncut tracts would need to be left undisturbed to maintain the historical processes that underwrite the genetic structure of the species. Our genetic data for mahogany and the data for other tropical trees suggest that processes resulting in widespread population declines (such as unregulated extraction) and habitat fragmentation will have long-term consequences for inbreeding and population genetic viability in tropical timber tree species.

Acknowledgments. Funding for seed collections in Bolivia was provided by grants to R.E.G. and M.D.L. from USAID (Bolfor Project). We thank Richard Rice for his support. Carmen Miranda, of the Bolivian Academy of Sciences, gave us permission to make collections in the Reserve. Sabina Staab and the staff of the Reserve assisted with logistical support. Chofo and Valentín García and Rosendo were our field assistants and climbers, without whom many of these collections would have been impossible. Bolivian Mahogany, S.A., allowed us to conduct fieldwork in their timber concession. Funding for laboratory genetic analysis was provided by the International Institute of Tropical Forestry (IITF), USDA Forest Service. Funding from IITF also allowed M.D.L. and R.E.G. to attend the symposium in San Juan, Puerto Rico, in October 1996. We thank Susan Sommer for assistance in the laboratory and in data entry, and we thank André Rodríguez, Carrie Myers, Polly Hicks, Suma Rao, and Paula Tuttle for laboratory help. The Department of Natural Resources at the Ohio Agricultural Research and Development Center of The Ohio State University provided greenhouse space and plant care. The Instituto do Homem e Meio Ambiente da Amazônia (IMAZON) in Belém, Brasil, provided facilities for computer data analysis. Additional financial support was provided to M.D.L. by The College of Wooster.

Literature Cited

Barrett, S.C.H., and Kohn, J.R. 1991. Genetic and evolutionary consequences of small population size in plants: implications for conservation. In *Genetics and Conservation of Rare Plants*, eds. D.A. Falk and K.E. Holsinger, pp. 3–30. Oxford University Press, Oxford.

Bawa, K.S. 1976. Breeding of tropical hardwoods: an evaluation of underlying bases, current status, and future prospects. In *Tropical Trees: Variation, Breeding, and Conservation*, eds. J. Burley and B.T. Styles, pp. 43–59. Academic Press, London.

Bawa, K.S., and Ashton, P.S. 1991. Conservation of rare trees in tropical rain forests: a genetic perspective. In *Genetics and Conservation of Rare Plants*, eds. D.A. Falk and K.E. Holsinger, pp. 62–71. Oxford University Press, Oxford.

Bawa, K.S., and Krugman, S.L. 1991. Reproductive biology and genetics of tropical trees in relation to conservation and management. In *Rain Forest Regeneration and Management*, eds. A. Gómez-Pompa, T.C. Whitmore, and M. Hadley, pp. 119–136. Man and the Biosphere Series, Vol. 6. UNESCO, Paris.

Boshier, D.H., Chase, M.R., and Bawa, K.S. 1995a. Population genetics of *Cordia alliodora* (Boraginaceae), a Neotropical tree. 2. Mating system. *American Journal of Botany* **82**:476–483.

Boshier, D.H., Chase, M.R., and Bawa, K.S. 1995b. Population genetics of *Cordia alliodora* (Boraginaceae), a Neotropical tree. 3. Gene flow, neighborhood, and population substructure. *American Journal of Botany* **82**:484–490.

Brown, A.H.D., and Schoen, D.J. 1992. Plant population genetic structure and biological conservation. In *Conservation of Biodiversity for Sustainable Development*, eds. O.T. Sandlund, K. Hindar, and A.H.D. Brown, pp. 88–104. Scandinavian University Press, Oslo.

Chase, M.R., Boshier, D.H., and Bawa, K.S. 1995. Population genetics of *Cordia alliodora* (Boraginaceae), a Neotropical tree. 1. Genetic variation in natural populations. *American Journal of Botany* **82**:468–475.

Chase, M.R., Moller, C., Kessell, R., and Bawa, K.S. 1996. Distant gene flow in tropical trees. *Nature* (London) **383**:398–399.

de Paiva, J.R., Kageyama, P.Y., and Vencovsky, R. 1994. Genetics of rubber tree (*Hevea brasiliensis* (Willd. ex Adr. de Juss) Müll. Arg.). 1. Genetic variation in natural populations. *Silvae Genetica* **43**:307–312.

Eguiarte, L.E., Pérez-Nasser, N., and Piñero, D. 1992. Genetic structure, outcrossing rate and heterosis in *Astrocaryum mexicanum* (tropical palm): implications for evolution and conservation. *Heredity* **69**:217–228.

Ellstrand, N.C., and Elam, D.R. 1993. Population genetic consequences of small population size: implications for plant conservation. *Annual Review of Ecology and Systematics* **24**:217–242.

Frankel, O.H. 1983. The place of management in conservation. In *Genetics and Conservation: A Reference for Managing Wild Animal and Plant Populations*, eds. C.M. Schoenwald-Cox, S.M. Chambers, B. MacBryde, and L. Thomas, pp. 1–14. Benjamin/Cummings, Menlo Park, CA.

Frankel, O.H., and Soulé, M.E. 1981. *Conservation and Evolution*. Cambridge University Press, Cambridge.

Gibson, J.P., and Wheelwright, N.T. 1996. Mating system dynamics of *Ocotea tenera* (Lauraceae), a gynodioecious tropical tree. *American Journal of Botany* **83**:890–894.

Gullison, R.E., Panfil, S.N., Strouse, J.J., and Hubbell, S.P. 1996. Ecology and management of mahogany (*Swietenia macrophylla* King) in the Chimanes Forest, Bení, Bolivia. *Botanical Journal of the Linnean Society* **122**:9–34.

Hall, P., Orrell, L.C., and Bawa, K.S. 1994. Genetic diversity and mating system in a tropical tree, *Carapa guianensis* (Meliaceae). *American Journal of Botany* **81**:1104–1111.

Hamrick, J.L. 1989. Isozymes and analyses of genetic structure of plant populations. In *Isozymes in Plant Biology*, eds. D.E. Soltis and P.S. Soltis, pp. 87–105. Dioscorides Press, Portland.

Hamrick, J.L., and Godt, M.J.W. 1990. Allozyme diversity in plant species. In *Plant Population Genetics, Breeding, and Genetic Resources*, eds. A.H.D. Brown, M.T. Clegg, A.L. Kahler, and B.S. Weir, pp. 43–63. Sinauer, Sunderland, MA.

Hamrick, J.L., and Godt, M.J.W. 1996. Conservation genetics of endemic plant species. In *Conservation Genetics: Case Histories from Nature*, eds. J.C. Avise and J.L. Hamrick, pp. 281–304. Chapman & Hall, New York.

Hamrick, J.L., and Loveless, M.D. 1987. The influence of seed dispersal mechanisms on the genetic structure of plant populations. In *Frugivores and Seed Dispersal*, eds. A. Estrada and T.H. Fleming, pp. 211–223. Junk, The Hague.

Hamrick, J.L., and Loveless, M.D. 1989. The genetic structure of tropical tree populations: associations with reproductive biology. In *The Evolutionary Ecology of Plants*, eds. J.H. Bock and Y.B. Linhart, pp. 129–146. Westview, Boulder, CO.

Hamrick, J.L., and Murawski, D.A. 1990. The breeding structure of tropical tree populations. *Plant Species Biology* **5**:157–165.

Hamrick, J.L., and Murawski, D.A. 1991. Levels of allozyme diversity in populations of uncommon Neotropical tree species. *Journal of Tropical Ecology* **7**:395–399.

Hamrick, J.L., Godt, M.J.W., Murawski, D.A., and Loveless, M.D. 1991. Correlations between species traits and allozyme diversity: implications for conservation biology. In *Genetics and Conservation of Rare Plants*, eds. D.A. Falk and K.E. Holsinger, pp. 75–86. Oxford University Press, Oxford.

Kanowski, P., and Boshier, D. 1997. Conserving the genetic resources of trees *in situ*. In *Plant Genetic Conservation: The In Situ Approach*, eds. N. Maxted, B.V. Ford-Lloyd, and J.G. Hawkes, pp. 207–219. Chapman & Hall, London.

Lamb, F.B. 1966. *Mahogany of Tropical America. Its Ecology and Management*. University of Michigan Press, Ann Arbor.

Ledig, F.T. 1986. Conservation strategies for forest gene resources. *Forest Ecology and Management* **14**:77–90.

Ledig, F.T. 1988. The conservation of diversity in forest trees. *BioScience* **38**:471–479.

Lee, H.Y. 1967. Studies in *Swietenia* (Meliaceae): observations on the sexuality of the flowers. *Journal of the Arnold Arboretum* **48**:101–104.

Liengsiri, C., Yeh, F.C., and Boyle, T.J.B. 1995. Isozyme analysis of a tropical forest tree, *Pterocarpus macrocarpus* Kurz., in Thailand. *Forest Ecology and Management* **74**:13–22.

Loveless, M.D. 1992. Isozyme variation in tropical trees: patterns of genetic organization. *New Forests* **6**:67–94.

Loveless, M.D., and Hamrick, J.L. 1984. Ecological determinants of genetic structure in plant populations. *Annual Review of Ecology and Systematics* **15**:65–95.

Loveless, M.D., and Hamrick, J.L. 1987. Distribución de la variación en especies de árboles tropicales. *Revista de Biología Tropical* **35**(Suplemento 1):165–175.

Loveless, M.D., Hamrick, J.L., and Foster, R.B. 1998. Population structure and mating system in *Tachigali versicolor*, a monocarpic Neotropical tree. *Heredity* **81**:134–143.

Murawski, D.A. 1995. Reproductive biology and genetics of tropical trees from a canopy perspective. In *Forest Canopies*, eds. M.D. Lowman and N.M. Nadkarni, pp. 457–493. Academic Press, San Diego.

Murawski, D.A., and Bawa, K.S. 1994. Genetic structure and mating system of *Stemonoporus oblongifolius* (Dipterocarpaceae) in Sri Lanka. *American Journal of Botany* **81**:155–160.

Murawski, D.A., and Hamrick, J.L. 1991. The effects of the density of flowering individuals on the mating system of nine tropical tree species. *Journal of Heredity* **67**:167–174.

Murawski, D.A., Hamrick, J.L., Hubbell, S.P., and Foster, R.B. 1990. Mating systems of two bombacaceous trees of a Neotropical moist forest. *Oecologia* (Berlm) **82**:501–506.

Murawski, D.A., Nimal Gunatilleke, I.A.U., and Bawa, K.S. 1994. The effects of selective logging on inbreeding in *Shorea megistophylla* (Dipterocarpaceae) from Sri Lanka. *Conservation Biology* **8**:997–1002.

National Research Council. 1991. *Managing Global Genetic Resources: Forest Trees*. National Academy Press, Washington, DC.

Nei, M. 1973. Analysis of gene diversity in subdivided populations. *Proceedings of the National Academy of Sciences* of the United States of America **70**:3321–3323.

Nei, M. 1987. *Molecular Evolutionary Genetics*. Columbia University Press, New York.

Newton, A.C., Leakey, R.R.B., and Mesén, J.F. 1993. Genetic variation in mahoganies: its importance, capture and utilization. *Biodiversity and Conservation* **2**:114–126.

Newton, A.C., Cornelius, J.P., Baker, P., Gillies, A.C.M., Hernández, M., Ramnarine, S., Mesén, J.F., and Watt, A.D. 1996. Mahogany as a genetic resource. *Botanical Journal of the Linnean Society* **122**:61–73.

O'Malley, D.M., and Bawa, K.S. 1987. Mating system of a tropical rain forest tree species. *American Journal of Botany* **74**:1143–1149.

O'Malley, D.M., Buckley, D.P., Prance, G.T., and Bawa, K.S. 1988. Genetics of Brazil nut (*Bertholletia excelsa* Humb. & Bonpl.: Lecythidaceae). 2. Mating system. *Theoretical and Applied Genetics* **76**:929–932.

Ritland, K. 1990. A series of FORTRAN computer programs for estimating plant mating systems. *Journal of Heredity* **81**:235–237.

Rodan, B.D., Newton, A.C., and Verissimo, A. 1992. Mahogany conservation: status and policy initiatives. *Environmental Conservation* **19**:331–338.

Shaw, D.V., Kahler, A.L., and Allard, R.W. 1981. A multilocus estimator of mating system parameters in plant populations. *Proceedings of the National Academy of Sciences* of the United States of America **78**:1298–1302.

Snook, L.K. 1996. Catastrophic disturbance, logging, and the ecology of mahogany (*Swietenia macrophylla* King): grounds for listing a major tropical timber species in CITES. *Botanical Journal of the Linnean Society* **122**:35–46.

Stacy, E.A., Hamrick, J.L., Nason, J.D., Hubbell, S.P., Foster, R.B., and Condit, R. 1996. Pollen dispersal in low-density populations of three Neotropical tree species. *American Naturalist* **148**:275–298.

Styles, B.T., and Khosla, P.K. 1976. Cytology and reproductive biology of Meliaceae. In *Tropical Trees: Variation, Breeding, and Conservation*, eds. J. Burley and B.T. Styles, pp. 61–67. Academic Press, London.

Verissimo, A., Barreto, P., Tarifa, R., and Uhl, C. 1995. Extraction of a high-value natural resource in Amazonia: the case of mahogany. *Forest Ecology and Management* **72**:39–60.

Wendell, J.F., and Weeden, N.F. 1989. Visualization and interpretation of plant isozymes. In *Isozymes in Plant Biology*, eds. D.E. Soltis and P.S. Soltis, pp. 5–45. Dioscorides Press, Portland.

Wilson, E.O., ed. 1988. *Biodiversity*. National Academy Press, Washington, DC.

3. Twenty Mahogany Provenances Under Different Conditions in Puerto Rico and the U.S. Virgin Islands

Sheila E. Ward and Ariel E. Lugo

Abstract. The Neotropical mahoganies (*Swietenia*, family Meliaceace) show great ecological and geographic amplitude. A provenance study, begun by the USDA Forest Service in the 1960s, included 20 genetic sources of the three species of mahogany from México, Central America, and the Caribbean. Trees were established at 14 sites ranging from dry to wet forests in Puerto Rico and the U.S. Virgin Islands. We assessed relative amounts of environmental and genetic variation in growth and size traits, survival; insect susceptibility; the segregation of genetic variation among species and among and within populations; and how provenance and species characteristics changed across the different environments. For size and growth traits by 14 years of age, this study showed the following results:

- Mahogany species had about 10 times more genetic variation among species than among provenances within species. Cluster analysis did not group provenances strictly by species, however.
- In big-leaf mahogany, the mean estimated environmental variance was 79%, the mean estimated genetic

variance was 11%, and the mean estimated genetic-by-environment interaction was 8% of the total phenotypic variance. In Pacific coast mahogany, these estimates were 41%, 15%, and 29%, respectively. Because of differing sampling designs, the same comparisons could not be made for small leaf mahogany.

- Big-leaf mahogany showed greater similarity in estimated environmental variances within and among (means of 33% vs. 46% of the total phenotypic variance) plantation sites in Puerto Rico and the Virgin Islands than did Pacific coast mahogany (within-site at 33% vs. among-site at 7%).

- For both big-leaf and Pacific coast mahogany, the estimated genetic variations among and within provenances were similar, between 6% and 9%.

- Significant correlations between environmental gradients at sites of origin and measures of growth and size suggested the action of natural selection in differentiating populations.

- Across all study sites, big-leaf mahogany always showed the greatest mean growth by year 14 (height = 10.4 m, dbh = 12.8 cm, total volume per tree = 114,700 cm^3). Small-leaf mahogany (height = 5.9 m, dbh = 7.8 cm, total volume per tree = 45,500 cm^3) and Pacific coast mahogany (height = 6.4 m, dbh = 8.0 cm, total volume per tree = 49,000 cm^3) were not different in most of the measured variables. All species showed maximum diameter and height growth rate in years 2 to 4, diminishing thereafter. The basal-area growth rates of big-leaf mahogany continued to increase with time; for all species, total volume growth rates continued to increase with time.

- All three species and several populations showed distinct responses to different life zones and soil types. Big-leaf mahogany died earliest in the subtropical dry life zone, grew least by age 14 in the lower montane wet zone, and grew best in the subtropical moist and subtropical wet zones. Pacific coast mahogany grew best and survived longest in the subtropical dry and moist zones and grew least and died soonest in the lower montane wet zone. Small-leaf mahogany lived longest in the dry zone, but other differences were not significant because of the small sample size.

- In both big-leaf and Pacific coast mahogany, 95% of the variation in shoot borer attack was caused by the estimated large- and small-scale environmental effects.

- Provenances became more genetically distinct as distance increased; the source populations came from 160 to 1600 km apart.

The results of our study suggested that genetic differentiation among provenances contributed significantly to the ecological breadth observed in the species of mahogany. Although most of the variation was caused by environmental effects, phenotypic plasticity was not adaptive for the traits examined and thus possibly did not contribute to the species ecological range. These findings suggested a strategy for conserving the genetic variation in mahogany based on both the genetic and the ecological distances among targeted populations.

Keywords: Meliaceae, *Swietenia* spp., Provenance study, Growth, Hierarchical analysis, Genetic variation, Environmental variation, Genetic-by-environment interaction, Puerto Rico, Conservation genetics

Introduction

The conservation status of mahogany in the Neotropics has become a matter of concern. The three species of mahogany have broad ecological tolerance (Lamb 1966; Styles 1981), growing across several life zones (*sensu* Holdridge 1967) from dry to wet forest (Webb et al. 1980; Styles 1981), and on a range of soil types (Styles 1981). Both genetic variation and phenotypic plasticity (ability of the same genotype to respond to different environments) may play a role in the ecological breadth of a species (Bradshaw 1965). Knowledge of the patterns and relative importance of genetic and plastic variation in these species is critical for developing strategies for conserving and managing them.

The patterns of environmental and genetic components of phenotypic variation result from and influence basic processes affecting the phenotype. Some species show narrow genetic variation but significant adaptive plasticity in a range of environments (Brown and Marshall 1981). On the other hand, plasticity may be an unavoidable consequence of environmental influences (Bradshaw 1965). Adaptive genetic differentiation may extend the ecological amplitude of the species beyond that resulting from plastic response to the environment. However, although most species have shown some degree of genetic differentiation among populations (Hamrick and Godt 1989), these differences are not necessarily adaptive to local environments (McGraw and Antonovics 1983; McGraw 1987; Rice and Mack 1991). Responses of genetic sources to environments (the genetic-by-environment interaction) can be examined for evidence of adaptation to local conditions. Patterns of genetic variation and the genetic-by-environment interaction also indicate potential to adapt to future environmental change.

Is more variation found at the local or at a wider geographic scale? The spatial scale with the most plastic variation (among or within sites) indicates where the influence of the environment affects plant traits, such as growth, the most. Genetic variation is partitioned among species and among and within populations. The three species of mahogany appear to interbreed freely and produce viable offspring (Whitmore and Hinojosa 1977), raising questions about the degree of differentiation among them (Helgason et al. 1996). High rates of outcrossing in the genus *Swietenia*, among the highest found (Loveless and Gullison, Chapter 2, this volume), may influence the distribution of genetic variation among and within populations and species.

Phenotypic patterns of variation have implications for management and conservation. These patterns may indicate whether controlling environmental variation on the large scale (site selection) or local scale (silvicultural management), or genetic variation via artificial selection, will maximize desirable traits such as growth rate. If most of the genetic variation is among rather than within populations, more emphasis should be placed on sufficient representative populations to conserve genetic variation.

Different type of traits may reveal contrasting patterns of variation. Most work on conservation genetics uses molecular markers assumed to be neutral, but neutral variation does not adequately reflect the patterns and amounts of variation seen in adaptive traits (Hedrick et al. 1996). Quantitative traits related to fitness may be more relevant to conserving the evolutionary potential of a species. Growth traits, although extremely plastic, are closely linked with fitness and thus are useful for examining adaptation and evolutionary potential.

Genetic or plastic variation may influence the apparent broad tolerance of mahogany species for a variety of environmental factors. Species tolerate pHs up to 8.5 and elevations from sea level to 900 m (Lamb 1966). The genus normally grows in climates with a dry season. Mahogany species have been found to perform poorly under cool, moist conditions (Marrero 1950b), under low soil pH, and in heavy clay soils with poor drainage (Marrero 1950b). Canopy position (F. Wadsworth, International Institute of Tropical Forestry, personal communication), shoot borer attack, soil (Marrero 1950b; Medina and Cuevas, Chapter 7, this volume), planting method, soil drainage, degree of land degradation and accompanying nutrient depletion (Marrero 1950b; Medina et al., Chapter 8, this volume; Wang and Scatena, Chapter 12, this volume), and overhead shade (Geary et al. 1973) are all believed to affect mahogany growth. On a local scale, a variety of environmental factors may affect mahogany growth, with complex interactions among them, and with different factors taking precedence on different sites (Geary et al. 1973). On sites that are poor or degraded, factors may act synergistically to worsen the effects of insects, disease, or competitors (Marrero 1948). Topographical position and slope may affect growth because they can affect drainage (Marrero 1950b; Wadsworth et al., Chapter 17, this volume), or interact with shoot borer attack (Weaver and Bauer 1986). Testing a broad range of genetic sources across a spectrum of environments may provide insight into the processes affecting the ecological amplitude of the species. In this chapter, we emphasize the interactions of soil type and climatic zone with genetic source to examine the potential adaptiveness of different genetic sources and plasticities to environmental conditions.

A provenance or population study of mahogany, begun in the 1960s by the USDA Forest Service in Puerto Rico (Barres 1963), provides a large database for assessing the patterns of variation in the genus. This study was designed as a multiple common-garden experiment, with genetic sources from México, Central America, and the Caribbean, replicated across an array of environments in Puerto Rico and the Virgin Islands. The Puerto Rico environments selected for this experiment encompassed a variety of life zones and soil types, potentially paralleling the range of environments of the samples. Because different genetic sources and environments were used in the same experimental design, this study affords a baseline on mahogany growth and the factors affecting it under different conditions.

We began by partitioning phenotypic variation in size, growth rates, longevity, and shoot borer attack into environmental and genetic components in a nested hierarchy: environmental variation on the large and local scale, and genetic variation among species, among provenances within species, and among mother trees (families) within provenances. Our questions were these: Which genetic and environmental components were significant? What was their relative importance? How did they change through time? Included in these hierarchical comparisons was an assessment of the relative amount of genetic variation among and within species.

We also assessed evidence for adaptive strategies including specialization to different habitats and adaptive plasticity. The interactions between genetic scales (species and provenances) and environmental conditions (life zone and soil type) were used to determine the relative performance of genetic sources across environments. The relation between environmental factors at the sites of origin and provenance performance were examined for potential factors selecting for different kinds of growth patterns.

Adaptive plasticity was sought in the relationship between trait variability and fitness parameters.

Methods

Species and Study Sites

Species

All three species of mahogany were included in this study. The native range of small-leaf mahogany is in the Caribbean from southern Florida, through the Bahamas, Cuba, and Hispañola to Jamaica. Pacific coast mahogany grows mainly on the west coast of Central America from México to Costa Rica. Big-leaf mahogany has the most extensive range, from central México into Bolivia and Brazil. The ranges of Pacific coast and big-leaf mahogany overlap, and they probably interbreed naturally (Whitmore and Hinojosa 1977). The range of small-leaf mahogany is discontinuous with that of the other two species, and thus it does not interbreed naturally, although it does hybridize with big-leaf mahogany when the two are planted nearby (Whitmore and Hinojosa 1977).

Provenance Collection and Study Sites

Sample sites were distributed as evenly as possible in the designated geographic area (Barres 1963). Seeds from 12 provenances of big-leaf mahogany and 7 provenances of Pacific coast mahogany were harvested in México and Central America in 1964 and 1965 (Table 3.1). In 1964, a seed source of small-leaf mahogany from St. Croix, U.S. Virgin Islands, was included in this study; this species is naturalized, but not native, to St. Croix. The original source of the St. Croix mahogany is unknown.

At each collection site, trees of obviously inferior quality were excluded from sampling (Barres 1963). The usual method of seed harvest was to collect at least five mature capsules from the targeted mother trees (Barres 1963). For each provenance harvested, seed was separated by mother tree, except the seed from St. Croix small-leaf mahogany, which was bulked in one lot. The 1964 collections were designated as SEEDGROUP 1, and the 1965 collections as SEEDGROUP 2 (Table 3.1). For each SEEDGROUP, seed was sown into soil-filled plastic bags, grouped by mother tree, and germinated in a common nursery. Seedlings were planted in 14 sites in Puerto Rico and St. Croix, including the 11 sites used in this analysis, encompassing a range of climatic and soil conditions (Table 3.2). Several plantation sites were on degraded lands, which may have affected soil fertility and thus tree growth and longevity (Geary et al. 1973).

SEEDGROUPs 1 and 2 planted out respectively in 1965 and 1966, were usually established in adjacent blocks at each location (Barres 1963; Geary

Table 3.1. Sites of Origin for the Mahogany Provenances in This Study

Seed Source Code	Country of Origin	Geographic Coordinates		Mother Trees Harvested (number)	Collection Date	Median Elevation (m)	Length of Dry Season (months)
		North Latitude	West Longitude				
Big-leaf mahogany							
MMA[a]	México	20°30'	97°20'	16	Jan 1964	25	7
MMB[a]	México	17°40'	94°55'	16	Feb 1964	90	5
MMC[a]	México	16°55'	91°35'	16	Feb 1964	640	3
MMD[a]	México	18°25'	89°25'	16	Feb 1964	110	6
MME[a]	México	19°30'	88°05'	16	Feb 1964	30	7
MGA[a]	Guatemala	17°00'	89°40'	16	Mar 1964	210	4
MHA[b]	Honduras	15°50'	86°00'	7	Apr 1965	35	3
MNA[b]	Nicaragua	13°40'	84°10'	16	Feb 1965	15	2
MNB[b]	Nicaragua	11°50'	83°55'	16	Feb 1965	15	2
MPA[b]	Panamá	7°35'	80°40'	15	Dec 1964	545	4.5
MPB[b]	Panamá	8°50'	78°00'	16	Dec 1964	60	3.5
Pacific coast mahogany							
HMA[a]	México	18°05'	102°15'	16	Jan 1964	180	6
HMB[a]	México	22°20'	105°20'	16	Jan 1964	60	8
HME[a]	México	16°20'	94°00'	16	Feb 1964	195	6
HGA[b]	Guatemala	14°15'	90°50'	16	Feb 1965	45	5
HSA[b]	El Salvador	13°40'	88°55'	16	Feb 1965	535	6
HNA[b]	Nicaragua	12°10'	86°40'	16	Feb 1965	150	6
HCA[b]	Costa Rica	12°10'	85°30'	16	Jan 1965	90	7
Small-leaf mahogany							
WIM[a]	This provenance was harvested from individuals naturalized to St. Croix, U.S. Virgin Islands						

a Provenance is member of SEEDGROUP1.
b Provenance is member of SEEDGROUP2.

Table 3.2. Plantation Sites in Puerto Rico and the U.S. Virgin Islands

Location	Soil Order	Elevation (m)	Date of Planting[a] SEED-GROUP 1	SEED-GROUP 2
Subtropical dry forest				
Coamo	Mollisol	250	65.1	66.0
Estate Thomas[b,c]	Mollisol	90	65.8	Not planted
Subtropical moist forest				
Cambalache	Oxisol	50	64.8	66.0
Guajataca[c]	Oxisol	230	64.8	66.0
Tract 105[d]	Inceptisol	500	64.9	66.0
Subtropical wet forest				
Las Marías[c]	Ultisol	220	64.9	65.9
El Verde[d]	Oxisol	100	64.9	66.0
Guavate[c]	Ultisol	600	64.8	66.0
Maricao	Inceptisol	730	64.9	65.9
Matrullas	Ultisol	750	64.9	66.0
Lower montane wet forest				
Guilarte[e]	Ultisol	1050	64.9	65.9

[a] Date as 1/10 year.
[b] Plantation site in St. Croix, U.S. Virgin Islands.
[c] Sites thinned in 1974.
[d] In the Luquillo Experimental Forest.
[e] Guilarte SEEDGROUP (SG) 1 was not measured in 1979 and therefore was thus not included in this study.

1969; Geary et al. 1973). Each seed group block consisted of 32 subblocks. Each subblock consisted of nine single-tree plots in a square, with a spacing of 2.4 m between trees. Subblocks and trees within subblocks were arranged randomly, but trees were not randomly assigned to subblocks. The number of mother trees per provenance was usually 16, and the number of seedlings available from each mother tree varied (Table 3.1; Geary et al. 1973). Even distribution of progeny groups from each provenance over the study sites was attempted. At each site in Puerto Rico and the U.S. Virgin Islands, an average of about two progeny per mother tree per provenance was planted.

Site preparation before planting included cleaning the area of ground and overhead vegetation, girdling, and herbicide application. Girdled trees died within 2 or 3 years. Intrusive vegetation was repeatedly cleared between planting and 1971, and some sites had more overhead shade than others. At some sites, forked trees (usually assumed to be due to shoot borer damage) were pruned. The plantation sites were never fertilized, but one site (Coamo SEEDGROUP 1) was watered for the first 4 months after planting because of drought. Seedlings that died were usually replaced within 4 months of the initial planting (Geary et al. 1973). The trees at all sites were remeasured at various intervals, but not all sites were measured every time.

Data Collection

At each measurement, total height (HT) and diameter at breast height (dbh) (DBH; 1.37 m) were recorded for all trees. The number of attacks made by the mahogany shoot borer during the previous year was also assessed for most sites between 1966 and 1971. Basal area (BA) was calculated as $\pi(DBH/2)^2$. Total wood volume (TV) was calculated by the formula ($TV = 0.017065 + [3.554 \times 10^{-5}][DBH^2 \times HT]$) used by Bauer and Gillespie (1990) for young hybrid mahogany. The yearly rates of increase in height (δHT), dbh (δDBH), basal area (δBA), and total wood volume (δTV) were calculated based on the differences between consecutive measurement dates. Longevity in years was calculated as the midpoint of the interval between when the tree was last recorded as alive and when it was considered to have died. Trees alive at year 14 were assigned a longevity of 14. The total shoot borer attack was the cumulative number of attacks for the first 5 years.

Measurement years for this study were 1966, 1967, 1968, 1969, 1970, 1971, and 1979 because they had the most complete representation of sites measured. Because SEEDGROUP 1 was planted a year before SEED-GROUP 2, SEEDGROUP 1 measurements were delayed a year between 1966 and 1970 so that measurement intervals could be compared with SEEDGROUP 2. Because both SEEDGROUPs 1 and 2 were measured in 1979, the 1979 measurements for SEEDGROUP 1 were not delayed.

Trees with missing measurements were removed from the analysis of size and growth rates. Of the 5606 trees for which data were recorded, about 2900 had complete measurements for 1966 to 1971 and 1979, and thus were included in the analysis of size and growth rates for the current study. About 3330 trees were included in the analysis of total shoot borer attacks. For longevity, 5550 trees were analyzed. On several sites, the smaller trees from each provenance were selectively removed in 1974 (see Table 3.2), and on two sites landslides removed some trees. These trees were excluded from analyses for all traits.

Analysis

Overview

The partitioning of phenotypic variation into environmental and genetic components at different hierarchical levels was examined through analysis of variance (ANOVA) and variance component analysis. The analyses of size parameters HT, DBH, BA, and TV, and growth rates $\delta HT, \delta DBH, \delta BA$, and δTV included repeated measures analysis of variance and variance component comparisons across years. Repeated measures analysis of variance (ANOVAR) allowed us to examine changes through time in the significance of different effects. Longevity and rate of shoot borer attack were examined by using analysis of variance (ANOVA) and variance component comparisons. Least square means comparisons were used to examine significant interaction effects from ANOVA and to relate environmental variation (at both sites of origin in México and Central America, and planting sites in the Caribbean) to variation in size and growth rate parameters, survival, and shoot borer attack. Correlations were used to investigate potential selection pressures and adaptive strategies.

ANOVAR and ANOVA Experimental Design

The same model was used for all analyses of variance (ANOVA) and repeated measures analyses (ANOVAR). Phenotypic variance has several components (Falconer 1981):

$$V_P = V_E + V_G + V_{G*E} + V_e$$

where V_P is the variance of the phenotype, V_E is the variance among environments, V_G is the genetic variance, V_{G*E} is the environment-by-genotype interaction, and V_e is the variance within environments (Falconer 1981; Table 3.3). For this study, $V_{SEEDGROUP}$ was a composite measure including aspects of genetic and environmental variance.

Table 3.3. Variance Components Definitions

Variance Component	Description of Effect	Random or Fixed Effect
LOCATION	Sites of plantations in Puerto Rico	Random
SUBBLOCK (LOCATION SEEDGROUP)	Subdivision of plantings within seed groups at each site consisting of nine single-tree plots	Random
SPECIES	Species of mahogany	Fixed
ORIGIN (SEEDGROUP)	Provenances of origin	Random
MOTHER (LOCATION SEEDGROUP ORIGIN)	Family membership	Random
SEEDGROUP	Provenances segregated into two groups based on year of harvest and planting, and area of planting in a PR location	Random
SPECIES*SEEDGROUP	Species-by-location interaction	Random
LOCATION*SEEDGROUP	Location-by-seed-group interaction	Random
LOCATION*SPECIES	Location-by-species interaction	Random
LOCATION*ORIGIN	Location-by-origin interaction	Random
LOCATION*MOTHER	Location-by-mother-tree interaction	Random
ERROR	Variation unexplained by the other effects	Random

- V_E: In this multiple common-garden study, the *LOCATION* effect was equivalent to variation among environments or variation among sites planted in Puerto Rico. It thus included differences in climate and soils among sites.
- V_G: Genetic effects were organized at several hierarchical levels; they included the amount of variation among species (*SPECIES*), the amount of variation among sampled populations within species and seed group (*ORIGIN*), and variation within population of origin among mother trees (*MOTHER*). Variation among *MOTHER*s (or among families) also included an environmental component resulting from a maternal effect that could not be removed. Because of sampling design, *MOTHER* could not be determined either for all species together or for small-leaf mahogany.
- V_{G*E}: The variance resulting from the environment-by-genetic interaction could also be broken down into several levels, including location-by-species interaction (*LOCATION*SPECIES*) and location-by-population-of-origin interaction (*LOCATION*ORIGIN*), and the location-by-mother-tree interaction (*LOCATION*MOTHER*). *LOCATION*MOTHER* could not be determined either for all species together or for small-leaf mahogany.
- $V_{e\ (total)}$: Variance resulting from small-scale environmental effects (V_e) included variance resulting from subblocks in each location in Puerto Rico (*SUBBLOCK*) and the error term (*ERROR*). The variance from error consisted of genetic and environmental variation in families of mother trees.
- $V_{SEEDGROUP}$: The model for this study included a SEEDGROUP main effect and the interactions of seed group with location and species (*SEEDGROUP, SEEDGROUP*LOCATION, SEEDGROUP* SPECIES*). The SEEDGROUP main effect included genetic and environmental components as discussed in Appendix A.

Composite variance components were calculated as detailed in Table 3.4 and Appendix A. The voluminous results of all analyses were deposited in the files of the International Institute of Tropical Forestry in Puerto Rico. Throughout this chapter, asterisks (*) indicate analyses available from the authors on request.

Results

ANOVAR, ANOVA, and Variance Component Analysis

For all species together, by year 14, the largest mean variance components for size and growth parameters were those resulting from *SPECIES* and *ERROR* effects (Fig. 3.1). This finding indicated that most of the variance was partitioned among species—or, as part of the error term—remained

Table 3.4. Aggregate Components of Variation[a]

Aggregate Components	Description	Subcomponents
V_P	Total phenotypic variation	All
$V_{FAMILIES\ (TOTAL\ ESTIMATED)}$	Total genetic variation within and among families	$4*V_{MOTHER}$ [b]
$V_{G\ (TOTAL\ ESTIMATED)}$	Total genotypic variation	$V_{SPECIES} + V_{ORIGIN} + 4*V_{MOTHER}$ [b]
$V_{G*E\ FAMILY\ (TOTAL\ ESTIMATED)}$	Total genetic-by-environment interaction among families	$2*V_{LOCATION*MOTHER}$ [b]
$V_{G*E\ (TOTAL\ ESTIMATED)}$	Total genotype-by-environment interaction	$V_{LOCATION*SPECIES} + V_{LOCATION*MOTHER}$
V_E	Large-scale environmental effects from location of plantations in Puerto Rico	$V_{LOCATION}$
$V_{e\ (WITHIN\ SUBBLOCKS\ TOTAL\ ESTIMATED)}$	Total environmental variation within subblocks	$V_{ERROR} - 3*V_{MOTHER} - V_{LOCATION*MOTHER}$ [b]
$V_{e\ (TOTAL\ ESTIMATED)}$	Total small-scale environmental effects	$V_{SUBBLOCK} + V_{ERROR} - 3*V_{MOTHER}$ [b]
$V_{E+e\ (TOTAL\ ESTIMATED)}$	Environmental effects Total environmental effects	$V_{LOCATION} + V_{e\ (total\ estimated)}$
$V_{SEEDGROUP\ (TOTAL)}$	Total seedgroup effects	$V_{SEEDGROUP} + V_{SEEDGROUP*SPECIES} + V_{LOCATION*SEEDGROUP}$

[a] For descriptions of specific effects, see Methods and Table 3.3.
[b] See Appendix for explanation of genetic variance estimation.

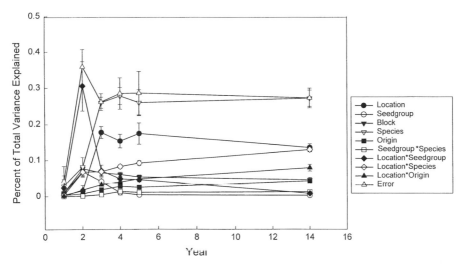

Figure 3.1. Percentage of the total variance explained by variance components for different effects (Table 3.3) in the years the three mahogany species were measured. Effects have been averaged across all size and growth rate parameters. *Error bars* indicate standard error.

within families and small-scale environmental effects. For big-leaf mahogany, the largest mean variance components for size and growth parameters were for the *LOCATION* and *ERROR* (Fig. 3.2), and for Pacific coast mahogany, the largest were caused by the *LOCATION*ORIGIN* interaction and *ERROR* (Fig. 3.3).

Total Environmental Versus Total Genetic Effects

Comparing total environmental ($V_{E+e\ (TOTAL\ ESTIMATED)}$) and genetic variation ($V_{G\ (TOTAL\ ESTIMATED)}$) and the total environmental by genetic interaction ($V_{G*E\ (TOTAL\ ESTIMATED)}$) for all species together was impossible because of confounded genetic and environmental effects in *ERROR* that could not be separated. Because of the amount of difference among species in mean size and growth by year 14, however, the genetic effects were larger than the large-scale environmental effects (Table 3.5). For big-leaf mahogany, $V_{E+e\ (TOTAL\ ESTIMATED)}$ accounted for about 80%, and $V_{G*E\ (TOTAL\ ESTIMATED)}$ accounted for 11% and $V_{G\ (TOTAL\ ESTIMATED)}$ 8% of the total variance (Table 3.6) in these traits. For Pacific coast mahogany, $V_{E+e\ (TOTAL\ ESTIMATED)}$ accounted for less (41%), and $V_{G*E\ (TOTAL\ ESTIMATED)}$ (29%) and $V_{G\ (TOTAL\ ESTIMATED)}$ (15%) accounted for more, of the total variation than in big-leaf mahogany (Table 3.7). In both species, $V_{E+e\ (TOTAL\ ESTIMATED)}$ dominated the variation in total shoot borer attack (more than 90%) and longevity (more than 65%). For small-leaf mahogany, these comparisons were not available because of the difference in sampling strategy (Table 3.8).

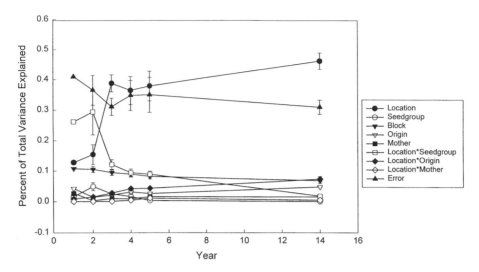

Figure 3.2. Percentage of the total variance explained by different effects (Table 3.3) for big-leaf mahogany at the years of measurement. Effects have been averaged across all size and rate parameters. *Error bars* indicate standard error.

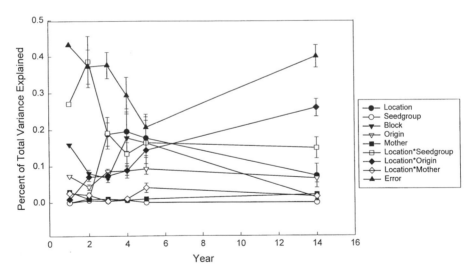

Figure 3.3. Percentage of the total variance explained by different effects (Table 3.3) for Pacific coast mahogany at the years of measurement. Effects have been averaged across all size and rate parameters. *Error bars* indicate standard error.

Table 3.5. Variance Components (as Proportions of Total Variance) and Standard Deviation (SD) for the Mean of Height and Growth Measurements (HT, δHT, DBH, δDBH, BA, δBA, TV, and δTV), Longevity, and Total Shoot Borer Attack by Age 5 for All Mahogany Species Together[a]

| Source of Variation | Mean for Height and Growth Measurements by Year | | | | | | | | | | | Longevity | Total Shoot Borer Attack |
	Year 1[b]	Year 2	SD	Year 3	SD	Year 4	SD	Year 5	SD	Year 14	SD		
LOCATION	0.004	0.015	0.043	0.178	0.046	0.154	0.053	0.175	0.083	0.136	0.030	0.000	0.537
SG	0.009	0.069	0.042	0.041	0.031	0.010	0.019	0.004	0.007	0.003	0.007	0.000	0.006
SUBBLOCK(LOCATION*SG)	0.010	0.077	0.024	0.065	0.021	0.061	0.018	0.053	0.017	0.046	0.019	0.024	0.030
SPECIES	0.029	0.082	0.075	0.260	0.043	0.278	0.067	0.260	0.101	0.272	0.064	0.076	0.049
ORIGIN(SG*SPECIES)	0.003	0.007	0.006	0.018	0.004	0.028	0.010	0.025	0.009	0.042	0.005	0.038	0.007
SG*SPECIES	0.000	0.001	0.002	0.005	0.006	0.013	0.017	0.011	0.012	0.012	0.012	0.039	0.000
LOCATION*SG	0.024	0.305	0.193	0.071	0.044	0.048	0.021	0.046	0.017	0.008	0.009	0.048	0.033
LOCATION*SPECIES	0.003	0.067	0.024	0.068	0.017	0.083	0.015	0.092	0.019	0.130	0.018	0.197	0.063
LOCA*ORIGI(SG*SPECI)	0.001	0.018	0.005	0.032	0.007	0.040	0.013	0.048	0.020	0.079	0.027	0.043	0.012
Error	0.042	0.359	0.138	0.262	0.065	0.285	0.123	0.286	0.169	0.273	0.077	0.536	0.262
Aggregate variance components													
V$_{SEEDGROUP (TOTAL)}$	0.033	0.375		0.116		0.072		0.061		0.023		0.087	0.040

[a] For descriptions of specific effects, see Methods and Table 3.3.
[b] Year 1 is for HT only because effects were not determined for other traits.

Table 3.6. Variance Components (as Proportions of Total Variance) and Standard Deviation (SD) for the Mean of Height and Growth Measurements (HT, δHT, DBH, δDBH, BA, δBA, TV, and δTV), Longevity, and Total Shoot Borer Attack by Age 5 for Big-Leaf Mahogany

Source of Variation	Mean for Height and Growth Measurements by Year											Longevity	Total Shoot Borer Attack
	Year 1	Year 2	SD	Year 3	SD	Year 4	SD	Year 5	SD	Year 14	SD		
LOCATION	0.128	0.154	0.092	0.388	0.080	0.365	0.130	0.379	0.139	0.463	0.076	0.111	0.670
SG	0.016	0.049	0.036	0.025	0.022	0.016	0.028	0.003	0.006	0.000	0.001	0.000	0.000
SUBBLOCK(LOCATION*SG)	0.107	0.105	0.029	0.095	0.025	0.090	0.027	0.082	0.023	0.068	0.026	0.043	0.034
ORIGIN(SG)	0.042	0.014	0.008	0.021	0.008	0.031	0.016	0.026	0.014	0.048	0.002	0.014	0.011
MOTHER(SG*ORIGIN)	0.026	0.003	0.005	0.010	0.003	0.009	0.004	0.016	0.004	0.015	0.004	0.010	0.004
LOCATION*SG	0.262	0.294	0.212	0.121	0.045	0.094	0.035	0.089	0.029	0.018	0.012	0.044	0.028
LOCATION*ORIGIN(SG)	0.008	0.015	0.005	0.028	0.009	0.042	0.012	0.043	0.014	0.073	0.025	0.023	0.005
LOCA*MOTHE(SG*ORIGI)	0.000	0.000	0.001	0.001	0.004	0.004	0.005	0.011	0.009	0.004	0.004	0.033	0.006
Error	0.410	0.365	0.139	0.311	0.080	0.349	0.140	0.350	0.163	0.310	0.066	0.722	0.241
Aggregate variance components													
V_{FAMILY} (TOTAL ESTIMATED)	0.105	0.011		0.039		0.035		0.063		0.060		0.041	0.015
V_{G} (TOTAL ESTIMATED)	0.147	0.025		0.060		0.066		0.089		0.108		0.055	0.027
V_{G*E} FAMILY (TOTAL ESTIMATED)	0.000	0.001		0.002		0.009		0.022		0.009		0.065	0.012
V_{G*E} (TOTAL ESTIMATED)	0.008	0.016		0.029		0.051		0.065		0.082		0.088	0.018
V_{E}	0.128	0.154		0.388		0.365		0.379		0.463		0.111	0.670
V_{e} WITHIN SUBBLOCKS (TOTAL ESTIMATED)	0.331	0.357		0.281		0.318		0.292		0.261		0.659	0.223
V_{e} (TOTAL ESTIMATED)	0.439	0.462		0.376		0.407		0.374		0.329		0.702	0.258
V_{E+e} (TOTAL ESTIMATED)	0.567	0.616		0.764		0.772		0.754		0.792		0.813	0.927
$V_{SEEDGROUP}$ (TOTAL)	0.278	0.343		0.146		0.110		0.092		0.018		0.044	0.028

[a] For descriptions of specific effects, see Methods and Table 3.3.
[b] Year 1 is for HT only because effects were not determined for other traits.

Table 3.7. Variance Components (as Proportions of Total Variance) and Standard Deviation (SD) for the Mean of Height and Growth Measurements (HT, δHT, DBH, δDBH, BA, δBA, TV, and δTV), Longevity, and Total Shoot Borer Attack by Age 5 for Pacific Coast Mahogany[a]

Source of Variation	Year 1[b]	Mean for Height and Growth Measurements by Year										Longevity	Total Shoot Borer Attack
		Year 2	SD	Year 3	SD	Year 4	SD	Year 5	SD	Year 14	SD		
LOCATION	0.000	0.011	0.009	0.188	0.090	0.195	0.152	0.177	0.137	0.073	0.093	0.217	0.464
SG	0.000	0.006	0.011	0.004	0.008	0.006	0.018	0.000	0.000	0.000	0.000	0.025	0.000
SUBBLOCK(LOCATION*SG)	0.159	0.080	0.028	0.067	0.032	0.178	0.225	0.164	0.202	0.013	0.024	0.009	0.053
ORIGIN(SG)	0.073	0.042	0.022	0.085	0.021	0.087	0.033	0.093	0.042	0.066	0.033	0.087	0.005
MOTHER(SG*ORIGIN)	0.029	0.010	0.005	0.010	0.007	0.007	0.004	0.010	0.006	0.022	0.003	0.019	0.000
LOCATION*SG	0.271	0.387	0.198	0.191	0.122	0.134	0.090	0.165	0.114	0.150	0.082	0.051	0.023
LOCATION*ORIGIN(SG)	0.009	0.071	0.032	0.073	0.029	0.088	0.056	0.144	0.103	0.260	0.067	0.088	0.037
LOCATION*MOTHER(SG*ORIGIN)	0.025	0.021	0.020	0.004	0.006	0.010	0.013	0.041	0.037	0.016	0.023	0.000	0.000
Error	0.433	0.373	0.133	0.376	0.102	0.294	0.142	0.207	0.101	0.400	0.087	0.505	0.419
Aggregate variance components													
V_FAMILY (TOTAL ESTIMATED)	0.118	0.038		0.040		0.027		0.039		0.087		0.075	0.000
V_G (TOTAL ESTIMATED)	0.191	0.081		0.125		0.114		0.131		0.153		0.162	0.005
V_G*E FAMILY (TOTAL ESTIMATED)	0.051	0.041		0.008		0.021		0.081		0.033		0.000	0.000
V_G*E (TOTAL ESTIMATED)	0.060	0.112		0.081		0.109		0.225		0.293		0.088	0.037
V_E	0.000	0.011		0.188		0.195		0.177		0.073		0.217	0.464
V_e WITHIN SUBBLOCKS (TOTAL ESTIMATED)	0.319	0.323		0.343		0.263		0.137		0.318		0.449	0.419
V_e (TOTAL ESTIMATED)	0.478	0.404		0.410		0.442		0.301		0.332		0.458	0.471
V_E+e (TOTAL ESTIMATED)	0.478	0.415		0.598		0.637		0.479		0.405		0.675	0.935
V_SEEDGROUP (TOTAL)	0.271	0.393		0.196		0.140		0.165		0.150		0.076	0.023

[a] For descriptions of specific effects, see Methods and Table 3.3.
[b] Year 1 is for HT only because effects were not determined for other traits.

Table 3.8. Variance Components (as Proportions of Total Variance) and Standard Deviation (SD) for the Mean of Height and Growth Measurements (HT, δHT, DBH, δDBH, BA, δBA, TV, and δTV), Longevity, and Total Shoot Borer Attack by Age 5 for Small-Leaf Mahogany[a]

Source of Variation	Mean for Height and Growth Measurements by Year											Longevity	Total Shoot Borer Attack
	Year 1[b]	Year 2	SD	Year 3	SD	Year 4	SD	Year 5	SD	Year 14	SD		
Var(LOCATION)	0.286	0.924	0.147	0.480	0.310	0.411	0.271	0.673	0.148	0.748	0.149	0.286	0.531
Var(Error)	0.714	0.076	0.147	0.520	0.310	0.589	0.271	0.327	0.148	0.252	0.149	0.714	0.469

[a] For descriptions of specific effects, see Methods and Table 3.3.
[b] Year 1 is for HT only because effects were not determined for other traits.

Large-Scale Versus Small-Scale Environmental Variation

Large-scale environmental variation (V_E or *LOCATION*) occurred among sites, and small-scale variation ($V_{e\ (TOTAL\ ESTIMATED)}$) occurred within sites among and within subblocks (see Table 3.4). For all species together and for each species individually, *LOCATION* effects were always highly significant ($p < 0.0001$; Tables 3.9, 3.10, 3.11, 3.12). For all species together and for big-leaf mahogany alone (Tables 3.9, 3.10), the *SUBBLOCK* effects were always highly significant ($p < 0.0001$; Table 3.10). For Pacific coast mahogany, *SUBBLOCK* effects were usually less significant (Table 3.11).

For all species together, for mean variance components of size and growth parameters by age 14, *ERROR* was greater than *LOCATION*. Because of several genetic components that could not be separated from *ERROR* when all species were considered together (see Appendix A), large-scale and small-scale environmental variation could not be compared.

For small-leaf mahogany alone, V_L was the only effect considered. For mean size and growth parameters by year 14, it accounted for an average of 75% of the total variation (Table 3.8). For this species, the *ERROR* term included *SUBBLOCK*, genetic, and genetic-by-environment interaction effects.

For big-leaf mahogany alone, in mean size and growth parameters by year 14, the large-scale environmental variation (V_E) was somewhat larger (46%) than the estimated small-scale environmental variation ($V_{e\ (TOTAL\ ESTIMATED)}$; 33%, Table 3.6). For Pacific coast mahogany, $V_{e\ (TOTAL\ ESTIMATED)}$ was much larger (33%) than V_E (7%) in these parameters (Table 3.7). Both species showed more $V_{e\ (TOTAL\ ESTIMATED)}$ than V_E for longevity (Tables 3.6, 3.7). In total shoot borer attack by age 5, V_E was larger than $V_{e\ (TOTAL\ ESTIMATED)}$ in big-leaf mahogany (Table 3.6), but V_E and $V_{e\ (TOTAL\ ESTIMATED)}$ were similar for Pacific coast mahogany (Table 3.7).

Genetic-by-Environment Interactions

The genetic-by-environment interactions ($V_{G*E\ (TOTAL\ ESTIMATED)}$) included *LOCATION*SPECIES, LOCATION*ORIGIN*, and *LOCATION* MOTHER* effects. For all species together, *LOCATION*SPECIES* and *LOCATION*ORIGIN* effects were always highly significant ($p < 0.0001$; Table 3.9). For size and growth parameters by age 14, these components accounted for 21% of the total variation, with *LOCATION*SPECIES* accountable for the greater part (see Table 3.5). For longevity and shoot borer attack as well, *LOCATION*SPECIES* accounted for more of the variation than did *LOCATION*ORIGIN*.

In both big-leaf (Table 3.10) and Pacific coast (Table 3.11) mahogany, *LOCATION*ORIGIN* effects were always highly significant ($p < 0.0001$; Table 3.9) and *LOCATION*MOTHER* effects were never significant. For big-leaf mahogany alone, in size and growth parameters by year 14, the average $V_{G*E\ (TOTAL\ ESTIMATED)}$ (8%; Table 3.6) was less than for Pacific coast

Table 3.9. Significance of Effects for Repeated Measures Analysis of Variance (ANOVAR) for HT, δHT, DBH, δDBH, BA, δBA, TV, and δTV and Analysis of Variance (ANOVA) for Longevity and Total Shoot Borer Attack by Age 5 for All Mahogany Species

Source	HT	δHT	DBH	δDBH	BA	δBA	TV	δTV	Longevity[a]	Shoot Borer[a]
Tests of hypotheses for among-subject effects: probability of a greater *F* statistic										
LOCATION	0.0001	0.0001	0.0001	0.0001	0.0001	0.0001	0.0001	0.0001	0.0001	0.0001
SG	0.0001	0.3623	0.2128	0.4873	0.0788	0.6011	0.2731	0.6557	0.1171	0.0858
SUBBLOCK(LOCATION*SG)	0.0001	0.0001	0.0001	0.0001	0.0001	0.0001	0.0001	0.0001	0.0001	0.0001
SPECIES	0.0001	0.0001	0.0001	0.0001	0.0001	0.0001	0.0001	0.0001	0.0006	0.0001
ORIGIN(SG*SPECIES)	0.0001	0.0001	0.0001	0.0001	0.0001	0.0001	0.0001	0.0001	0.0001	0.0001
SG*SPECIES	0.0009	0.0001	0.0073	0.0001	0.0992	0.0112	0.0903	0.0002	0.0616	0.6828
LOCATION*SG	0.0092	0.0001	0.0331	0.0044	0.1357	0.0464	0.0440	0.0001	0.9113	0.6019
LOCATION*SPECIES	0.0001	0.0001	0.0001	0.0001	0.0001	0.0001	0.0001	0.0001	0.0001	0.0001
LOCA*ORIGI(SG*SPECI)	0.0001	0.0001	0.0001	0.0001	0.0001	0.0001	0.0001	0.0001	0.0001	0.0001
Univariate tests of hypotheses for within-subject effects: probability for greater *F* statistic (adjusted by Greenhouse–Geisser epsilon)[b]										
TIME	0.0001	0.0001	0.0001	0.0001	0.0001	0.0001	0.0001	0.0001		
TIME*LOCATION	0.0001	0.0001	0.0001	0.0001	0.0001	0.0001	0.0001	0.0001		
TIME*SG	0.0001	0.0001	0.0001	0.0001	0.0001	0.0001	0.0218	0.0108		
TIME*SUBBLOCK(LOCATION*SG)	0.0001	0.0001	0.0001	0.0001	0.0001	0.0001	0.0001	0.0001		
TIME*SPECIES	0.0001	0.0001	0.0001	0.0001	0.0001	0.0001	0.0001	0.0001		
TIME*ORIGIN(SG*SPECIES)	0.0001	0.0001	0.0001	0.0001	0.0001	0.0001	0.0001	0.0001		
Greenhouse–Geisser epsilon	0.2961	0.5629	0.3222	0.5746	0.2061	0.5610	0.2010	0.2333		
Univariate tests of hypotheses for within-subject effects: probability for greater *F* statistic (adjusted by Greenhouse–Geisser epsilon)[b]										
TIME	0.0001	0.0219	0.0001	0.0001	0.0001	0.0001	0.0001	0.0001		
TIME*SG*SPECIES	0.0001	0.0876	0.0001	0.0005	0.0001	0.0001	0.0002	0.0001		
TIME*LOCATION*SG	0.1720	0.9502	0.0001	0.4455	0.0067	0.0638	0.0443	0.0001		
TIME*LOCATION*SPECIES	0.0001	0.0001	0.0001	0.0001	0.0001	0.0001	0.0001	0.0001		
TIME*LOCA*ORIGI(SG*SPECIES)	0.0001	0.0001	0.0001	0.0001	0.0001	0.0001	0.0001	0.0001		
Greenhouse–Geisser epsilon	0.6532	0.8677	0.7196	0.7899	0.6780	0.8992	0.6683	0.7962		

[a] No within-subject effects related to time were included for longevity and shoot borer attack because these effects were calculated only for ANOVAR.
[b] The Greenhouse–Geisser epsilon is given to indicate the lack of sphericity of the ANOVAR tests (see Methods).

Table 3.10. Significance of Effects for Repeated Measures Analysis of Variance (ANOVAR) for HT, δHT, DBH, δDBH, BA, δBA, TV, and δTV and Analysis of Variance (ANOVA) for Longevity and Total Shoot Borer Attack by Age 5 for All Provenances of Big-Leaf Mahogany

Source	HT	δHT	DBH	δDBH	BA	δBA	TV	δTV	Longevity[a]	Shoot Borer[a]
Tests of hypotheses for among-subjects effects: probability of a greater F statistic										
LOCATION	0.0001	0.0001	0.0001	0.0001	0.0001	0.0001	0.0001	0.0001	0.0001	0.0001
SG	0.0001	0.0001	0.6845	0.0818	0.0182	0.1352	0.6538	0.4366	0.8995	0.3487
SUBBLOCK(LOCATION*SG)	0.0001	0.0001	0.0001	0.0001	0.0001	0.0001	0.0001	0.0001	0.0001	0.0001
ORIGIN(SG)	0.0001	0.0001	0.0001	0.0001	0.0001	0.0001	0.0001	0.0001	0.0001	0.0001
MOTHER(SG*ORIGIN)	0.0007	0.0130	0.0001	0.0001	0.0006	0.0001	0.0108	0.0024	0.0006	0.0070
LOCATION*SG	0.9986	0.9380	0.9972	0.9179	0.9477	0.9592	0.6895	0.5451	0.7414	0.9970
LOCATION*ORIGIN(SG)	0.0001	0.0001	0.0001	0.0001	0.0001	0.0001	0.0001	0.0001	0.0001	0.0001
LOCA*MOTHE(SG*ORIGI)	0.9794	0.9300	0.9982	0.9395	0.9981	0.9821	0.9516	0.8393	0.8864	0.9204
Univariate tests of hypotheses for within-subject effects: probability for greater F statistic (adjusted by Greenhouse–Geisser epsilon)[b]										
TIME	0.0001	0.0001	0.0001	0.0001	0.0001	0.0001	0.0001	0.0001		
TIME*LOCATION	0.0001	0.0001	0.0001	0.0001	0.0001	0.0001	0.0001	0.0001		
TIME*SG	0.0010	0.0001	0.0001	0.0001	0.0002	0.0001	0.2157	0.2127		
TIME*SUBBLOCK(LOCATION*SG)	0.0001	0.0001	0.0001	0.0001	0.0001	0.0001	0.0001	0.0001		
TIME*ORIGIN(SG)	0.0001	0.0001	0.0001	0.0001	0.0001	0.0001	0.0001	0.0001		
TIME*MOTHER(SG*ORIGIN)	0.0003	0.2711	0.0001	0.0351	0.0014	0.0001	0.0178	0.0210		
Greenhouse–Geisser epsilon	0.3047	0.7153	0.3337	0.5435	0.2060	0.5426	0.2010	0.2323		
Univariate tests of hypotheses for within-subject effects: probability for greater F statistic (adjusted by Greenhouse–Geisser epsilon)[b]										
TIME	0.0001	0.0001	0.0001	0.0001	0.0001	0.0001	0.0001	0.0001		
TIME*LOCATION*SG	0.0001	0.0001	0.9995	0.9988	0.0001	0.0001	0.9990	0.9999		
TIME*LOCATION*ORIGIN(SG)	0.0001	0.0001	0.0001	0.0001	0.0001	0.0001	0.0001	0.0001		
TIME*LOCA*MOTHE(SG*ORIGIN)	0.9917	0.9993	0.9587	0.9995	0.9866	0.9954	0.9844	0.9745		
Greenhouse–Geisser epsilon	0.6457	0.8563	0.7006	0.7052	0.6721	0.8447	0.6525	0.8124		

[a] No within-subject effects related to time were included for longevity and shoot borer attack because these effects are calculated only for ANOVAR.
[b] The Greenhouse–Geisser epsilon is given to indicate the lack of sphericity of the ANOVAR tests (see Methods).

Table 3.11. Significance Levels of Effects for Repeated Measures Analysis of Variance (ANOVAR) for HT, δHT, DBH, δDBH, BA, δBA, TV, and δTV and Analysis of Variance (ANOVA) for Longevity and Total Shoot Borer Attack by Age 5 for All Provenances of Pacific Coast Mahogany

Source	HT	δHT	DBH	δDBH	BA	δBA	TV	δTV	Longevity[a]	Shoot Borer[a]
Tests of hypotheses for among-subjects effects: probability of a greater F statistic										
LOCATION*SG	0.0001	0.0001	0.0001	0.0001	0.0001	0.0001	0.0001	0.0001	0.0001	0.0001
SG	0.0003	0.0362	0.0029	0.0106	0.0156	0.0114	0.0652	0.0715	0.2418	0.8385
SUBBLOCK(LOCATION*SG)	0.0001	0.0001	0.0190	0.0005	0.0365	0.0036	0.0595	0.0096	0.1047	0.7709
ORIGIN(SG)	0.0001	0.0001	0.0001	0.0001	0.0001	0.0001	0.0001	0.0001	0.0001	0.0004
MOTHER(SG*ORIGIN)	0.0551	0.1183	0.0266	0.0496	0.0097	0.0103	0.1236	0.0988	0.0016	0.6603
LOCATION*SG	0.2070	0.0268	0.7833	0.0800	0.8410	0.5381	0.5535	0.2919	1.0000	0.9982
LOCATION*ORIGIN(SG)	0.0001	0.0001	0.0001	0.0001	0.0001	0.0001	0.0001	0.0001	0.0001	0.0007
LOCA*MOTHE(SG*ORIGI)	0.1588	0.6879	0.6418	0.8267	0.7826	0.8424	0.4732	0.3934	0.6462	0.9948
Univariate tests of hypotheses for within-subject effects: probability for greater F statistic (adjusted by Greenhouse–Geisser epsilon)[b]										
TIME	0.0001	0.0001	0.0001	0.0001	0.0001	0.0001	0.0001	0.0001		
TIME*LOCATION	0.0001	0.0001	0.0001	0.0001	0.0001	0.0001	0.0001	0.0001		
TIME*SG	0.0466	0.0001	0.0377	0.0255	0.0791	0.0821	0.1253	0.1107		
TIME*SUBBLOCK(LOCATION*SG)	0.0001	0.0001	0.0402	0.0125	0.0689	0.0001	0.0901	0.0368		
TIME*ORIGIN(SG)	0.0001	0.0001	0.0001	0.0002	0.0001	0.0001	0.0001	0.0001		
TIME*MOTHER(SG*ORIGIN)	0.0578	0.4657	0.0081	0.7529	0.0083	0.0091	0.1418	0.1754		
Greenhouse–Geisser epsilon	0.2738	0.7751	0.3551	0.5605	0.2085	0.7335	0.2015	0.2616		
Univariate tests of hypotheses for within-subject effects: probability for greater F statistic (adjusted by Greenhouse–Geisser epsilon)[b]										
TIME	0.0001	0.0001	0.0001	0.0001	0.0001	0.0001	0.0001	0.0001		
TIME*LOCATION*SG	0.0001	0.0001	0.8021	0.9997	0.9721	0.9978	0.9253	0.9894		
TIME*LOCATION*ORIGIN(SG)	0.0001	0.0001	0.0001	0.0001	0.0001	0.0001	0.0001	0.0001		
TIME*LOCA*MOTHE(SG*ORIGIN)	0.9986	1.0000	0.9993	0.9989	0.9969	1.0000	0.7517	0.9885		
Greenhouse–Geisser epsilon	0.6407	0.7996	0.6822	0.7555	0.6201	0.8079	0.5825	0.7705		

[a] No within-subject effects related to time were included for longevity and shootborer attack because these effects are calculated only for ANOVAR0.

[b] The Greenhouse–Geisser epsilon is given to indicate the lack of sphericity of the ANOVAR tests (see Methods).

Table 3.12. Significance of Effects for Repeated Measures Analysis of Variance (ANOVAR) for HT, δHT, DBH, δDBH, BA, δBA, TV, and δTV and Analysis of Variance (ANOVA) for Longevity[a] and Total Shoot Borer Attack by Age 5 for Small-Leaf Mahogany

Source	HT	δHT	DBH	δDBH	BA	δBA	TV	δTV	Longevity[a]	Shoot Borer[a]
Tests of hypotheses for among-subjects effects: probability of a greater F statistic										
LOCATION	0.0001	0.0001	0.0001	0.0001	0.0001	0.0001	0.0001	0.0001	0.0001	0.0001
Univariate tests of hypotheses for within-subject effects: probability for greater F statistic (adjusted by Greenhouse–Geisser epsilon)[b]										
TIME	0.0001	0.0001	0.0001	0.0029	0.0001	0.0001	0.0001	0.0001		
TIME*LOCATION	0.0001	0.0001	0.0001	0.0182	0.0001	0.0001	0.0001	0.0001		
Greenhouse–Geisser epsilon	0.2693	0.7348	0.3749	0.4609	0.2107	0.5712	0.2021	0.2733		

[a] No within-subject effects related to time were included for longevity and shootborer attack because these effects are calculated only for ANOVAR.
[b] The Greenhouse–Geisser epsilon is given to indicate the lack of sphericity of the ANOVAR tests (see Methods).

mahogany ($V_{G*E \ (TOTAL \ ESTIMATED)}$ = 29%; Table 3.7). For these traits in both species, most of V_{G*E} was due to the *LOCATION*ORIGIN* interaction. In Pacific coast mahogany, $V_{G*E \ (TOTAL \ ESTIMATED)}$ was reduced for both longevity (9%) and shoot borer attack (4%) compared to the other parameters, with the *LOCATION*ORIGIN* interaction accountable for the greater part (see Table 3.7). In big-leaf mahogany, $V_{G*E \ (TOTAL \ ESTIMATED)}$ was also small for both longevity (9%) and shoot borer attack (2%), with *LOCATION*ORIGIN* and *LOCATION*MOTHER* interactions accountable for similar amounts (Table 3.6).

Genetic Effects

The hierarchical levels of total genetic effects ($V_{G \ (TOTAL \ ESTIMATED)}$) included variation among *SPECIES*, variation among population of *ORIGIN* within species, and variation within populations ($V_{FAMILY \ (TOTAL \ ESTIMATED)}$: estimated as $4*V_{MOTHER}$, see Appendix; Table 3.4). For all species together, *SPECIES* and *ORIGIN* were almost always highly significant ($p < 0.0001$; Table 3.9), and *SPECIES* was usually 10 times greater than *ORIGIN* (see Table 3.5).

For big-leaf and Pacific coast mahogany, *ORIGIN* was always highly significant ($p < 0.0001$; Tables 3.10, 3.11). *MOTHER* was usually very significant in big-leaf mahogany ($p < 0.001$; Table 3.10) and usually significant in Pacific coast mahogany ($p < 0.05$; Table 3.11). The mean estimated total genetic effects ($V_{G \ (TOTAL \ ESTIMATED)}$) were more similar for big-leaf (11%; Table 3.6) and Pacific coast (15%; Table 3.7) mahogany in size and growth parameters by age 14 than for longevity (16% in big-leaf vs. 5.5% in Pacific coast mahogany; Tables 3.6, 3.7). Both V_{ORIGIN} and $V_{FAMILY \ (TOTAL \ ESTIMATED)}$ appeared similar for most traits in both species (Tables 3.6, 3.7). Total genetic effects for total shoot borer attack for both species were very low (less than 3%; Tables 3.6, 3.7).

Total Seed-Group Effects

Total seed-group effects ($V_{SEEDGROUP \ (TOTAL)}$) included SEEDGROUP, SEEDGROUP*SPECIES, and SEEDGROUP*LOCATION. These factors included confounded environmental and genetic aspects, as discussed in the Methods and Appendix sections. For all species together, these effects were often less significant ($p < 0.01$) than the other effects (see Table 3.9). For size and growth parameters by year 14, $V_{SEEDGROUP \ (TOTAL)}$ accounted for an average of 2% of the variation in all species together (Table 3.5) and also in big-leaf mahogany (Table 3.6). In Pacific coast mahogany for these parameters, $V_{SEEDGROUP \ (TOTAL)}$ was 15% of the total variation (Table 3.7). For total shoot borer attack, $V_{SEEDGROUP \ (total)}$ accounted for 3 to 4% of the total phenotypic variation over all species and within species in individually (Tables 3.5, 3.6, 3.7). Among and within species, $V_{SEEDGROUP \ (TOTAL)}$ accounted for 4% to 9% of total variation for longevity (Tables 3.5, 3.6, 3.7).

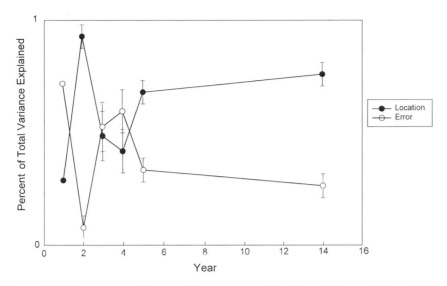

Figure 3.4. Percentage of the total variance explained by different effects (Table 3.3) for small leaf mahogany at the years of measurement. Effects have been averaged across all size and rate parameters. *Error bars* indicate standard error.

Changes with Time

From ANOVAR, significant within-subject interactions with time between years 1 and 14 (Tables 3.9, 3.10, 3.11, 3.12) indicated that the variation from main effects and interactions changed through time. For all species together, the interactions of *SEEDGROUP*, *SEEDGROUP*SPECIES*, and *LOCATION*SEEDGROUP* with time were less often highly significant at $p < 0.0001$ than the other main effects and interactions.

Changes in variance components through time revealed more detailed trends. Mean variance components across all size and growth parameters indicated that, over all species, seed-group effects (*SEEDGROUP*, *SEEDGROUP*SPECIES*, and *SEEDGROUP*LOCATION*) decreased with time, to a mean of 2% of total variance by year 14 (Fig. 3.1, Table 3.5). By species, big-leaf mahogany (Fig. 3.2, Table 3.6) showed an increase in largescale environmental effects. Pacific coast mahogany showed an increase in *LOCATION*ORIGIN* (Fig. 3.3, Table 3.7) and a decrease in *LOCATION*SEEDGROUP* interaction effects through time. Small-leaf mahogany showed no trend over time in variance components (Fig. 3.4, Table 3.8).

Clustering of Provenances

Growth and size traits from year 5 showed clear separation of most provenances of big-leaf and Pacific coast mahogany, and small-leaf clustered with

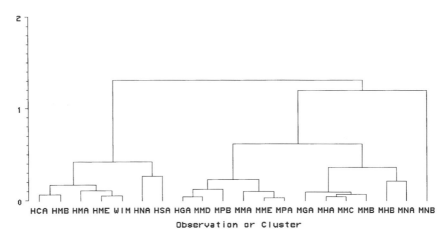

Figure 3.5. Tree of similarity among provenances from cluster analysis based on least square means of growth and size parameters from year 5. The *y* axis indicates relative separation of joined groupings. See Table 3.1 for key to provenances.

Pacific coast mahogany (Fig. 3.5). The exception was the Pacific coast mahogany provenance HGA from Guatemala, which other studies (Geary et al. 1973) found to be more similar in leaf morphology and growth characteristics to big-leaf mahogany. Traits in year 14 showed a stronger separation of big-leaf provenances MMB, MMC, MNA, and MNB than of all the others grouped together (Fig. 3.6). Most of the differences in pattern of provenance separation between years 5 and 14 appeared to result from changes in *TV* and *δTV*.*

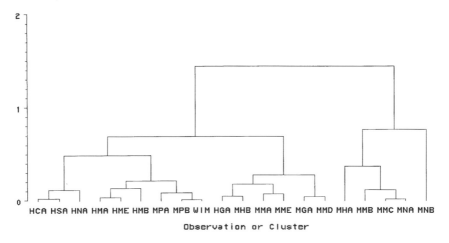

Figure 3.6. Tree of similarity from among provenances from cluster analysis based on least square means of growth and size parameters from year 14. The *y* axis indicates relative separation of joined groupings. See Table 3.1 for key to provenances.

Comparison of Genetic and Geographic Distances

The genetic distance between populations was based on all quantitative traits used in this study (see the Appendix) and indicated relative differences between population pairs. For big-leaf mahogany alone, the correlation between genetic and geographic distance was significant ($r^2 = 0.40$, $p < 0.05$), as well as for Pacific coast mahogany ($r^2 = 0.50$, $p < 0.05$).

Shoot Borer Attack

Shoot borer attack was less likely to differ significantly among provenances under different environmental conditions than the other parameters (Figs. 3.7c, 3.8c, 3.9c, 3.10c, 3.11c, 3.12c, *). In big-leaf mahogany, 69% of the phenotypic variance in total shoot borer attack was explained by environmental variance among locations in Puerto Rico (Table 3.6), and, in Pacific coast mahogany, this proportion was 51% (Table 3.7). Shoot borer attack correlations with HT in year 5 differed among locations, and tended to be positive. Correlations between cumulative attack in year 5 and HT in year 14 also differed among locations and could be positive.*

Least Square Means Analysis

Differences Among Species

Differences discussed in the following sections are significant at $p \leq 0.05$, unless otherwise indicated. Averaged across sites, the mean tree size and growth rates were always significantly greater for big-leaf than for small-leaf and Pacific coast mahogany (Fig. 3.13a–h, *). Small-leaf and Pacific coast mahogany were not different for size and growth parameters (Figs. 3.13a–h, 3.14a, *), but small-leaf mahogany died sooner than did Pacific coast mahogany (Fig. 3.14b, *). Small-leaf mahogany showed lower total borer attack than the other two species (Figs. 3.14c, 3.15c, 3.16c, *). All species showed maximum height and diameter growth rate in years 2 to 4, diminishing thereafter (Fig. 3.13b,d; not tested for significance). For big-leaf mahogany, the basal-area growth rate continued to increase with time (Fig. 3.13g; not tested) for all species, and total volume growth rates continued to increase with time (Fig. 3.13h; not tested).

Differences Among Provenances of Origin

Averaged across sites, size and growth parameters varied and were not significantly different among many provenances. In Pacific coast mahogany, provenance HGA (Table 3.1) was usually in the range of the performance of big-leaf mahogany in size, growth, and longevity (Figs. 3.7a,b, 3.8a,b, 3.9a,b, 3.10a,b, 3.11a,b, 3.12a,b, *). At year 14, provenance HGA consistently performed well, and provenances HAS and HNA were usually at or near the bottom (Fig. 3.8, *). Provenances HGA and HCA lived longest

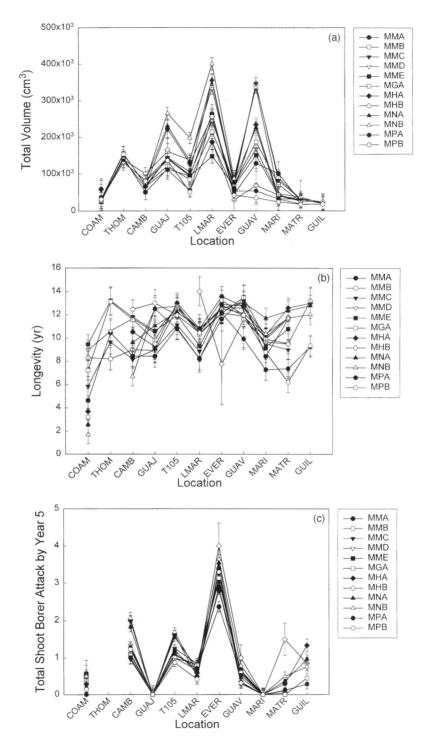

Figure 3.7a–c. Total volumes (TV) in year 14 (**a**), longevity (**b**), and total shoot borer attack by year 5 (**c**), based on least square means for the provenances of big-leaf mahogany across the study sites. Least square means are preferable to simple means (see Appendix). *Error bars* indicate standard error. See Table 3.2 for key to plantation sites. *Lines* connect points for ease of reading the graph.

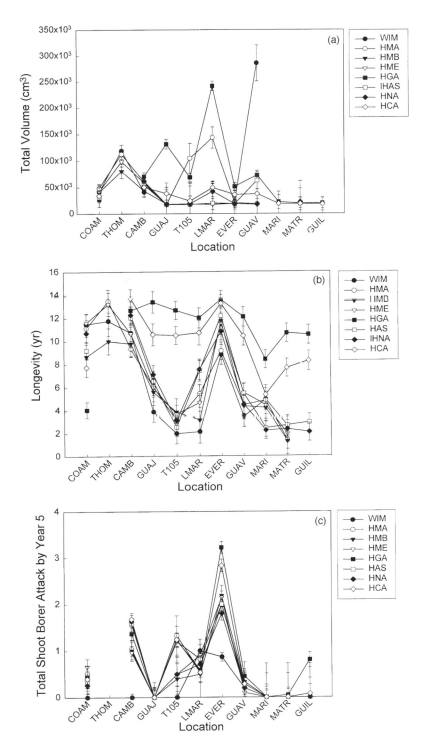

Figure 3.8a–c. Total volumes (TV) in year 14 (**a**), longevity (**b**), and total shoot borer attack by year 5 (**c**), based on least square means for provenances of small-leaf and Pacific coast mahogany across the study sites. Least square means are preferable to simple means (see Appendix). *Error bars* indicate standard error. See Table 3.2 for the key to plantation sites. *Lines* connect points for ease of reading the graph.

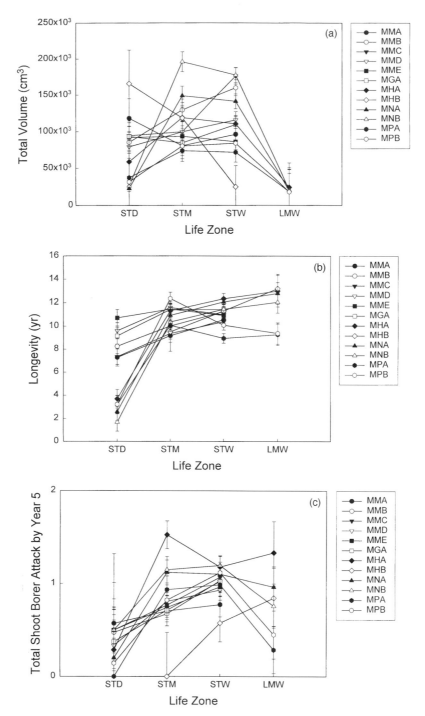

Figure 3.9a–c. Total volumes (TV) in year 14 (**a**), longevity (**b**), and total shoot borer attack by year 5 (**c**), based on least square means for the provenances of big-leaf mahogany across the life zones in this study. STD: subtropical dry, STM: subtropical moist, STW: subtropical wet, LMW: subtropical lower montane wet. Least square means are preferable to simple means (see Appendix). *Error bars* indicate standard error. *Lines* connect points for ease of reading the graph.

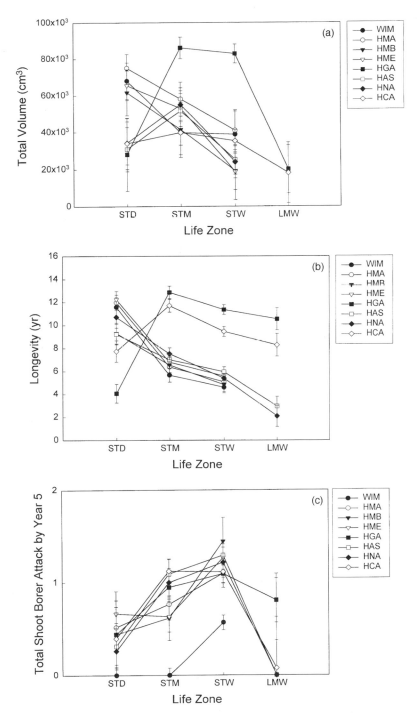

Figure 3.10a–c. Total volumes (TV) in year 14 (**a**), longevity (**b**), and total shoot borer attack by year 5 (**c**), based on least square means for the provenances of small-leaf and Pacific coast mahogany across the life zones in this study. Least square means are preferable to simple means (see Appendix). *Error bars* indicate standard error. *Lines* connect points for ease of reading the graph.

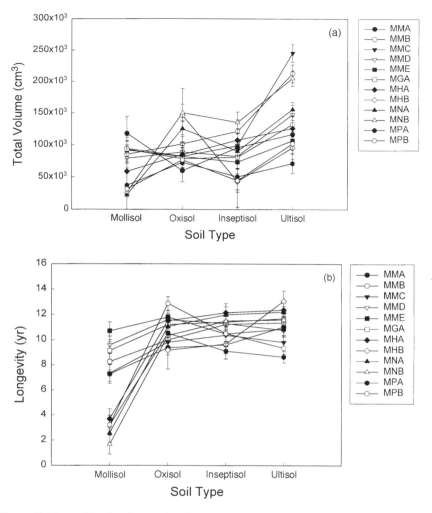

Figure 3.11a–c. Total volumes (TV) in year 14 (**a**), longevity (**b**), and total shoot borer attack by year 5 (**c**), based on least square means for provenances of big-leaf mahogany across the soil types in this study. Least square means are preferable to simple means (see Appendix). *Error bars* indicate standard error. *Lines* connect points for ease of reading the graph.

$(p < 0.001, *)$. No significant differences were found among provenances in shoot borer attack.

For big-leaf mahogany in year 14, the provenance MNB was always at the top, and MNA, MMB, and MMC were usually near the top for size and growth rate traits; provenances MPA and MPB performed poorly across most traits in most locations (Fig. 3.7a, 3.9a, 3.11a, *). No provenances were significantly different in longevity (Fig. 3.7b, 3.9b, 3.11b, *), but provenances MMA, MPA, and MPB had fewer shoot borer attacks than did MNB.

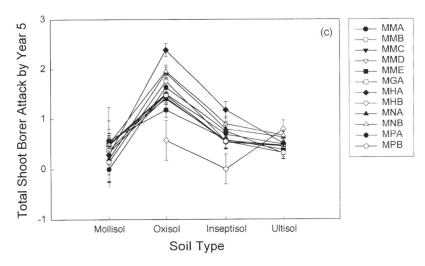

Figure 3.11a–c. *Continued.*

Location of Plantations in Puerto Rico

Averaged across all species at year 14, three sites—Guilarte, Matrullas, and Maricao—usually performed poorest in size and growth. Las Marias, Guavate, Guajataca, and Estate Thomas in St. Croix usually performed

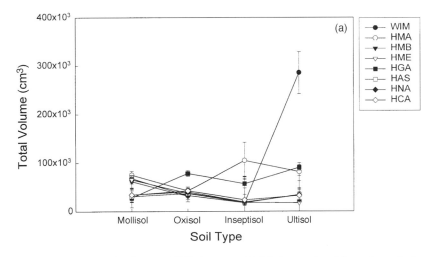

Figure 3.12a–c. Total volumes (TV) in year 14 (**a**), longevity (**b**), and total shoot borer attack by year 5 (**c**), for provenances of small-leaf and Pacific coast mahogany across the soil types in this study. Least square means are preferable to simple means (see Appendix). *Error bars* indicate standard error. One unusually large small-leaf mahogany tree remained in Ultisol soil at Guavate. *Lines* connect points for ease of reading the graph.

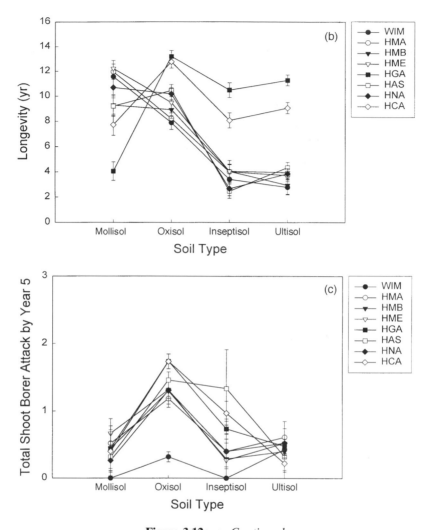

Figure 3.12a–c. *Continued.*

best. Longevity was best at El Verde, Estate Thomas, and Cambalache, and poorest at Matrullas, Maricao, and Coamo. Shoot borer attack was highest at El Verde (LEF; see Table 3.2), Cambalache, and Tract 105 (LEF; Table 3.2) and lowest at Matrullas, Maricao, and Guajataca.*

Sites were also characterized by life zone, elevation, and soil type. Soil orders and climate in Puerto Rico sites were to some extent confounded: the only sites that were in the subtropical dry life zone were also the only sites with Mollisol soils (Table 3.2); the other soil types were found at more than one life zone. The greatest longevity was found at subtropical moist and subtropical wet forest sites, and the greatest shoot borer attack was at

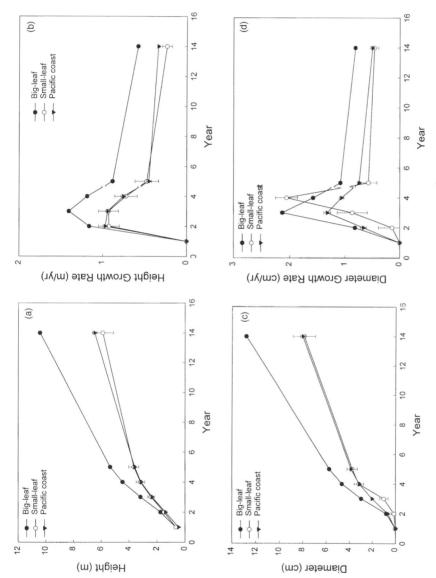

Figure 3.13a–h. Size and growth rate parameters through time for the three mahogany species based on least square means: (**a**) height, (**b**) height growth rate, (**c**) diameter, (**d**) diameter growth rate, (**e**) basal area, (**f**) basal area growth rate, (**g**) total volume, and (**h**) total volume growth rate. Least square means are preferable to simple means (see Appendix). *Error bars* indicate standard error.

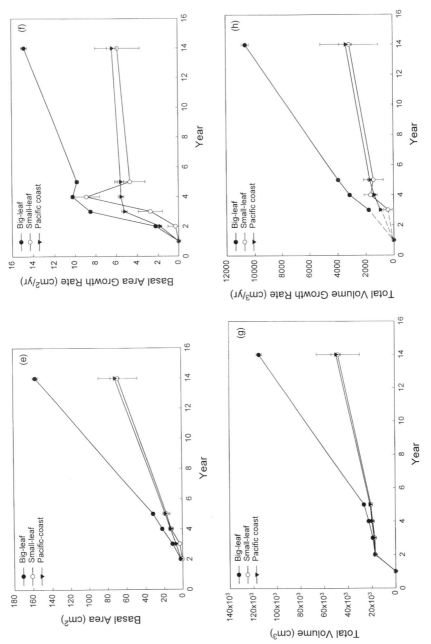

Figure 3.13a–h. *Continued.*

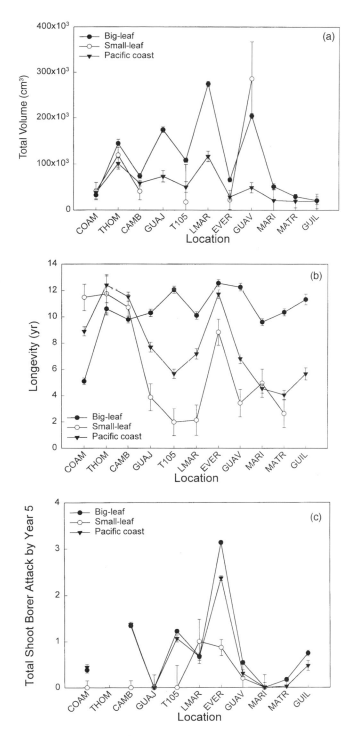

Figure 3.14a–c. Total volumes (TV) in year 14 (**a**), longevity (**b**), and total shoot borer attack by year 5 (**c**), based on least square means for the three mahogany species across the study sites. Least square means are preferable to simple means (see Appendix). *Error bars* indicate standard error. See Table 3.2 for the key to plantation sites. One unusually large individual of small-leaf mahogany remained at Guavate. *Lines* connect points for ease of reading the graph.

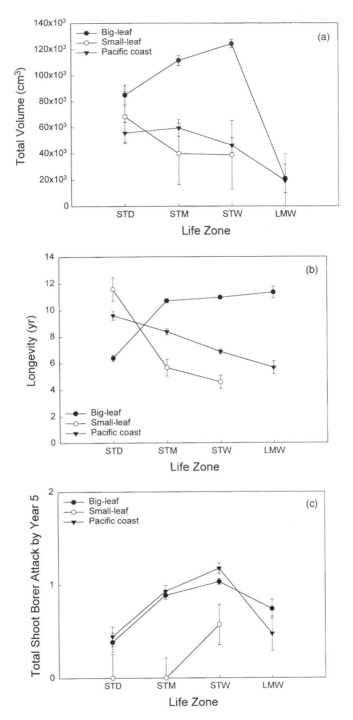

Figure 3.15a–c. Total volumes (TV) in year 14 (**a**), longevity (**b**), and total shoot borer attack by year 5 (**c**), based on least square means for the three mahogany species across the life zones in this study. Least square means can result in negative estimates of means, but they are preferable to simple averaging (see Methods). *Error bars* indicate standard error. Least square means are preferable to simple means (see Appendix). *Error bars* indicate standard error. *Lines* connect points for ease of reading the graph.

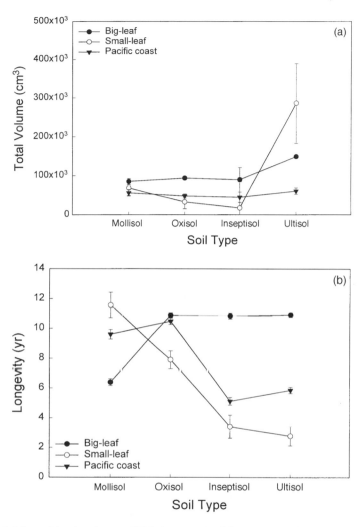

Figure 3.16a–c. Total volumes (TV) in year 14 (**a**), longevity (**b**), and total shoot borer attack by year 5 (**c**), based on least square means for species across the soil types in this study. Least square means are preferable to simple means (see Appendix). *Error bars* indicate standard error. One unusually large small-leaf mahogany tree remained in Ultisol soil at Guavate. *Lines* connect points for ease of reading the graph.

the subtropical wet forest sites. The poorest size and growth performance was at the lower montane wet forest site (Guilarte), and the lowest longevity and least shoot borer attack was at the subtropical dry forest sites.* By year 14, trees generally grew better on Ultisols than on the other soil types. Longevity was greatest on the Oxisols and least on Mollisols. Shoot borer attack was greatest on Oxisols and least on Mollisols.*

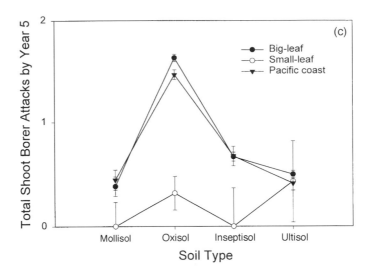

Figure 3.16a–c. *Continued.*

Interactions Between Species or Provenances and
Environmental Factors in Puerto Rico

Further investigation of location effects showed that more of the variance in size and growth traits by year 14 was due to life zone than to soil type (Table 3.13). Probably because of the imbalance in distribution of locations among life zone and soil categories, however, the life zone effects were seldom significant and soil-type effects were sometimes significant.* The interactions of life zone and soil type with species and provenance of origin were usually significant, but they accounted for much less of the variation (Table 3.13, *).

The differential performance of species under varying environmental conditions at the Puerto Rico plantation sites was also assessed. Big-leaf mahogany performed better in size and growth in the subtropical moist and subtropical wet sites (Fig. 3.15a, *), and showed the poorest longevity in the subtropical dry sites (Fig. 3.15b, *). Because of the small sample size, small-leaf mahogany seldom showed significant differences in size and growth among sites, but these were generally better in the dry sites (Fig. 3.15a, *), and the trees lived longest there (Fig. 3.15b, *). Pacific coast mahogany showed generally better performance and lived longer in the dry and moist sites (Fig. 3.15a,b, *). More shoot borer attack was seen in both big-leaf and Pacific coast mahogany in the wet than in the dry sites (Fig. 3.15c, *).

The differential performance of species across soil types was not as marked as across life zones. Big-leaf mahogany performed best across several parameters on Ultisols (Fig. 3.14a, *) and less well on Mollisols and Inceptisols; it died soonest on Mollisols (Fig. 3.16b, *). Small-leaf mahogany lived longest on Mollisols and Oxisols and shortest on Ultisols (Fig. 3.16b, *). Pacific coast mahogany grew best in diameter on Mollisols (*). This species lived longest on Mollisols and Oxisols and shortest on Inceptisols and Ultisols (Fig. 3.16b, *). Both big-leaf and Pacific coast mahogany had the highest shoot borer attack on Oxisols (Fig. 3.16c, *).

Correlations between plantation site elevation and size were usually negative for all three species.* Strong negative correlations in big-leaf ($r = -0.44$, $p < 0.0001$) and Pacific coast mahogany ($r = -0.43$, $p < 0.0001$) were found between shoot borer attack and elevation. Site elevation also showed negative correlations with longevity in big-leaf ($r = -0.43$, $p < 0.0001$) and Pacific coast mahogany ($r = -0.40$, $p < 0.0001$).*

Within big-leaf and Pacific coast mahogany, the ordering of provenance performance and longevity changed across locations, life zone, and soil type in Puerto Rico (Figs. 3.7a,b, 3.8a,b, 3.9a,b, 3.10a,b, 3.11a,b, 3.12a,b, *). Because the pairwise comparisons of least square means used the Scheffe *post hoc* criteria, however, they may have been too restrictive (Milliken and Johnson 1992). Thus, no significant differences in total volume were found between big-leaf provenances at age 14 in different life zones.* Total shoot borer attack also showed no differences in these comparisons.* Under the Scheffe criteria, several provenances showed significant differences for longevity comparisons among different life zones and soil types.* In particular, several of the provenances from big-leaf mahogany in SEEDGROUP 2 showed poorer longevity in the subtropical dry life zone

Table 3.13. Variance Components for the Mean of Height and Growth Measurements (HT, δHT, δBH, δDBH, BA, δBA, TV, and δTV) for Year 14, Longevity, and Total Shoot Borer Attack by Age 5, with LIFE ZONE and SOIL Considered as Separate Effects for All Species of Mahogany

Source of Variation	Mean Variance Component, Year 14	SD, Year 14	Longevity	Total Shoot Borer Attack
LIFE ZONE	0.2328	0.0976	0.0000	0.0000
SOIL	0.1393	0.0158	0.0550	0.2589
SG	0.0013	0.0036	0.0000	0.0000
BLOCK(LIFEZ*SOIL*SG)	0.0066	0.0126	0.0231	0.2143
SPECIES	0.0701	0.0321	0.0459	0.0000
ORIGIN(SG*SPECIES)	0.0009	0.0015	0.0185	0.0000
SG*SPECIES	0.0031	0.0040	0.0420	0.0000
LIFE ZONE*SG	0.0135	0.0079	0.0100	0.0000
SOIL*SG	0.0004	0.0008	0.0022	0.0170
LIFE ZONE*SPECIES	0.0272	0.0146	0.0128	0.1118
SOIL*SPECIES	0.0586	0.0201	0.1124	0.0805
LIFE*ORIGI(SG*SPECI)	0.0168	0.0084	0.0045	0.0346
SOIL*ORIGI(SG*SPECI)	0.0139	0.0094	0.0231	0.0133
ERROR	0.4156	0.1093	0.6504	0.2697

(confounded with Mollisol soil type) than in the other life zones (or soil types) (Figs. 12b, 15b, *). Provenances from big-leaf mahogany in SEEDGROUP 1, however, showed no significant differences in longevity between the dry (or Mollisols) and other life zones (or soil types) (Figs. 3.9b, 3.11b, *) even when Estate Thomas (where only SEEDGROUP 1 was planted) was omitted from analysis.

In Pacific coast mahogany, HGA showed poorer longevity in the dry than in the moist and wet zones (Fig. 3.10b, *), and on Mollisols than on the other soil types (Fig. 3.12b, *), even with Estate Thomas excluded. The provenance HMA lived longer in the dry than in the wet zone (Fig. 3.10b, *), and several provenances showed greater longevity on Mollisols and Oxisols than on Inceptisols and Ultisols (Fig 3.12b), even with Estate Thomas excluded. Total volume at age 14 and total shoot borer attack showed few or no significant differences among provenances, by life zone or soil type.*

Potential Selection Pressures

Potential selection pressures were investigated using correlations between performance measures and environmental conditions at sites of origin and plantation sites in Puerto Rico (see the Appendix). In plantation sites in the subtropical dry life zone, big-leaf and Pacific coast mahogany showed small (usually $r < 0.30$) but significant positive correlations between dry season length at site of origin and performance for some size and growth parameters in Puerto Rico. Similarly, in plantation sites in Puerto Rico in the subtropical moist, subtropical wet, and lower montane wet life zones, both species had small (usually $-0.30 < r < 0$) but significant negative correlations between dry season length at site of origin and performance measures. Pacific coast mahogany had larger and more significant negative correlations.* By life zone, this species showed more significant negative correlations than did big-leaf mahogany between elevation at site of origin and performance for at least some size and growth parameters in Puerto Rico.*

Again, for both species, more significant positive correlations were found between latitude at site of origin and performance for most size and growth parameters in subtropical dry and lower montane wet, than in subtropical wet and subtropical moist life zones.* The same results were obtained using the difference between the latitude at site of origin and the latitude of Puerto Rico.*

Adaptive Strategies

A positive correlation between the coefficient of variation of a trait (CV; see Appendix) as an indicator of plasticity, and the mean value of fitness related traits imply that plasticity is adaptive (Schmitt 1993). In this study, both big-leaf and Pacific coast mahogany usually showed negative correlations between trait CVs and fitness-related traits.* In addition, no provenance displayed superior performance across all locations, life zones, or soil types (see Figs. 3.7, 3.8, 3.9, 3.10, 3.11, 3.12, *). Negative correlations between performance in different environments may indicate specialization to a given habitat. More evidence of specialization to wet and dry life zones was found for Pacific coast mahogany, with stronger negative correlations between performance in subtropical dry and the wetter life zones than for big-leaf mahogany.*

Discussion

Hierarchical Comparisons

Hierarchical comparisons included environmental effects among and within plantation sites, and genetic and genetic-by-environment interactions among species, among provenances, and within-provenance variance components. Variance components should be treated as causes of variation among individuals, not as causal factors of the individual phenotype (Sultan 1987). For all species together, by year 14, the largest mean variance components for size and growth were those resulting from SPECIES and ERROR (see Fig. 3.1). Thus, most of the variance was partitioned among species or as part of the error term, and remained within families and small-scale environmental effects. For both big-leaf and Pacific coast mahogany in size and growth by year 14, the mean local-scale environmental variance was among the largest variance components, but big-leaf mahogany showed more large-scale environmental variation and Pacific coast mahogany showed more genetic-by-environment interaction than did big-leaf mahogany (Figs. 3.2, 3.3).

Total Environmental Versus Total Genetic Effects

For big-leaf mahogany longevity, shoot borer attack, and size and growth at age 14, the mean total environmental effects ($V_{E+e \ (TOTAL \ ESTIMATED)}$) accounted for more variance than the mean total genetic ($V_{G \ (TOTAL \ ESTIMATED)}$) and genetic-by-environment interaction effects together (V_{G*E}; see Table 3.6). In Pacific coast, $V_{E+e \ (TOTAL \ ESTIMATED)}$, and $V_{G*E \ (TOTAL \ ESTIMATED)}$ and $V_{G \ (TOTAL \ ESTIMATED)}$ together accounted for more similar amounts of variation (Table 3.7). More pronounced, $V_{E+e \ (TOTAL \ ESTIMATED)}$ accounted for 92.7% in big-leaf mahogany and 93.5% in Pacific coast mahogany of the variation in the 5-year cumulative shoot borer attack in the provenance trial. Average heritabilities for height ($h^2 = 0.28$) and diameter ($h^2 = 0.23$) in forest trees (Cornelius 1994) indicate that environmental variation is greater than that caused by genetic background in these traits. Other plant studies (Silander 1985; Ward 1996) found that the environmental variance for quantitative traits was usually greater than the genetic and genetic-by-environment effects.

In general, plants have a large amount of phenotypic plasticity (Bradshaw 1965). Fitness in plants is strongly correlated with biomass, which is extremely plastic (Sultan 1987). In consequence, most variations in longevity and reproduction in plants are induced by environmental rather than genetic differences (Sarukhan et al. 1984). By year 14, big-leaf mahogany showed a higher proportion of variation from $V_{E+e \ (TOTAL \ ESTIMATED)}$ in Puerto Rico than did Pacific coast mahogany (Tables 3.6, 3.7; Figs. 3.2, 3.3). Because of a more labile phenotype, big-leaf mahogany appeared to be more under the influence of the environment than did Pacific coast mahogany. The broad ecological tolerance of mahogany could be in part the result of adjustment of the phenotype. This adjustment may result from phenotypic plasticity in some traits producing stability in others. Miscellaneous genetic stocks of mahogany show a low failure rate when planted in different life zones in Puerto Rico (J. Francis, USDA Forest Service, Río Piedras, PR, personal communication). This result suggests that phenotypic adjustment to local conditions plays a significant role in niche breadth in the species.

In mahogany, the adjustment would probably result from plasticity in traits not measured in this study, because our evidence did not indicate that plasticity in size and growth rates was adaptive. At 14 years of age, plasticity was seldom positively related to fitness parameters.* Thus, plastic variation in these traits appeared to be caused by unavoidable environmental effects resulting from a limited capacity to stabilize the phenotype. Much of the phenotypic variation seen in other species is probably similarly unavoidable (Bradshaw 1965). In annual bluegrass, for example, plasticity in allocation and biomass parameters were not found to be adaptive (Ward 1996). The potential capacity for phenotypic adjustment in mahogany needs further investigation because it may result from other traits.

Large-Scale Versus Small-Scale Environmental Variation

The scale with the most plastic variation indicates where the environment most affects plant growth. We consider the local scale of the microhabitat to be the immediate vicinity of an organism and related to its size (for example, the local environment would be much smaller in seedlings than in adult plants). Large-scale environmental differences include gross differences of climate and soil type. For size and growth of big-leaf mahogany by year 14, the large-scale environmental variation (V_E) was somewhat larger than the estimated small-scale environmental variation ($V_{e \ (TOTAL \ ESTIMATED)}$; see Table 3.6). For Pacific coast mahogany, however, $V_{e \ (TOTAL \ ESTIMATED)}$ was much larger than V_E for size and growth (Table 3.7). Longevity and shoot borer attack showed other patterns. Ward (1996) found more phenotypic variance resulting from trait variation within than among environments in annual bluegrass, using treatments simulating contrasting environments. Silander (1985) also found more environmental variance within than among environments in *Spartina patens* across three habitats. Ward (1996) concluded that although adaptation to geographically separated environments may be largely due to genetic differentiation, and that adaptive plasticity may be more important at the smaller scale because it is where the same genetic background may be more likely to encounter different environments. However, our analysis did not examine the adaptiveness of plasticity on the local scale, and found no evidence of adaptiveness for large scale plasticity for the traits examined in mahogany.

Although plasticity may be adaptive for some species, such environmentally induced variation reduces the rate of response to selection by reducing heritability of traits. Phenotypic plasticity at the macroenvironmental scale can retard differentiation among populations as a response to selection. Although the local scale affects plant growth, genetic adaptation may not be possible because of limits to local adaptation imposed by microhabitat unpredictability. Plasticity would not affect evolutionary change resulting from genetic drift. The relative roles of large- and small-scale environmental variation in niche width need to be further investigated in plants.

Genetic-by-Environment Interactions

The genetic-by-environment interactions represent differences among genetic sources in the capacity to adjust to environmental change. The genetic-by-environment variance components provide a rough measure of the heritability of plasticity (Scheiner and Lyman 1989) and thus the potential for plasticity to respond to selection (Via and Lande 1985). For size and growth traits by year 14, the LOCATION*SPECIES interaction was about twice as large as the LOCATION*ORIGIN interaction but lower than the genetic component attributable to species (see Table 3.5). This finding indicated that these species have distinct responses to large-scale environmental variation.

For most species, the amount of genetic variance is greater than the genotype-by-environment interaction (Scheiner 1989). In contrast, in Pacific coast mahogany, by age 14 in size and growth, $V_{G*E\ (TOTAL\ ESTIMATED)}$ appeared greater than $V_{G\ (TOTAL\ ESTIMATED)}$ effects (see Table 3.7). The total genetic-by-environment interaction was greater for Pacific coast (Table 3.7) in most traits than for big-leaf mahogany (Table 3.6), indicating that Pacific coast mahogany has greater potential to evolve a plastic response to these different environmental conditions. For both species, the $V_{G*E\ FAMILY\ (TOTAL\ ESTIMATED)}$ appeared smaller than the LOCATION*ORIGIN interactions, indicating that less genetic variation for plasticity remains within than among populations. The size of genetic-by-environment interactions was partially related to the range of environments across which these species were planted in Puerto Rico. This range resulted in a greater differential response of different populations and families within populations than if a narrower range of environments had been used.

Genetic Variation

Variation Among Species

The mahogany genus has been viewed as consisting of three poorly defined species (Helgason et al. 1996). All three have been found to overlap in chromosome number and morphological traits such as leaf size (Helgason et al. 1996), and the species apparently hybridize readily when growing together (Whitmore and Hinojosa 1977). In our study, variance component analysis (see Table 3.5) indicated about 10 times more variance among species than among populations in size and growth parameters by age 14. The particular populations used for comparisons may accentuate the differences among species. In this study, however, the criteria for population selection were neutral, because they are based on even coverage of the selected parts of the species distributions, not on matching particular species characteristics. Two of the Pacific coast provenances, HGA and HCA, had growth characteristics intermediate between it and big-leaf mahogany (Geary et al. 1973), indicating that the criteria for population selection were neutral. The provenances HGA and HCA occurred in Guatemala and Costa Rica, where the natural ranges of the two species overlap.

Cluster analyses (Figs. 3.5, 3.6) at different ages provided a picture of the relationships among provenances. Size and growth traits from year 5 showed greater similarity among small-leaf mahogany and Pacific coast mahogany provenances than among big-leaf mahogany. The exceptional provenance was HGA of Pacific coast mahogany, which Geary et al. (1973) found to be more similar in leaf morphology and growth characteristics to big-leaf specimens. In year 14, the greater separation of big-leaf provenances MMB, MMC, MNA, and MNB from all other provenances of the three species combined may be related to growth patterns, especially in total volume and volume growth rate. By year 14, the spacing between trees on the study sites was probably starting to close. With canopy closure, the differences between the faster- and slower-growing provenances probably became more pronounced.

Our analysis is based on quantitative traits influenced by a multitude of genes. Because of the close association between size, growth, and fitness, these traits also have probably been subjected to natural selection. Genetic markers, such as those revealed by isozyme electrophoresis and DNA analysis, are considered to be less subject to selection than are quantitative traits (Muona 1990), and to reflect evolutionary relationships among groups (Petit et al. 1997). Abrams (1995) found that the three species and the hybrid of big-leaf and small-leaf mahogany could be distinguished with RAPDs (random polymorphic DNA) markers; however, they did not assess the relative differentiation among populations within species. On the other hand, Helgason et al. (1996) found that DNA from all three species tested with RAPDs did not clearly differentiate them, which may have been an artifact of including hybrids in their analysis.

The results of our study indicated that the relations among growth parameters for these species were complex and depended on both the traits and the timeframe considered. The inconclusiveness of trait distinction among species, from both DNA analysis and quantitative traits, points to the need for further clarification of the actual differentiation among these species. The potential hybrid zone between neighboring populations of big-leaf and Pacific coast mahogany, in particular, should be sampled and genetic markers compared for potential gene flow. The more distinct the gene pools of these species are, the greater the imperative to maintain sufficient gene banks of each of them.

Variation Among Populations

For both big-leaf and Pacific coast mahogany, the estimates of variation among populations based on size and growth traits by year 14 from our study (Tables 3.6, 3.7, Figs. 3.2, 3.3) are about half the total genetic variance detected within species. This fraction is greater than more general estimates for tropical species based on isozyme data (11%; Loveless 1992). Using isozymes, Loveless and Guillison (Chapter 2, this volume) found almost no differentiation among populations of big-leaf mahogany from Bolivian Amazonia in relatively uniform habitat and separated by distances of 50 to 200km. In a survey of big-leaf mahogany populations from México to Panamá, based on RAPD markers, Gillies et al. (1999) also found a much smaller fraction of total genetic variation among (20%) than within (80%) populations.

Two factors probably contribute to the apparent lack of agreement between the results found by our study and those based on genetic markers. In trees, quantitative traits, such as those considered in our study, show more genetic differentiation among populations than do genetic markers (Muona 1990), perhaps because they are more likely to be subject to selection (Bonnin et al. 1996). Distance among populations sampled probably also contributed to discrepancils between our results and those of Loveless and Gullison (Chapter 2, this volume). The populations in our study were separated by 160 to 1600km (Barres 1963). A comparison of genetic distance and geographic distance between populations indicated that, for both species, as geographic distance between populations increased, populations could be expected to be more genetically distinct.* The consequences of this effect for conservation are discussed later. If gene flow decreases with fragmentation, the genetic differentiation seen among the populations used in our study is also likely to be accentuated through time as the habitats at the sites of origin of these populations have become more fragmented.

Both genetic effects and genetic-by-environment interactions among populations were significant for the provenances used in our study (Tables 3.9, 3.10, 3.11). These results indicated genetic differentiation among populations in the face of the high gene flow rates measured by other investigators (Loveless and Gullison, Chapter 2, this volume), at least on a geographic scale. High rates of gene flow could be expected to retard differentiation; it increases the uncertainty about the environmental conditions that a given genotype will experience. Therefore, if gene flow is high between environments, selection must be strong in each generation to maintain differentiation. A high gene flow rate would be expected to encourage selection for phenotypic adjustment to a range of environments. The high gene flow also suggested that genetic drift was not important in the differentiation of local populations. Alternatively, natural barriers to gene flow may exist in some parts of the range of mahogany that reduce gene flow and contribute to the genetic differentiation seen among some of these populations.

The lower proportion of genetic differentiation and genetic-by-environment variation among populations of big-leaf than of Pacific coast mahogany may indicate higher gene flow among big-leaf populations. Populations of Pacific coast mahogany may be more specialized and isolated, and big-leaf mahogany may be more selected to survive across an array of habitats.

Family-Level Variation

Family level, or mean genetic variation within provenances ($V_{FAMILY\ (TOTAL\ ESTIMATED)}$), accounted for 6% of total phenotypic variation for big-leaf mahogany (see Table 3.6) and 9% for Pacific coast mahogany in size and growth traits by age 14 (Table 3.7), and somewhat less in longevity and shoot borer attack. For comparison, Ward (1996) found that 16% of the total phenotypic variation, or about 40% of the total genetic variation, was at the family level in populations of annual bluegrass, a highly inbreeding species.

Changes in Variance Components over Time

Variance components for different traits may assume different patterns of change over time. Big-leaf mahogany showed a greater increase in large-scale environmental effects (see Table 3.6, Fig. 3.2), Pacific coast mahogany showed a greater increase in LOCATION*ORIGIN interaction effects through time (Table 3.7, Fig. 3.3), and small-leaf mahogany (Table 3.8, Fig. 3.4) showed no trend. During the early years of a trial, provenance effects usually increase and family-level effects decrease with age. Later, with increased competition between trees, family-level effects may again become more pronounced (Xie and Ying 1996). We did not find these trends in our analyses.

We found a general decline in the total SEEDGROUP-related effects through time, indicating that the differences in population composition and in timing of harvest, planting, and measurement became less important with time. These changes help to justify the lagging of SEEDGROUP 1 by 1 year when we directly compared growth between the populations in SEEDGROUP 1 and SEEDGROUP 2.

Relation of Size and Growth Characteristics to Environmental Factors

The interactions between genetic levels (species and provenances) and the plantation sites in Puerto Rico showed the relative performance of genetic sources across environments. In turn, the relation between environmental factors at the sites of origin and provenance performance may indicate factors (or correlated factors) selecting for different growth patterns. Because size and growth rates are proportional to reproductive capacity in most plants, these traits also indicate potential fitness of individuals and provenances under different conditions. The degree of potentially adaptive change in size and growth rates across different environments provides some indication of the range of genetic diversity to consider preserving in these species and provenances.

Species-Scale Comparisons

In our study, big-leaf mahogany performed best and lived longest, and small-leaf mahogany had the least shoot borer attack.* Previous work based on the Puerto Rican provenance study also found that big-leaf mahogany showed better growth and longevity than do either of the other species across most plantation sites (Geary 1969; Geary et al. 1973; Glogiewicz 1986).

At the species scale, although soil type was more often significant than life zone, more of the variation among locations in size and growth appeared to be a result of life zone rather than soil type (see Table 3.13). The poor replication for certain categories of life zone and soil type and the great variation possible within a soil order, however, did not allow more general conclusions about the relative importance of soil and climate in the performance of these species and provenances. Effects were confounded between climate and soil type because the soils in the driest sites (Coamo and Estate Thomas) were both Mollisols, and wetter sites were a scattered array of the other three soil types. Thus, we could not differentiate the effects of climate and soil in the good performance of small-leaf and poor performance of big-leaf mahogany on dry sites.

By species, big-leaf mahogany performed best in moist and wet zones, Pacific coast mahogany in moist and dry zones, and small-leaf mahogany in dry zones (see Fig. 3.15). On dry sites, neither Pacific coast nor small-leaf mahogany was significantly greater in growth or size than big-leaf mahogany. Analyses based on year 5 alone (Geary et al. 1973) and a more limited data set in 1986 (Glogiewicz 1986) showed similar findings. Pacific coast and small-leaf mahogany showed a greater mean longevity than big-leaf mahogany on dry sites (Fig. 3.15). Some overlap between big-leaf and Pacific coast provenances was found on the moist and wet forest sites (Geary et al. 1973; this study).

All the species of mahogany performed poorly at Guilarte, which is in the lower montane wet life zone, and at a higher elevation than the normal range of mahogany. The three wettest sites [with annual rainfall of 2460 mm (mean), Maricao; 3000 mm (1999 only), Matrullas; and 2330 mm (1999 only), Guilarte, at the nearest stations] (NOAA 1999) appeared to fall outside the optimal range for the big-leaf provenances used in this study. Small-leaf mahogany is adapted to well-drained soils in areas where annual rainfall is between 800 and 1800 mm (Marrero 1950b); Marrero (1950a) recommended this species for drier, degraded sites.

Performance of species on different soil types generally corresponded with the expectations of Medina and Cuevas (Chapter 7, this volume). The relatively better performance of big-leaf mahogany on Ultisols than on Mollisols (Fig. 3.16a,b *) corresponds with their finding that this species is sensitive to drought and less tolerant of the calcium found in well-drained soils such as Mollisols (Bonnet Benitez 1983). We found that small-leaf mahogany lived longer on Mollisols than on Ultisols (Fig 3.16b, *). This finding agreed with their hypothesis of this species greater drought tolerance and sensitivity to the high iron or aluminum concentrations found in Ultisols. In our study, however, small-leaf mahogany also lived longer on Oxisols than on Ultisols. Oxisols are also high in iron and aluminum but tend to be better drained than Ultisols (Bonnet Benítez 1983). Although the performance of Pacific coast mahogany in this study was poorer on Inceptisols and Ultisols. Medina and Cuevas (Chapter 7, this volume) considered the growth patterns of this species to be less interpretable relative to soil characteristics.

Provenance-Scale Comparisons

Other studies (Geary et al. 1973; Glogiewicz 1986) have interpreted the two Nicaraguan provenances of big-leaf mahogany, MNA and MNB, to show the best growth. We found a more complex picture. In year 5, provenance performance depended on the growth or size trait considered*. By year 14, however, for all locations averaged together, MNB was always at the top, and MNA, MMB, and MMC were usually near the top in size and growth rate.* Provenances with the best growth in year 14 were not necessarily those that lived longest or had the least shoot borer attack.* Although Geary et al. (1973) found no evidence of geographic variation among provenances in susceptibility to borer attack in big-leaf mahogany, our analysis showed that MNB was attacked more often.* According to Geary et al. (1973), HGA (from Guatemala) and HCA (from Costa Rica) showed characteristics more similar to big-leaf mahogany (HGA more so than HCA) than did other Pacific coast mahogany provenances in height growth, longevity, and number of leaflets. We found HGA, but not HCA, to be more similar to big-leaf than Pacific coast mahogany in growth characteristics.

As indicated by significant LOCATION*ORIGIN interactions (see Tables 3.5, 3.6, 3.7), performance patterns of mahogany provenances differed across locations, climates, and soil types in Puerto Rico (Figs. 3.7, 3.8, 3.9, 3.10, 3.11, 3.12, *). The Scheffe criteria for establishing significance may have been too restrictive (Milliken and Johnson 1992) to detect differences in growth between provenances. However, we found that different provenances of both big-leaf and Pacific coast mahogany showed distinct patterns of longevity between life zones and soil types. Big-leaf mahogany provenances in SEEDGROUP 2 were significantly shorter lived in the subtropical dry life zone (confounded with Mollisol soil type) than in wetter life zones (or other soil types), but SEEDGROUP 1 provenances were not (Fig. 3.9b, *). This difference held even when Estate Thomas (where only SEEDGROUP 1 had been planted) was excluded. On the remaining subtropical dry site, Coamo, great differences in the conditions for planting existed between years for the two seed groups, but SEEDGROUP 1 had endured a bad drought (although it was watered for the first 4 months). This drought could have been expected to increase mortality in SEEDGROUP 1 but did not. Some of the difference between the seed groups is thus probably related to their genetic difference. Although the differences were not significant, MNA and MND of big-leaf mahogany were poor performers in size and growth parameters in the subtropical dry zone and at the top in the wetter zones (Fig. 3.16b, *). In Pacific coast mahogany, HGA had shorter longevity in the subtropical dry zone (and Mollisols) than in the subtropical moist and subtropical wet zones (and other soil types); several other provenances showed the opposite pattern (Fig. 3.10b, *).

Thus, some provenances of both Pacific coast and big-leaf mahogany showed specialization to different life zones. Geary et al. (1973) also detected that provenances of these two species, which performed better in the dry areas, were also poor performers in wetter areas. This differential response suggested extension of the phenotype beyond that possible by plasticity, and that the wide geographic range can be attributed in part to genetic differentiation among populations.

For both species, we found significant positive correlations between provenance performance in the subtropical dry life zone and dry season length at site of origin. In the wetter life zones, these correlations were negative and stronger in Pacific coast than in big-leaf mahogany.* These correlations indicated adaptive differentiation related to dry season length. Differences in correlations between species are probably caused by the greater genetic-by-environment interaction in Pacific coast than in big-leaf mahogany. A higher genetic-by-environment interaction may result from adaptation to more specialized environmental conditions.

Shoot Borer Attack

We found that more big-leaf than Pacific coast mahogany trees were attacked by shoot borers; small-leaf mahogany was relatively immune. In contrast, in Haiti where the small-leaf species is native, it is attacked more by shoot borers than is big-leaf mahogany (J. Whitmore, USDA Forest Service, Washington, DC, personal communication).

Within species, our analysis showed that the environment had much larger effects on shoot borer attack than did genetic background (see Tables 3.6, 3.7). Much of the phenotypic variation in shoot borer attack for both big-leaf and Pacific coast mahogany was explained by differences among plantation sites in Puerto Rico (Tables 3.6, 3.7). Part of this environmental effect appears to be related to differences among life zones: significantly more shoot borer attack was found in both big-leaf and Pacific coast mahogany in the subtropical wet sites than in the subtropical dry sites (Fig. 3.15c, *). Shoot borer attack also appeared to be heaviest at sites with Oxisol soils (Figs. 3.15c, 3.16c, *).

Geary et al. (1973) considered that much of the difference in growth among locations was associated with varying rates of shoot borer attack. Among locations, big-leaf mahogany growth was greatest where shoot borer attack was reduced (Geary et al. 1973). Shoot borer attack at a given site may have been affected by the initial amount of shading (Geary et al. 1973). Our results indicated that the relationship between attack and growth varied among sites.* Within sites, the shoot borer tended to attack taller trees but did not cause changes in provenance height ranks.* Provenances MNA and MNB experienced more attack than most others, and they were also among the taller ones. Pacific coast provenances did not differ significantly in rate of shoot borer attack.

Potential Correlates of Selection Pressures

Genetic variation in traits that can be related to environmental gradients has been interpreted as evidence for natural selection (Endler 1977). Our study showed weak, but significant, correlations between length of dry season, altitude, latitude, and difference in latitude from Puerto Rico, at site of origin, and performance among provenances in Puerto Rico.* These factors may not themselves have been the selective factors but rather correlates of unmeasured variables. Earlier, Geary et al. (1973) had concluded that differences among the seed sources used for our study in seedling characteristics, survival, shoot borer attack, growth rate, and leaf patterns could not be related to geographic patterns.

Within each species, we found that geographic distance between provenances was significantly correlated with genetic distance. Greater genetic separation with geographic distance probably resulted from a combination of progressively reduced gene flow and contrasting environmental conditions. In general, the degree of genetic differentiation among populations probably depends on the consistency and degree of difference in environmental conditions at different locations (McNeilly 1981; Platenkamp 1990, 1991).

Strategies

Species may exhibit different suites of adaptive traits that can be defined as strategies. These strategies include generalist adaptations to a range of environments and specialization to a given environment. A given genotype may have the properties of a generalist if it exhibits high fitness across environments. Consistent high fitness may result from adaptive plasticity. The current analysis showed no evidence for a generalist strategy. For big-leaf and Pacific coast mahogany, no provenance performed consistently well across all environments (see Figs. 3.7, 3.8, 3.9, 3.10, 3.11, 3.12, *). In addition, the correlation between provenance coefficients of variation (*CV*) and means for traits related to fitness were usually negative when significant.* This finding indicated that plasticity was not generally adaptive. Evidence of trade-offs in performance between wet and dry environments were found, however, especially for Pacific coast mahogany, which suggested specialization for these different conditions.

Conservation of Mahogany and Recommendations for Future Research

Relevance of the Puerto Rican Provenance Study to Natural Populations of Mahogany

The method of selecting provenances and mother trees for our study probably does not seriously limit the applicability of this study to mahogany conservation, at least in its native range in México and Central America. Provenances were selected to cover the geographic range evenly within the constraints of accessibility for sampling. Mother trees lacking obvious defects were chosen for seed harvest, which may have reduced within-provenance genetic variability and skewed trait means toward the high end. However, we encountered significant among-provenance genetic variation, indicating that important differences were not masked by how mother trees were selected.

Also of concern is the relevance of an *ex situ* provenance study to genetic conservation in the home range of the sampled populations. Provenance performance away from the environment of origin has been argued to be relevant to the home site, if the environmental conditions are sufficiently similar between those sites (Lindgren 1986). Puerto Rico lies within the same range of annual rainfall and latitude found in the mahogany habitats of México and Central America. Even if trait expression deviated in these provenances between Puerto Rico and their sites of origin, the detection of

significant genetic variation among sources planted in Puerto Rico indicated that among-provenance variation should be taken into account for
genetic conservation.

Plasticity and Genetic Conservation

Besides assuring overall genetic variation, conservation of genetic resources
would also maintain the genetic basis of plasticity. Genetic variation for
plasticity helps to buffer a species against future environmental change. In
mahogany, greater genetic-by-environment interactions among, rather than
within, populations indicated more of this buffering capacity existed at the
population level. In addition to potential for evolutionary change, the
genetic-by-environment interactions also indicated current differential
responses on the part of provenances to various environmental conditions.
In particular, the provenances in our study showed differential responses
to wet and dry life zones. Appropriate selection of populations for conservation would maintain genetic variation for plasticity in the face of spacial
and temporal environmental variation.

Genetic Impoverishment

Threats to genetic diversity in mahogany include selective harvest in
populations and loss of populations themselves. Harvest of the most desir-
ably formed trees at a given location has caused much debate about genetic impoverishment in the mahogany species. Genetic erosion is likely in
neutral rare alleles with even minimal harvest. Dysgenic selection can be
defined as the selective removal of desirable trees, resulting in the reduction of genetic frequencies for adaptive or marketable traits. Few data are
available on whether this actually occurs. Because of homogenization of the
local environment before planting, planting individuals at the same time,
and subsequent plantation maintenance, the small-scale environmental
variation at the plantation sites used in this study would have been reduced
compared to the natural environment. There, environmental effects on tree
growth form and size may be so strong that the influence of genetic variation is relatively small. The high gene flow and potentially low heritability
for most traits as expressed in the native environment may buffer against
loss of desirable genetic variation. Thus, most of the genetic variation may
remain in the unharvested stock on a given site. Within certain constraints,
these species may be relatively genetically resilient. The large environmental component of phenotypic variation may have retarded the effects
of selection, both natural and human, and thus the rate of genetic erosion
of desirable traits. Hard data are lacking for these speculations, however.

A more likely cause of genetic impoverishment in mahogany is the loss
of populations. Individual stands of mahogany in a relatively uniform environment are not genetically unique (Loveless and Gullison, Chapter 2, this
volume). Our study suggested that a different pattern prevails on a larger

geographic scale that incorporates the effects of natural selection. From our study, we see that probably half the genetic variation and the major part of the genetic-by-environment interaction are among rather than within populations. Eliminating mahogany in whole regions would reduce the genetic stock of these species, especially Pacific coast mahogany, which had greater LOCATION*ORIGIN interactions and perhaps more genetic variation among populations than did big-leaf mahogany.

Population Selection

A standard used for designating populations to be preserved in conservation biology is to select distinct lineages based on the degree of genetic divergence among populations (Meffe and Carroll 1994; Vogler and Desalle 1994). We believe that a functional aspect, based on adaptation to different environments and quantitative trait expression, should also be incorporated. In our study, we found that none of the populations used for this study showed superior performance across all environmental conditions in Puerto Rico. Rather, their relative performance varied across environmental conditions and showed apparent adaptive differences relative to the subtropical dry and the wet life zones. We also saw that the correlation between genetic variation and geographic distance was between 40% and 50% in the populations surveyed. Thus, the more populations are separated geographically and ecologically, the more genetically distinct they are likely to be.

These findings argue for conserving populations that cover the range of potentially adaptive variation. This strategy could be integrated with the more usual approach based strictly on measures of genetic divergence (usually determined by neutral genetic markers; Petit et al. 1997). A marker analysis could be used to characterize genetic lineages among mahogany populations. Setting priorities among populations for preservation should be based not only on the uniqueness of descent, but also on the range of environments inhabited by the species, including any information on adaptive traits that is available, such as the quantitative traits of this study.

The high gene flow rate in big-leaf mahogany showed that not every intervening population between those at environmental and geographic extremes is needed to preserve genetic variation. With the loss of intermediate populations, however, gene flow may be reduced, and inbreeding may be increased. Pacific coast mahogany showed more differentiation among populations than does big-leaf mahogany; thus, the Pacific coast species may require a modified strategy to maximize genetic conservation. This need might entail shorter geographic distances between populations targeted for conservation.

Preserve Design

Two strategies to consider for maintaining genetic variation for mahogany are population preservation *in situ* and gene bank holdings. Given the fast-changing nature of the Latin American landscape, the most practical approach for *in situ* preservation of genetic variation in mahogany would emphasize the reserves already in existence. Incorporating this reserve design for mahogany into a biome preservation strategy for co-occurring species would make efficient use of land and monetary resources. Such a preserve system holds intact the evolutionary and ecological dynamic in which each population is embedded and helps to maintain the genetic character of the species. The findings of this study suggested the implementing of a preserve system to cover the geographic range at regular intervals and different environmental conditions. Are existing reserves large enough to maintain a locally viable mahogany population under natural disturbance and recovery regimes? Do these reserves cover the range of environments in which mahogany grows?

For areas of the species range where such reserves do not exist and cannot be put in place, a rescue operation would be important to implement, that is, by collecting genetic variants from existing populations for *ex situ* protection in a gene bank. One problem with this approach is that the high rates of outcrossing seen in big-leaf mahogany act against maintaining distinct genetic sources if they are planted together in the same site. Keeping these sources distinct by planting them on a reserve near native habitat, such as by underplanting existing forest at wide spacing, may be possible. *Ex situ* conservation may also result in the dominance of a few genetic sources that grow faster or flower earlier (S. Krugman, USDA Forest Service, Washington, DC, personal communication).

Further Research

The original plan of Puerto Rican mahogany genetic trials was to compare the performance of provenances across the ranges of the three species. Completing the sampling of the species ranges in the Caribbean and in South America would provide information on the total environmental and genetic variation exhibited in the genus. Populations, especially those at the extremes of the species tolerance ranges, should be examined.

In conservation genetics, many of the arguments about amounts and distributions of genetic variation for preservation, and avoidance of genetic erosion, have been based on genetic markers. Genetic markers are more efficiently sampled for obtaining information on genetic variation than are quantitative traits. The relation between markers and variation in quantitative traits is extremely variable, however (Hedrick et al. 1996), including those traits linked to fitness. Establishing a relation between genetic markers and quantitative traits in mahogany would help integrate distinctness of descent and trait expression across environments into a genetic conservation strategy. We intend to follow up on this provenance study with genetic marker analysis of trees in the extant plantations.

Our study has been framed in terms of genetic variation relevant to fitness. Assessing the components of phenotypic variation in marketable traits, such as wood quality, would assist in preserving genetic diversity in these characteristics as well. Although several authors have suggested that most variation in mahogany wood quality is a result of environmental factors (e.g., Lamb 1966; Chudnoff and Geary 1973), this conclusion needs to be assessed more rigorously. In other species, wood quality is known to have a genetic component (Zoble and Talbert 1984; Cornelius 1994).

For both fitness-related traits and the characteristics people want in mahogany, effects of genetic variation on the phenotype at the local environmental scale remain in question. In this study at the regional scale, genetic differences among populations were partially related to distance and to environmental variables. At the local scale, the high proportion of environmental variation suggested that genetic variation may be partially masked by the local environment. Information about the relative influence of the genotype and environment at the local scale in the natural environment would go a long way toward resolving the confusion on dysgenic selection.

Acknowledgments. This work was done in cooperation with the University of Puerto Rico. We acknowledge H. Barres and C.B. Briscoe for originally conceiving and implementing this study, and all those who have tended the plots, recorded data, and arranged the records over the years. Special thanks go to S. Chin and E. Muns of Gallaudet University for long hours in preparing this data set for analysis and to M. Alayón for assistance in producing the manuscript. Thanks to W. Arendt, J. Francis, J. Hamrick, S. Krugman, E. Medina, G. Moreno, R. Ostertag, B. Parresol, J. Parrotta, F. Wadsworth, L. Whitmore, and three anonymous reviewers whose comments contributed to improving this manuscript.

Literature Cited

Abrams, J.S. 1995. *Taxonomic Study of* Swietenia *spp. Using Restriction Fragment Length Polymorphic DNA and Morphological Traits*. Thesis, Department of Plant and Soil Science, Alabama Agricultural and Mechanical University, Normal.

Barres, H. 1963. A provenance study of *Swietenia*. Study Plan 2421 (unpublished). U.S. Department of Agriculture, Forest Service, Institute of Tropical Forestry, Río Piedras, PR.

Bauer, G.P., and Gillespie, A.J.R. 1990. *Volume Tables for Young Plantation Grown Hybrid Mahogany* (Swietenia macrophylla × S. mahagoni) *in the Luquillo Experimental Forest of Puerto Rico*. Research Paper SO 257. U.S. Department of Agriculture, Forest Service, Southern Forest Experimental Station, New Orleans, LA.

Bonnet Benítez, J.A. 1983. *Evaluación de la Nueva Clasificación Taxonomica de los Suelos de Puerto Rico*. Publicación 147. Estación Experimental Agrícola, Universidad de Puerto Rico. Río Piedras, PR.

Bonnin, I., Prosperi, J.M., and Olivieri, I. 1996. Genetic markers and quantitative trait variation in *Medicago trunculatus* (Leguminosae): a comparative analysis of population structure. *Genetics* **143**:1795–1805.

Bradshaw, A.D. 1965. Evolutionary significance of phenotypic plasticity in plants. *Advances in Genetics* **13**:115–155.

Brown, A.H.D., and Marshall, D.R. 1981. Evolutionary changes accompanying colonization in plants. In *Evolution Today*, eds. G.C.E. Scudder and J.L. Reveal, pp. 351–363. Proceedings of the Second International Congress of Systematics and Evolutionary Biology. Hunt Institute for Botanical Documentation, Carnegie-Mellon University, Pittsburgh, PA.

Chudnoff, M., and Geary, T.F. 1973. On the heritability of wood density in *Swietenia macrophylla*. *Turrialba* **23**:359–362.

Cornelius, J. 1994. Heritabilities and additive genetic coefficients of variation in forest trees. *Canadian Journal of Forest Research* **24**:372–379.

Dodd, D.H., and Schultz, R.F., Jr. 1973. Computational procedures for estimating magnitude of effects for some analysis of variance designs. *Psychological Bulletin* **79**:391–395.

Endler, J.A. 1977. *Geographic Variation, Speciation, and Clines*. Princeton University Press, Princeton.

Falconer, D.S. 1981. *Introduction to Quantitative Genetics*. Longman, New York.

Fleiss, J.L. 1969. Estimating the magnitude of experimental effects. *Psychological Bulletin* **72**:273–276.

Freund, R.J., Littel, R.C., and Spector, P.C. 1986. *SAS System for Linear Models*. SAS Institute, Cary, NC.

Geary, T.F. 1969. Adaptability of Mexican and Central American provenances of *Swietenia* in Puerto Rico and St. Croix. OF FTB 69 2/19. Food and Agricultural Organization of the United Nations, Rome, Italy.

Geary, T.F., Barres, H., and Ybarra-Coronado, R. 1973. *Seed Source Variation in Puerto Rico and Virgin Islands Grown Mahoganies*. Research Paper ITF-17. U.S. Department of Agriculture, Forest Service, Institute of Tropical Forestry, Río Piedras, PR.

Gillies, A.C.M., Navarro, C., Lowe, A.J., Newton, A. C., Hernández, M., Wilson, J., and Cornelius, J. Genetic diversity in Mesoamerican populations of mahogany (*Swietenia macrophylla*), assessed using RAPDs. *Heredity* **83**:722–732.

Glass, G.V., Peckham, P.D., and Sanders, J.R. 1972. Consequences of failure to meet assumptions underlying the fixed effects analyses of variance and covariance. *Review of Educational Research* **42**:237–288.

Glogiewicz, J.S. 1986. *Performance of Mexican, Central American, and West Indian Provenances of* Swietenia *Grown in Puerto Rico*. Thesis, Faculty of Forestry, College of Environmental Science and Forestry, State University of New York, Syracuse.

Hamrick, J.L., and Godt, M.J.W. 1989. Allozyme diversity in plant species. In *Plant Population Genetics, Breeding, and Genetic Resources*, eds. A.H.D. Brown, M.T. Clegg, A.L. Kahler, and B.S. Weir, pp. 43–63. Sinauer, Sunderland, MA.

Hedrick, P.W., Savolainen, O., and Holter, L. 1996. Molecular and adaptive variation: a perspective for endangered plants. In *Southwestern Rare and Endangered Plant Species: Proceedings of the Second Conference*, eds. J. Maschinski and H.D. Hammond, pp. 92–102. General Technical Report RM-283. U.S. Department of Agriculture, Forest Service, Rocky Mountain Forest and Range Experiment Station, Fort Collins, CO.

Helgason, T., Russell, S.J., Monro, A.K., and Vogel, J.C. 1996. What is mahogany? The importance of a taxonomic framework for conservation. *Botanical Journal of the Linnean Society* **122**:47–59.

Holdridge, L.R. 1967. *Life Zone Ecology*. Tropical Science Center, San José, Costa Rica.

Huber, D.A., White, T.L., and Hodge, G.R. 1994. Variance component estimation techniques with forest genetic architecture through computer simulation. *Theoretical and Applied Genetics* **88**:236–242.

Lamb, F.B. 1966. *Mahogany of Tropical America. Its Ecology and Management*. The University of Michigan Press, Ann Arbor.

Lindgren, K. 1986. Can we utilize provenance test results from other countries for choice of lodgepole pine provenances in Sweden? In *Provenances and Forest Tree Breeding for High Latitudes*, ed. D. Lindgren, pp. 219–234. Proceedings of the Frans Kempe symposium in Umea, June 10–11, 1986, Swedish University of Agricultural Sciences. Department of Forest Genetics and Plant Physiology, Umea, Sweden.

Loveless, M.D. 1992. Isozyme variation in tropical trees: patterns of genetic organization. *New Forests* **6**:67–94.

Manley, B.F.J. 1986. *Multivariate Statistical Methods: A Primer*. Chapman & Hall, New York.

Marrero, J. 1948. Forest planting in the Caribbean National Forest: past experience as a guide for the future. *Caribbean Forester* **9**:85–148.

Marrero, J. 1950a. Reforestation of degraded lands in Puerto Rico. *Caribbean Forester* **11**:3–15.

Marrero, J. 1950b. Results of forest planting in the insular forests of Puerto Rico. *Caribbean Forester* **11**:107–147.

McGraw, J.B. 1987. Experimental ecology of *Dryas octapetala* ecotypes. IV. Fitness response to reciprocal transplanting in ecotypes with differing plasticities. *Oecologia* (Berlin) **73**:465–468.

McGraw, J.B., and Antonovics, J. 1983. Experimental ecology of *Dryas octopetala* ecotypes. I. Ecotypic differentiation and life-cycle stages of selection. *Journal of Ecology* **71**:879–897.

McNeilly, T. 1981. Ecotypic differentiation in *Poa annua*: interpopulational differences in response to competition and cutting. *New Phytologist* **93**:539–547.

Meffe, G.K., and Carroll, C.R. 1994. *Principles of Conservation Biology*. Sinauer, Sunderland, MA.

Milliken, G.A., and Johnson, D.E. 1992. *Analysis of Messy Data. Vol. I: Designed Experiments*. Van Nostrand Reinhold, New York.

Muona, O. 1990. Population genetics in forest tree improvement. In *Plant Population Genetics, Breeding, and Genetic Resources*, eds. A.H.D. Brown, M.T. Clegg, A.L. Kahler, and B.S. Weir, pp. 282–298. Sinauer, Sunderland, MA.

NOAA (National Oceanographic and Atmospheric Administration). 1999. Climatological data: Puerto Rico and Virgin Islands **45**(13):4–5. National Climate Data Center, Asheville, NC.

Petit, R.J., El Mousadik, A., and Pons, O. 1997. Identifying populations for conservation on the basis of genetic markers. *Conservation Biology* **12**:844–855.

Platenkamp, G.A. 1990. Phenotypic plasticity and genetic differentiation in the demography of the grass *Anthoxanthum odoratum*. *Journal of Ecology* **78**:772–788.

Platenkamp, G.A. 1991. Phenotypic plasticity and population differentiation in seeds and seedlings of the grass *Anthoxanthum odoratum*. *Oecologia* (Berlin) **88**:515–520.

Potvin, C., Lechowicz, M.J., and Tardif, S. 1990. The statistical analysis of ecophysiological response curves obtained from experiments involving repeated measures. *Ecology* **71**:1389–1400.

Rice, K.J., and Mack, R.N. 1991. Ecological genetics of *Bromus tectorum*. III. The demography of reciprocally sown populations. *Oecologia* (Berlin) **88**:91–101.

Salter, K.C., and Fawcett, R.F. 1985. A robust and powerful rank test of treatment effects in balanced incomplete block designs. *Communications in Statistics—Simulations and Computations* **14**:807–828.

Salter, K.C., and Fawcett, R.F. 1993. The ART test of interaction: a robust and powerful rank test of interaction in factorial models. *Communications in Statistics—Simulations and Computations* **22**:137–153.

Sarukhan, J., Martínez-Ramos, M., and Piñero, D. 1984. The analysis of demographic variability at the individual level and its population consequences. In *Perspectives on Plant Population Ecology*, eds. R. Dirzo and J. Sarukhan, pp. 83–106. Sinauer, Sunderland, MA.

SAS Institute. 1990. *SAS/STAT User's Guide, Version 6*, Vol. 2. SAS Institute Inc., Cary, NC.

Scheiner, S.M. 1989. Genetics and evolution of phenotypic plasticity. *Annual Review of Ecology and Systematics* **24**:35–68.

Scheiner, S.M., and Goodnight, C.J. 1984. The comparison of phenotypic plasticity and genetic variation in populations of the grass *Danthonia spicata*. *Evolution* **38**:845–855.

Scheiner, S.M., and Lyman, R.F. 1991. The genetics of plasticity. I. Heritability. *Journal of Evolutionary Biology* **2**:95–107.

Schmitt, J. 1993. Reaction norms of morphological and life-history traits to light availability in *Impatiens capensis*. *Evolution* **47**:1654–1668.

Seaman, J.W., Jr., Walls, S.C., Wise, S.E., and Jaeger, R.G. 1994. Caveat emptor: rank transformations and interactions. *Trends in Ecology and Evolution* **9**:261–263.

Silander, J.A. 1985. The genetic basis of the ecological amplitude of *Spartina patens*. II. Variance and correlation analysis. *Evolution* **39**:1034–1052.

Styles, B.T. 1981. Swietenioideae. In *Meliaceae*, ed. T.D. Pennington, pp. 359–406. Flora Neotropica Monograph Vol. 28. New York Botanical Garden, Bronx.

Sultan, S. 1987. Evolutionary implications of phenotypic plasticity in plants. *Evolutionary Biology* **21**:127–178.

Underwood, A.J., and Petraitis, P.S. 1993. Structure of intertidal processes at different locations: how can local processes be compared? In *Species Diversity in Ecological Communities*, eds. R.E. Ricklefs and D. Schluter, pp. 39–51. University of Chicago Press, Chicago.

Vaughan, G.M., and Corvallis, M.C. 1969. Beyond tests of significance: estimating strength of effects in selected ANOVA designs. *Psychological Bulletin* **72**:204–213.

Via, S., and Lande, R. 1985. Genotype-environment interaction and the evolution of phenotypic plasticity. *Evolution* **39**:505–522.

Vogler, A.P., and Desalle, R. 1994. Diagnosing units of conservation management. *Conservation Biology* **8**:354–363.

Ward, S.E. 1996. Aspect of phenotypic, genetic, and plastic variation in golf course populations of annual bluegrass (*Poa annua* L.). Dissertation, University of California, Davis.

Weaver, P.L., and Bauer, G.P. 1986. Growth, survival, and shoot borer damage in mahogany plantings in the Luquillo Forest in Puerto Rico. Turrialba **36**:509–522.

Webb, D.B., Wood, P.J., and Smith, J. 1980. *A Guide to Species Selection for Tropical and Subtropical Plantations*. Tropical Paper 15. Commonwealth Forestry Institute, Oxford.

Westfall, P.H. 1987. A comparison of variance component estimates of arbitrary underlying distributions. *Journal of the American Statistical Association* **82**:866–873.

Whitmore, J.L., and Hinojosa, G. 1977. *Mahogany* (Swietenia) *Hybrids*. Research Note ITF-23. U.S. Department of Agriculture, Forest Service, Institute of Tropical Forestry, Río Piedras, PR.

Winer, B.J., Brown, D.R., and Michels, K.M. 1991. *Statistical Principles in Experimental Design*. McGraw-Hill, San Francisco.

Xie, C.-Y., and Ying, C.C. 1996. Heritabilities, age-age correlations, and early selection in lodge pole pine (*Pinus contorta* ssp. *latifolia*). *Silvae Genetica* **45**:101–107.

Zoble, B., and Talbert, J. 1984. *Applied Forest Tree Improvement*. Wiley, New York.

Appendix A

Data Distributions and Transformations

Data distributions were nonnormal, with bimodal distributions for the early years for DBH-based measurements (*DBH, δDBH, BA, δBA, TV, δTV*), because of a high proportion of zero DBHs from rounding errors. Error variances were also heterogeneous. No power family or other parametric transformation could normalize the distribution of these data and homogenize the error. Analysis of variance, used for most comparisons, is robust against most violations of assumptions except for inequality of variance. If variances are unequal in an unbalanced design, the significance levels for the ANOVA tests can be greatly affected (Glass et al. 1972).

For these reasons, we performed an aligned ranks transformation for both Analysis of variance (ANOVA) and Repeated analysis of variance (ANOVAR) (Salter and Fawcett 1985, 1993). This transformation allows for testing of interaction and main effects (Seaman et al. 1994). First, the variables were ranked, and then SAS General Linear Models Procedure (SAS 6.12 Proc GLM; SAS Institute 1990) was performed for only the main effects. The residuals from this step of analysis essentially had the main effects removed and were ranked again by time and reanalyzed to test for interactions.

Details of ANOVA and ANOVAR Analysis

When all species were considered together (see Tables 3.5 and 3.9), ERROR included some genetic variance from MOTHER effects, which were not separated out at this level, and the G*E interaction for MOTHER*LOCATION. For each species taken individually, ERROR included genetic variation from variation among fathers and among progeny. The maternal effects estimate $1/4\ V_a + V_{ec}$, where *ec* is maternal effects from a common environment during seed development (Falconer 1981). The variation among MOTHERs provided an outside estimate of $1/4\ V_a$, if maternal effects can be assumed to be 0. The additive genetic variance remaining within families (or subsumed into the error variance) would be $1/4\ V_a$ from paternal variation, and $1/2\ V_a$ from variation among offspring (Falconer 1981). The term $4(V_{MOTHER})$ was used to estimate total additive genetic variance at the family scale (Table 3.4). The term

$2(V_{LOCATION*MOTHER})$ was used to estimate the total genetic-by-environment interaction at the family scale after Scheiner and Goodnight (1984). The term $V_e - 3(V_{MOTHER}) - (V_{LOCATION*MOTHER})$ was used to remove from ERROR the estimated additive genetic variance from fathers and within family variation, and estimated genetic environment interaction from fathers (see Falconer 1981; Scheiner and Goodnight 1984; Table 3.4). Remaining in V_e would be the nonadditive genetic variance, which can be strong for growth characteristics (Zoble and Talbert 1984). Nonadditive genetic variance cannot be passed on to the next generation because it is the result of particular combinations of genes in the offspring (Falconer 1981).

The experiment used a randomized incomplete block design. The design was unbalanced because of tree mortality and uneven replication by MOTHERs across sites. Also, multiple levels of nesting were present for different terms. The SUBBLOCKS were nested within LOCATION and SEEDGROUP; the ORIGIN was nested within SPECIES and SEED-GROUP; and the MOTHER tree was nested within SEEDGROUP, SPECIES, and population of ORIGIN (see Table 3.3).

Random variables included LOCATION, SUBBLOCKS, SEED-GROUP, ORIGIN, and MOTHER effects and all interactions (Table 3.3). These effects were considered to be random because they were samples of a larger group of levels for each effect. Only SPECIES was considered to be a fixed effect because at least one population from all three mahogany species was included in this study. When fixed effects are considered, the interpretation must not be extrapolated. The SEEDGROUP effects included the following confounded environmental effects: probable clima-tological effects from different years of seed harvest, planting, and slightly different sites in each location in Puerto Rico. In addition, SEEDGROUP included a genetic effect because different provenances were contained in each SEEDGROUP. In the repeated measures analysis, SEEDGROUP effects decreased over time. This decrease helped to justify aggregating both SEEDGROUPS in one analysis to allow comparisons across all sites of origin.

For this study, fruits were collected from openly pollinated trees. An open-pollinated system is usually assumed to result in half-sib offspring (J. Hamrick, University of Georgia, personal communication). In such a system, maternal effects cannot be separated from genetic effects, or dominance from additive genetic effects (Falconer 1981). Additive genetic effects are the components of variation on which selection can act (Falconer 1981).

The analytical tool SAS 6.12 Proc GLM was used for ANOVAR to analyze the size and growth rate parameters. We used type III sums of squares because they are considered desirable for making inferences about main effects, even if the interaction effects are significant (Freund et al. 1986). Approximate F tests on aligned transformed data were used from

SAS Proc GLM version 6.12 with the Random statement and Test option (SAS Institute 1990). The analyses were performed similarly for longevity and shoot borer attack, by using ANOVA. For the ANOVAR, Mauchly's criteria for covariance structure of measurements between dates were usually outside the accepted limits. For this reason, effects in ANOVAR were tested with univariate tests and F statistics modified by the Greenhouse–Geisser method, which results in more conservative tests than the Huynh–Feldt modified F statistics (SAS Institute 1990). The Greenhouse–Geisser epsilon was also given to indicate the lack of sphericity of the ANOVAR tests. The applicability of the modified F statistics when sphericity tests depart from one is a point of debate (Potvin et al. 1990; K. Keirnan, SAS Institute, Cary, NC, personal communication).

We used variance component analyses, ANOVAR, and ANOVA over all species together, and then for each individual species. These analyses were needed because seed was not partitioned by mother trees for small-leaf mahogany. Analyses of all species together included species effects but excluded mother effects. Analyses for big-leaf and Pacific coast mahogany included mother effects but excluded among-species comparisons. Small-leaf mahogany comprised only one provenance and lacked mother-tree definition. Because of the limited number of degrees of freedom, analysis for this species included only LOCATION effects.

Variance Component Analysis

In ANOVA with large sample sizes, the individual effects may be extremely significant, but they contribute only marginally to explaining the total variance. Thus, variance component analysis has been developed to determine the relative contributions of different effects to total variance (Fleiss 1969; Underwood and Petraitis 1993), including models with both fixed and random effects (Vaughan and Corvallis 1969; Dodd and Schultz 1973). Variance components were calculated by using the same model as for the significance tests, but on untransformed data. To determine the components of variance for all species together, the actual and expected mean sums of squares were used for each effect from SAS GLM, according to the formulas provided by Winer et al. (1991). The aggregate components of variance composed of various genetic and environmental effects were also calculated (see Table 3.4). Because heights at time zero were lacking, and all initial DBHs were rounded to zero, variance components for $\delta HT1$, $DBH1$, $\delta DBH1$, $BA1$, $\delta BA1$, $TV1$, and $\delta TV1$ were set to zero. Negative variance components for individual effects were also set to zero, after the SAS Institute (1990). Mean variance components for all size and growth parameters together were calculated for each time interval. Standard deviations were calculated for these means. Confidence intervals were not calculated for variance components because of the unbalanced experimental design.

Within each species, all effects were judged random, so variance components were determined using restricted maximum likelihood (SAS 6.12 PROC VARCOMP). This method is relatively robust both for unbalanced designs (Huber et al. 1994) and for departures from normality (Westfall 1987). Species differentiation was also examined by using cluster analysis on growth and size parameters at years 5 and 14. The parameters were averaged for each population, then clustered by similarity using the unweighted pair group method based on arithmetic means Unweighted Paired Group Method Arithmetic (UPGMA; PROC CLUSTER, SAS 6.12). This analysis allowed a comparison between previous provenance assignment to species and actual provenance similarity based on the traits examined.

The geographic and genetic distances of provenances were also compared over all provenances of big-leaf and Pacific coast mahogany, and within species. Small-leaf mahogany was excluded from these comparisons because the population was not from the native range of the species. All size and growth parameters from years 1 to 14, total shoot borer attack, and longevity were used to calculate Mahalnobis distances between provenance pairs (PROC DISCRIM SAS 6.12). Mahalnobis distances estimate group similarities, taking into account correlations among the parameters considered (Manley 1986); here they provided an estimate of genetic distances between provenances. Geographic distance was calculated from latitude and longitude coordinates. These distances were then correlated to each other.

Interaction Analysis

The LOCATIONs in Puerto Rico were classified by soil type, elevation, and life zone; ORIGINs were classified by the site of provenance origin in México and Central America. Species and provenance performance across sites in Puerto Rico was compared by least square means (LSMEANS, PROC GLM, SAS 6.12). Least square means may provide negative estimates of some parameters, but this procedure is superior to simple averaging in that it allows comparisons based on unequal representation per subgroup in unbalanced experimental designs. In our comparisons, replication was unequal, especially among species. Performance across different Puerto Rican and Virgin Island locations, soil types, and life zones was also compared, by LSMEANS, for all species together. Potential elevation effects on measured traits were examined by using Pearson correlations between site elevation and trait expression. To examine genetic-by-environment interactions, LSMEANS comparisons were made of the interactions between species and provenance performances at different locations, life zones, and soil types in Puerto Rico. For all LSMEANS tests, the probabilities for significant differences were adjusted by Scheffe's criteria because these adjustments are conservative for *post hoc* comparisons.

To further differentiate the relative importance of soil type and life zone in LOCATION effects, additional ANOVAS were performed that replaced LOCATION with LIFE ZONE and SOIL type categories.

Analysis of Potential Selection Pressures and Adaptive Strategies

Length of dry season, altitude, latitude of sites of origin, and the difference between latitude at the site of origin and the latitude of Puerto Rico (or unmeasured associated factors) were examined as potential selection pressures using Pearson correlations with trait expression based on family means in Puerto Rico. The potential for adaptive strategies in these species was also examined. Evidence for generalist provenances was sought from LSMEAN plots for species or provenances consistently performing well across all locations, life zones, or soil types. The coefficient of variation (standard deviation/mean × 100) allows standardization of the standard deviation by the mean of a given parameter, and indicates its relative range of variation. Family means for parameters related to fitness were compared to family level coefficients of variation (after Schmitt 1993) by Pearson correlations, as a test for the adaptiveness of plasticity. Specialization by habitat was examined by Pearson correlations based on a family means between performance in subtropical lower montane sites and the wetter zones.

4. Hurricane Damage to Mahogany Crowns Associated with Seed Source

John K. Francis and Salvador E. Alemañy

Abstract. The International Institute of Tropical Forestry's mahogany provenance test was established with collections from México, Central America, and the Caribbean. In 1989, when the trees were 26 years old, Hurricane Hugo damaged three of the stands in Puerto Rico and St. Croix, U.S. Virgin Islands. Four months after the storm, tree diameters were measured, and the percentage of the crowns of trees removed by the storm was estimated. Nearly 4 years later, the percentage of the normal crown complement then present was estimated. Using analysis of covariance, with site and post-storm diameter as covariants, the effect of provenance (16 seed sources) was evaluated. The percentage of crown lost in the storm was significantly influenced by provenance. Percentage of full crown volume 4 years after the storm was significant, but it was probably more an effect of the original loss than of regrowth. Small-leaf mahogany and the northern sources of big-leaf mahogany were the most resistant to storm damage; Pacific coast mahogany and the southern sources of big-leaf mahogany were most affected. Crown recovery, measured by change in percentage of the estimated full crown complement over the 4-year period between measurements, was significantly affected by provenance. Those provenances suffering the greatest crown loss tended to make a greater percentage recovery, although they did not catch up with those provenances that had suffered less initially.

Keywords: Mahogany, Hurricanes, Crown damage, Provenances, *Swietenia*

Introduction

Seeds of big-leaf and Pacific coast mahogany were collected through México and Central America to identify the best seed sources (or provenances) of mahogany for Puerto Rico and the U.S. Virgin Islands. One source of small-leaf mahogany, St. Croix, Virgin Islands, was also included in the study, making a total of 20 provenances. Each provenance was represented by 16 mother trees. The seeds were planted in replicated trials in 13 locations in Puerto Rico and St. Croix, and the identities of each individual were maintained throughout the study. Establishment and early growth of the various plantations were reported by Geary et al. (1973). The provenance study was also used to study the heritability of wood density in big-leaf mahogany (Briscoe et al. 1963; Chudnoff and Geary 1973). After 21 years, Glogiewicz (1986) measured 3 of the plantations and reported their growth. As many of the seed sources in México and Central America become scarcer, the value of these trials increases.

Methods

On September 18, 1989, Hurricane Hugo (a category four storm) passed through the Virgin Islands and Puerto Rico, causing near-complete defoliation and severe breakage of small and large branches of the mahoganies in the provenance trial. During January and February 1990, 4 and 5 months after the passage of the storm, we measured three of the damaged plantations of the provenance study near El Verde and Cubuy in Puerto Rico, and Estate Thomas in St. Croix, Virgin Islands. The three sites, all on USDA Forest Service Land, receive, on average, about 2500, 1900, and 1150 mm annual rainfall, respectively. The maximum gust speeds affecting the three areas during Hurricane Hugo were estimated at 190, 140, and 240 km h^{-1}.

Table 4.1. Source of Mahogany Provenances Used in This Study

Species Code	Country	North Latitude	West Longitude
Pacific coast mahogany			
HCA	Costa Rica	10°40'	85°30'
HGA	Guatemala	14°15'	90°50'
HMA	México	18°05'	102°15'
HME	México	16°20'	94°00'
Big-leaf mahogany			
MGA	Guatemala	17°00'	89°40'
MHA	Honduras	15°50'	86°00'
MMA	México	20°30'	97°20'
MMB	México	17°40'	94°55'
MMC	México	16°55'	91°35'
MMD	México	18°25'	89°25'
MME	México	19°30'	88°05'
MNA	Nicaragua	13°40'	84°10'
MNB	Nicaragua	11°50'	83°55'
MPA	Panamá	7°35'	80°40'
MPB	Panamá	8°50'	78°00'
Small-leaf mahogany[a]			
WIM	St. Croix		

[a] Naturalized, probably from Jamaica.

After finding and identifying each individual tree, either by the tag from previous measurements or by map position, we measured the diameter at breast height (dbh) with a diameter tape. We estimated the volume of extant crown as a percentage (by 10% increments) of what would have been there without the storm, by using the trunk and remaining branch structure as indicators of crown dimension. Between September 1993 and January 1994 (nearly 4 years after the storm), we revisited the sites, measured the trees, and reevaluated the crowns by the same procedure.

Of the 20 provenances planted, 4 had suffered such high mortality over the years because of poor adaptability to the plantation site and as a result of the storm that they had to be eliminated from the analysis. The 16 provenances used in this analysis are summarized in Table 4.1. We evaluated the resulting data by analysis of covariance, by using the GLM procedure of the SAS statistical software package (SAS 1988). The dependent variables were percentage crown loss as measured shortly after the storm, extant crown 4 years later, and percentage net crown change between the two measurements. Independent variables were provenance and site (class variables) and diameter (dbh) in 1990 (a continuous variable). We calculated the least square means for each provenance and used a probability of $\alpha = 0.05$ in all tests of significance.

Results

When we tested the variances of percentage of estimated crown loss by the analysis of covariance, the general model, provenance, and the covariant dbh in 1990 proved to be highly significant; site (a class variable) was not significant. Ranking of least square means showed that small-leaf mahogany was the least damaged and Pacific coast mahogany provenances were ranked among big-leaf mahogany provenances (Table 4.2). Significant differences between individual provenances were tested by comparing least square means of each with those of every other provenance by t tests. Adjusted means of crown loss ranged from 40% for the small-leaf mahogany provenance to 72% for big-leaf mahogany from Panamá (MPB). Provenances at the extremes of the range of damages were significantly different from each other (Table 4.3).

We also evaluated the estimated percentages of crowns present in 1993–1994 by analysis of covariance. The model proved to be highly significant. The component variances of provenance, site, and dbh were also highly significant. Rankings of least square means also placed small-leaf mahogany with the most complete crown and Pacific coast mahogany and big-leaf mahogany provenances interspersed in the ranking (see Table 4.2).

Table 4.2. Least Square Means of Estimated Crown Loss After Hurricane Hugo in 1990 and Extant Crown in the 1993–1994 Measurements[a]

Provenance[b]	Extant Crown		Crown Recovery (%)
	1990	1993–1994	
WIM	60.5	70.3	9.8
MME	59.2	58.9	−0.3
MGA	55.0	60.5	5.5
MMB	54.2	54.8	0.6
HME	52.5	50.4	−2.1
MMC	49.2	51.4	2.2
MMA	47.7	55.7	8.0
HMA	47.7	51.9	4.2
MPA	44.3	46.2	1.9
MMD	42.5	56.2	2.2
HCA	41.1	49.6	8.5
HGA	39.0	44.1	5.1
MNB	38.6	53.8	15.2
MNA	36.6	58.2	21.6
MHA	31.2	55.7	24.5
MPB	28.0	40.2	12.2

[a] From analysis of covariance with provenance as the independent variable, dbh as continuous covariant, and site as a class covariant.
[b] For key to provenances, see Table 4.1.

Table 4.3. Comparison of Least Square Means of Percentage of Crown Loss Measured in 1990[a]

Provenance Codes[b,c]

	WIM	MME	MGA	MMB	HME	MMD	MMC	MMA	HMA	HPA	HCA	HGA	MNB	MNA	MHA	MPB
WIM	—															
MME	0.88	—														
MGA	0.52	0.49	—													
MMB	0.46	0.42	0.90	—												
HME	0.45	0.46	0.79	0.86	—											
MMD	0.35	0.28	0.70	0.80	0.99	—										
MMC	0.20	0.13	0.40	0.47	0.73	0.63	—									
MMA	0.16	0.11	0.32	0.39	0.63	0.52	0.85	—								
HMA	0.21	0.18	0.40	0.46	0.64	0.58	0.87	0.99	—							
HPA	0.12	0.09	0.23	0.27	0.47	0.36	0.59	0.72	0.76	—						
HCA	0.04	0.01	0.06	0.07	0.26	0.12	0.29	0.42	0.50	0.73	—					
HGA	0.02	0.00	0.02	0.03	0.17	0.05	0.16	0.27	0.36	0.56	0.77	—				
MNB	0.02	0.00	0.02	0.03	0.16	0.05	0.15	0.25	0.34	0.53	0.73	0.96	—			
MNA	0.01	0.00	0.01	0.01	0.10	0.02	0.08	0.15	0.24	0.39	0.54	0.74	0.79	—		
MHA	0.00	0.00	0.00	0.00	0.03	0.00	0.01	0.04	0.08	0.15	0.19	0.29	0.32	0.32	—	
MPB	0.00	0.00	0.00	0.00	0.02	0.00	0.01	0.02	0.04	0.09	0.11	0.17	0.19	0.27	0.69	—

Data are probability that *t* is greater than critical value.
[a] Adjusted for covariants dbh and site by *t* tests in all possible combinations.
[b] For provenance codes, see Table 4.1.
[c] Underlined values indicate significance at $\alpha = 0.05$.

Adjusted mean extant crowns in 1993–1994 ranged from 40% for big-leaf mahogany from Panamá (MPB) to 70% for small-leaf mahogany. The small-leaf mahogany differed significantly from all the other provenances, and some of the big-leaf and Pacific coast mahoganies with the least developed crowns differed significantly from those in the center of the range (Table 4.4).

Net change in percentage of extant crown (crown recovery), whose provenance means ranged from −2% to +24% (Table 4.2), was analyzed as a potentially more accurate measure of recovery. The total model and independent variable provenance were significant. Comparisons by t tests in all possible combinations of least square means of provenances is presented in Table 4.5. The percentage recovery was inversely related to the percentage of extant crown. In other words, those provenances that lost the greatest proportion of their crowns regained a greater proportion of that lost, but they failed to catch up with those provenances which had originally suffered less crown loss.

Discussion

Strong hurricanes such as Hugo cause severe damage to mahogany crowns. The slow buildup and long duration of hurricane-force winds resulted in defoliation, twig and limb breakage, trunk snap, and throwing of whole trees. The extent of breakage has more to do with the ease with which trees shed their leaves than with the strength of the wood in the limbs and trunks (Francis and Gillespie 1993). The profile and degree of saturation of soils may also have affected the extent of snap and throw. The distribution through the Caribbean of wind-resistant provenances was not incompatible with the hypothesis that differences in resistance to damage between provenances may have resulted from selection by frequent storms through recent geologic history. The most resistant provenances, namely, WIM, MME, MGA, and MMB, were small-leaf and big-leaf mahoganies from the frequent hurricane track in the Caribbean and the East Coast and Central México through Guatemala. The provenances most affected by Hurricane Hugo were big-leaf mahoganies from the southern portion of the range sampled (Honduras, Nicaragua, and Panamá), where storms are less frequent. The very worst provenance in both damage and extant crown 4 years after the hurricane was the most southern, from the isthmus of Panamá (MPB).

Differences in amount of crown present in 1993–1994 by provenance were mostly a result of damage sustained initially. The contribution of crown growth was only important in a few provenances. Tree trunk diameters influenced crown damage significantly because larger trees with larger diameters were normally taller and more exposed than smaller trees and consequently suffered more. Although the sites were different in mean

Table 4.4. Comparison of Least Square Means of Percentage of Crown in the 1993–1994 Measurement[a]

	WIM	MGA	MME	MNA	MMA	MHA	MMD	MMB	MNB	HMA	MMC	HME	HCA	MPA	HGA	MPB
							Provenance Codes[b,c]									
WIM	—															
MGA	0.04	—														
MME	0.02	0.66	—													
MNA	0.02	0.56	0.86	—												
MMA	0.01	0.29	0.47	0.60	—											
MHA	0.01	0.24	0.42	0.56	0.99	—										
MMD	0.01	0.27	0.47	0.61	0.91	0.90	—									
MMB	0.00	0.14	0.27	0.40	0.84	0.83	0.71	—								
MNB	0.00	0.11	0.21	0.21	0.68	0.66	0.40	0.68	—							
HMA	0.00	0.10	0.17	0.26	0.50	0.50	0.56	0.58	0.74	—						
MMC	0.00	0.03	0.06	0.10	0.35	0.31	0.23	0.35	0.58	0.93	—					
HME	0.00	0.08	0.13	0.19	0.38	0.38	0.30	0.44	0.58	0.82	0.87	—				
HCA	0.00	0.01	0.03	0.04	0.20	0.17	0.12	0.22	0.35	0.69	0.69	0.89	—			
MPA	0.00	0.01	0.01	0.01	0.09	0.07	0.05	0.09	0.01	0.37	0.33	0.53	0.53	—		
HGA	0.00	0.00	0.00	0.00	0.01	0.01	0.00	0.01	0.03	0.17	0.09	0.29	0.21	0.70	—	
MPB	0.00	0.00	0.00	0.00	0.00	0.00	0.00	0.00	0.00	0.04	0.02	0.10	0.05	0.29	0.40	—

Data are probability that t is greater than critical value.

[a] Adjusted for covariants dbh in 1990 and site by t tests in all possible combinations.

[b] For provenance codes, see Table 4.1.

[c] Underlined values indicate significance at $\alpha = 0.05$.

Table 4.5. Comparison of Least Square Means of Net Change of Extant Crown Between 1990 and the 1993–1994 Measurement[a]

Provenance Codes[a,c]

	MNA	MHA	MMC	MNB	HMA	MGA	MPA	MMD	MMB	HCA	MPB	HGA	MMA	MME	HME	WIN
MNA	—															
MHA	0.86	—														
MMC	0.10	0.15	—													
MNB	0.24	0.32	0.78	—												
HMA	0.00	0.00	0.00	0.00	—											
MGA	0.01	0.02	0.43	0.33	0.01	—										
MPA	0.87	0.93	0.61	0.71	0.04	0.42	—									
MMD	0.00	0.01	0.25	0.20	0.02	0.69	0.34	—								
MMB	0.00	0.01	0.23	0.19	0.02	0.66	0.33	0.95	—							
HCA	0.37	0.42	0.91	0.80	0.10	0.81	0.63	0.67	0.66	—						
MPB	0.15	0.19	0.65	0.54	0.08	0.99	0.48	0.82	0.80	0.84	—					
HGA	0.01	0.01	0.23	0.18	0.06	0.57	0.30	0.82	0.86	0.60	0.72	—				
MMA	0.00	0.00	0.12	0.10	0.07	0.37	0.25	0.59	0.64	0.51	0.59	0.81	—			
MME	0.00	0.00	0.04	0.04	0.07	0.21	0.21	0.41	0.46	0.44	0.50	0.67	0.86	—		
HME	0.03	0.03	0.22	0.18	0.13	0.48	0.27	0.41	0.46	0.53	0.62	0.83	0.97	0.93	—	
WIN	0.01	0.02	0.20	0.16	0.02	0.45	0.26	0.63	0.66	0.51	0.60	0.80	0.95	0.95	0.98	—

Data are probability that t is greater than critical value.
[a] Adjusted for covariants dbh and site by t tests in all possible combinations.
[b] For provenance codes, see Table 4.1.
[c] Underlined values indicate significance at $\alpha = 0.05$.

annual rainfall and exposure to wind velocity, the hurricane effects tended to overpower any differences in site. As explained by Francis and Gillespie (1993), in storms like this one the degree of damage does not increase much with increased wind speed above a certain threshold of maximum wind velocity and resulting crown damage.

Net change in percentage of crown volume over the 4-year period was relatively small and indicated a disappointingly slow rate of recovery (branch regrowth). Actual reductions in crown volumes were found in two provenances. Several trees that partially refoliated after the storm died afterward or slowly declined. The ranking of provenances in this study does not mean that the most hurricane-resistant provenances are better than others for all traits. In fact, the faster-growing provenances tended to suffer more in storms because they were taller and thus more exposed to damaging winds.

Acknowledgments. This study was performed in cooperation with the University of Puerto Rico.

Literature Cited

Briscoe, C.B., Harris, J.B., and Wyckoff, D. 1963. Variation of specific gravity in plantation-grown trees of big-leaf mahogany. *Caribbean Forester* **24**(2):67–74.

Chudnoff, M., and Geary, T.F. 1973. On the heritability of wood density in *Swietenia macrophylla. Turrialba* **23**(3):359–361.

Francis, J.K., and Gillespie, A.J.R. 1993. Relating gust speed to tree damage in Hurricane Hugo, 1989. *Journal of Arboriculture* **19**(6):368–373.

Geary, T.F., Barres, H., and Ybarra-Coronado, R. 1973. *Seed Source Variation in Puerto Rico and Virgin Islands Grown Mahoganies.* Research Paper ITF-17. U.S. Department of Agriculture, Forest Service, Institute of Tropical Forestry, Río Piedras, PR.

Glogiewicz, J.S. 1986. *Performance of Mexican, Central American, and West Indian Provenances of* Swietenia *Grown in Puerto Rico.* Thesis, State University of New York, Syracuse.

SAS. 1988. *SAS/STAT User's Guide.* Release 6.03. SAS Institute, Inc., Cary, NC.

5. A New Mesoamerican Collection of Big-Leaf Mahogany

Carlos Navarro, Julia Wilson, Amanda Gillies,
and Marvin Hernández

Abstract. In 1994, a project was funded by the European Union to assess the genetic diversity of a range of Central American and Caribbean tree species and to determine its implications for conservation, sustainable use, and management. We describe the germplasm-sampling strategy and field collection methods for sampling big-leaf mahogany populations across Meso-america from Mexico to Panamá. Our observations confirm the severe diminution of populations of this species.

Keywords: *Swietenia macrophylla*, Collection, Current distribution

Introduction

Members of the family Meliaceae, including some 50 genera and 1000 species, grow in the Americas, Africa, and Asia. The family is highly valued, chiefly for its high-quality timber and for the ease with which some species

103

can be grown in plantations in certain places. Eight genera have been described in the Neotropics: *Cabralea, Carapa, Cedrela, Guarea, Ruegea, Schmardea, Swietenia,* and *Trichilia.* Of these, *Swietenia* and *Cedrela* contain the most important timber species (Pennington and Styles 1975).

The natural range of big-leaf mahogany extends from Mexico to Brazil. In Central America, it grows in the humid areas of Guatemala (Petén), Belize, Honduras (La Mosquitia, Colón, Atlantida, Olancho), Nicaragua (Mosquitia), Costa Rica (Los Chiles, Upala), and Panamá (Darién, Azuero). Two other species are in this genus: Pacific coast mahogany naturally grows from western Mexico to Costa Rica, and small-leaf mahogany grows in the Caribbean.

At a workshop in Costa Rica (Matamoros and Seal 1996), both big-leaf and Pacific coast mahogany were described as critically endangered. Indeed, Costa Rican populations of both these species have declined by 80% in the past 50 years (Matamoros and Seal 1996). Big-leaf mahogany is more abundant in Costa Rica than is Pacific coast mahogany, mainly because of the populations remaining in the north of the country. Small-leaf mahogany has been even more intensively exploited; both it and Pacific coast mahogany are already listed in Appendix 3 of the Convention on International Trade in Endangered Species (CITES), and both are subject to close control in international trade (Patiño 1997). Selective logging and population fragmentation are also threatening big-leaf mahogany, not only in Costa Rica but also across much of its range in Mesoamerica.

Our project was intended to determine the current distribution of big-leaf mahogany and to collect material for herbarium, gene bank, and genetic diversity studies that would identify populations with particular value for conservation and breeding. A series of questions was formulated to determine the current state of genetic variation in the species (Wilson et al. 1995, 1997). The questions were about the distribution of variation in and among geographically distinct regions: the Yucatan Peninsula (Belize, Petén in Guatemala, and Quintana Roo in México); Central Zone (Honduras, Nicaragua, and northern Costa Rica); Guanacaste (northwestern Costa Rica); and Panamá. Factors included the influence of geographic distance, environmental variation, population size, logging history, and population protection on the distribution of variation within and between populations (Gillies et al. 1999).

Methods

Identifying Populations

Locations of extant populations and their degree of exploitation were determined by literature review, visits to herbaria, and consultations with personnel from government agencies and nongovernmental organizations in all the countries listed. We obtained collecting permits for each country.

To collect across the full range of genetic variation, we selected many different types of sites, using criteria based on geography and climate, including topography, geology, soil type, vegetation, and land use; and on socioeconomics, including population density, agricultural surveys, economic indicators, and information on the transportation infrastructure. This information was used to define sampling areas and estimate the likely extent of within-species variation, based on their heterogeneity. We also ascertained the best time for collecting seeds.

Determining the Collection Strategy

To reduce the likelihood of collecting siblings, we took pollination biology and seed dispersal into account when determining the minimum distance needed between trees and populations. The unisexual flowers of big-leaf mahogany are pollinated by bees, moths, and thrips (Styles and Khosla 1976; Patiño 1997). Although little is known about the details of pollen flow in this species, previous work in other species with similar pollinators shows that certain euglossine bees may forage over distances of several kilometers (Janzen 1971). Moreover, in a disturbed area of tropical dry forest in Guanacaste Province, Costa Rica, Frankie et al. (1976) recorded individuals of eight species of bees moving between trees 0.8 km apart. Long-distance pollen dispersal of wasps up to 10 km by wasps has also been recorded (Nason et al. 1996).

Seeds from our collection ranged from 2.0 to 2.8 cm wide, 5.5 to 8.9 cm long, and 0.2 to 0.5 g. The winged seeds are relatively heavy and tend to fall close to the maternal tree (median distance, 32–36 m) (Mayhew and Newton 1998), but they can travel distances up to 80 m (Gullison et al. 1996) or 100 m (Navarro, personal observation). Considering these factors, the minimum collecting distance between trees in the same population was set at 100 m, the distance of maximum flight recorded for the seeds. A maximum population area of 5 km diameter was designated, the maximum likely distance of dispersal of the pollen; populations in the same area were separated by at least 5 km.

Collecting Protocol

Within populations, mature trees were sampled along an approximate transect. All trees along or visible from the transect were sampled, until either a maximum sample size of 65 or the end of the transect was reached, with a minimum distance of 100 m between trees. The minimum sample size was 10. When trees grew in clumps, a maximum of five were selected per clump, maintaining the minimum distance between them. Tree climbers collected leaf material from all trees and seed, when available.

We recorded the following information for each tree: identification number, population name, collector's name, date of collection, country, department or province, local governmental-political division ("municipio" or "cantón"), address, owner; meteorological data (rainfall, temperature, number of dry months); cartographic map; slope, position (valley, slope, and so on), altitude, Global Positioning System (GPS) coordinates, Holdridge life zone; associated species; description of the tree (total height, diameter at breast height, height of main stem, height of first fork, presence of buttresses, tree form, and flowering, fruiting, and leaf condition); and a sketch of the tree.

Leaf samples for DNA analysis were preserved with silica gel (Chase and Hills 1991). Herbarium material was dried, mounted, and deposited in herbaria at the Tropical Agricultural Research and Higher Education Center (Centro Agronómico Tropical de Investigación y Enseñanza [CATIE]), the National Museum of Costa Rica, and the Pan American Agricultural School in Zamorano, Honduras. Collection numbers of spe-cimens are available from the first author on request. Seeds were dried for storage in the CATIE seed bank, and trials are in progress in Costa Rica to identify populations of particular value for *in situ* conservation or breeding.

Results

Seed Production

Seed production times were found to vary from November to March, according to geographic location (Table 5.1). Seeds generally reached maturity at the end of the rainy season.

Table 5.1. Variation in the Production Time of Big-Leaf Mahogany Seeds in Mesoamerica as Determined from Field Experience

Country	Seed Production Commences	Seed Production Ends
Panamá	November	December
Costa Rica		
Pacific lowlands	October	December
Pacific highlands	December	January
North humid areas	December	January
Nicaragua		
Atlantic north	January	February
Honduras		
Atlantic lowlands	February	March
Guatemala		
Petén	February	March
México	February	March
Belize		
Lowlands	February	March
Highlands	March	April

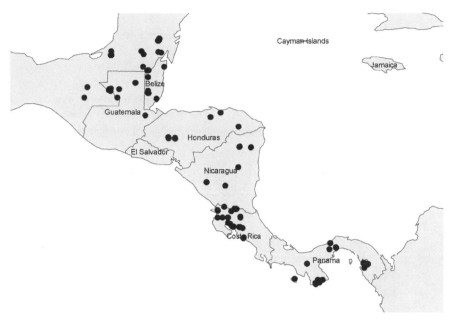

Figure 5.1. Areas of collection of herbarium material, seeds, and leaf material from mahogany trees for DNA studies.

Collections

Collections were made from 42 different populations, ranging from Mexico to Panama (Fig. 5.1). Numbers of trees sampled varied according to population size and accessibility. Leaf material was collected from 782 trees for DNA analysis, and herbarium specimens and seeds were collected from some of them (Table 5.2). Further details on the individual populations included in the collection are given in Table 5.3.

Table 5.2. Collections of Big-Leaf Mahogany Leaf Material for DNA Studies, Herbarium Samples, and Seed

Country	DNA Samples	Herbarium Samples	Seed Collections
México	100	40	63
Belize	109	36	3
Guatemala	138	18	18
Honduras	55	25	25
Nicaragua	64	21	38
Costa Rica	252	62	79
Panamá	64	39	30
Total	782	241	256

Table 5.3. Locations and Environment of Sampled Populations of Big-Leaf Mahogany

Country	Population	Latitude (°N)[a]	Longitude (°W)[a]	Altitude[b]	Sample Size	Precipitation (mm)[c]	Temperature (°C)[c]	Dry Months
México	Naranjal, Quintana Roo	19.36	88.46	50	15	1200	24	4
	Nuevo Becal, Campeche	18.80	89.32	150	46	1200	24	4
	San Felipe, Quintana Roo	18.74	88.35	50	20	1300	25	4
	Escarcega, Campeche	18.60	90.82	50	9	1100	25	5
	Laguna Kana, Quintana Roo	19.44	88.44	50	5	1200	24	3
	Madrazo, Quintana Roo	18.03	89.24	150	5	2000	26	4
Belize	Las Cuevas, Cayo	16.75	89.00	600	10	2900	22	3
	San Pastor, Cayo	16.70	88.97	600	26	2900	22	3
	New María, Cayo	16.82	89.00	600	13	2900	22	3
	Grano de Oro, Cayo	16.71	89.01	600	25	2900	22	3
	Río Bravo, Orange Walk	17.84	89.03	50	35	1700	27	3
Guatemala	Bethel, Petén	16.48	90.50	120	32	1800	25	3
	Tikal, Petén	17.22	89.61	250	56	1955	28	5
	La Tecnica, Petén	16.91	90.91	125	47	1800	27	4
	Bio-Itza, Petén	16.85	90.93	20	2	1800	25	3
Honduras	Corrales, Colón	15.51	85.94	650	12	3100	24	2
	Lancetilla, Atlántida	15.73	85.45	30	35	3278	25	3
	Mangas	15.51	85.94	680	1	3100	24	2
	Comayagua, Comayagua	14.46	87.68	500	5	1619	25	5
	Otoro, Siguatepeque	14.52	88.00	600	3	1103	25	6
Nicaragua	Terciopelo, Sahsa	14.00	83.93	60	26	2750	24	4
	Mukuwas, Bonanza	14.04	84.49	200	38	2750	24	4

Country	Location							
Costa Rica	Marabamba, Alajuela	10.94	84.63	45	67	2885	24	3
	Caño Negro, Alajuela	10.91	84.42	55	37	2885	24	3
	Santa Cecilia, Guanacaste	11.06	85.27	300	12	2585	26	4
	Upala, Alajuela	10.53	85.08	50	13	2558	25	4
	Pocosol, Guanacaste	10.53	85.35	270	37	1940	27	5
	Playuelas, Alajuela	10.92	84.69	35	4	2885	24	3
	San Emilio, Alajuela	10.97	84.77	30	64	2885	24	3
	Abangares, Guanacaste	10.06	84.49	50	6	1940	27	5
	Orotina, Puntarenas	9.55	84.29	250	1	2358	27	5
	Turrubares, Puntarenas	9.51	84.31	350	1	2358	27	5
	Chapernal, Puntarenas	10.07	84.82	50	10	1940	27	5
Panamá	Quintín, Darien	8.22	78.08	70	10	2500	26	4
	Punta Alegre, Darien	8.26	78.23	10	5	2500	26	4
	Tonosí, Los Santos	7.44	80.29	100	15	2500	25	4
	Gatúm, Gatúm	9.26	79.91	20	4	2500	25	4
	Paraiso, Paraiso	9.03	79.62	50	1	2500	25	4
	Balboa, Ancon	8.95	79.95	50	1	2500	26	2
	Summit, Ancon	9.06	79.64	50	3	2500	26	2
	Calabacito, Veraguas	8.24	81.08	50	1	2500	26	2
	Coiba, Veraguas	7.50	81.69	10	1	3500	25	4
	Cerro Hoya	7.32	80.59	500	23	1500	22	4

[a] Decimal degrees.
[b] Meters above sea level.
[c] Mean annual values.

Ecology

We found that big-leaf mahogany in a wide range of environments (Table 5.3): the altitudinal range was from 20 to 680 m, annual precipitation ranged from 1100 to 3500 mm, and rainy season duration varied from 6 to 10 months. Density of populations ranged from 0.1 trees ha^{-1} (for example, at Quintin, Punta Alegre) to 2.5 trees ha^{-1} (San Emilio, Río Bravo) (Gillies et al. 1999). Soil types and drainage varied enormously; for example, in Guatemala, the species was found on flooded soils with high yellow clay content at La Técnica, on vertisols with montmorillonite clays in flat lands at Tikal, and on Inceptisols on slopes of 70% in the Maya Biosphere Reserve. It also grew on volcanic tuba and rich soils in dry areas of Costa Rica. Big-leaf mahogany grows in both dry and humid regions, but it is most frequently found in flat areas with high humidity, sometimes flooded like the "julubales" in the north of Petén, Guatemala. It is also found in hilly areas and slopes such as those of Colón and Atlántida in Honduras where precipitation is high.

Associated Tree Species

Because of the wide range of environmental conditions in which we found big-leaf mahogany, many tree species were found to be associated with it, in several different forest types (Table 5.4).

Distribution

According to Lamb's (1966) map, the distribution of big-leaf mahogany in Mesoamerica was predominantly along the Pacific Coast of Panama and the Caribbean coasts of Nicaragua and countries farther north. Within these defined areas, we have plotted the current distribution of the species (Fig. 5.2), which is substantially smaller.

Discussion

Species Range

Ours is the most extensive seed collection in Mesoamerica. The Institute of Tropical Forestry in Puerto Rico collected seeds of big-leaf mahogany from 12 locations in Mexico and Central America in 1964–1965 (Boone and Chudnoff 1970). Our collection emphasizes the diversity of habitats in which this species grows.

This study highlights effects of exploitation on one of the most important commercial forest species of the Neotropics. In Mesoamerica, the area of natural forest containing important populations of mahogany has been reduced to one-third of the area described by Lamb (1966) (see Fig. 5.2). Not only has the area declined, but it has also become increasingly

Table 5.4. Tree Species Associated with Big-Leaf Mahogany

Country	Tree Species
Panamá	*Guaicum sanctum, Ceiba pentandra, Cordia alliodora, Bombacopsis quinata, Cedrela odorata, Ochroma lagopus, Chlorophora tinctoria, Cecropia obtusifolia, Spondias mombim*
Costa Rica	North associated with *Vochysia ferruginea, Vochysia guatemalensis, Tabebuia pentaphylla, Platymiscium pinnatum, Calophyllum brasiliense, Hyeronima alchorneoides, Tabebuia rosea, Cordia alliodora, Pithecellobium arboreum*
Nicaragua	*Calophyllum brasiliense, Cedrela odorata, Apeiba aspera, Cecropia obtusifolia, Ochroma lagopus, Sabal mexicana, Sapium nitium, Orbygnia cohune, Linociera dominguensis, Simarouba glauca, Spondias mombin, Bursera simarouba, Tabebuia* sp., *Pithecellobium, Brosimum arboreum, Pouteria* sp.
Honduras	*Terminalia amazonia, Guarea* sp., *Brosimum allicastrum, Dialium guianense, Dendropanax arboreum, Licania platypus, Vochysia ferruginea, Vochysia guatemalensis, Guarea grandifolia, Pithecellobium arboreum, Mosquitoxylon jamaicense, Virola koschnyi, Bombacopsis quinatum, Peltogyne purpurea, Cavanillesia platanifolia, Astronium graveolens, Anacardium excelsum, Xylopia frutescens, Aspidosperma megalocarpum, Chlorophora tinctoria, Ceiba pentandra, Hyeronyma alchorneoides*
Guatemala	*Cedrela odorata, Manilkara achras,* Cedrillo colorado, *Lonchocarpus castilloi (Brosimum), Calophyllum brasiliense, Coccoloba* sp., *Sabal mexicana, Bucida buseras, Bursera simarouba, Terminalia amazonia*
Belize	Las Cuevas and San Pastor Camp (Cayo District) associated with *Vitex gaumeri, Sabal mexicana, Mosquitoxylom jamaicense, Xylopia frutescens, Matayba oppositifolia, Terminalia amazonia, Pithecellobium arboreum, Zuelania guidonia, Coccoloba* sp., *Lysiloma demostachys, Orbygnia cohune, Trichospermun grewieaefolium, Ocotea veraguensis, Laethia thamnia, Lonchocarpus castilloi, Bernardia interrupta, Drypetes brownii, Calyptrantes* sp., *Nectandra interrupta, Inga edulis, Rehdera penninervia, Acacia angustissima, Cupania macrophylla,* quina, *Manilkara achras, Pouteria* sp., *Protium copal, Ficus tonduzii, Calophyllum brasiliense, Guettarda combsii, Spondias mombim, Metopium brownei, Pouteria* sp., *Achras zapota, Swartzia sacuayun, Pimienta dioica, Astronium graveolens, Quercus oleoides, Pinus oocarpa, Aspidosperma megalocarpum, Cordia dodecandra, Xylopia frutescens, Pouteria amygdalina, Nectandra, Vitex*

fragmented by agriculture, cattle farming, and harvest of the forest. Not all species have been adversely affected; for instance, species such as *Cedrela odorata* (Gillies et al. 1997) and *Cordia alliodora* are more abundant in grasslands and agroforestry systems than they are in natural forest.

Comparison of Figures 5.1 and 5.2 show that some of our collections were obtained outside the area defined by Lamb. The sources of Lamb's

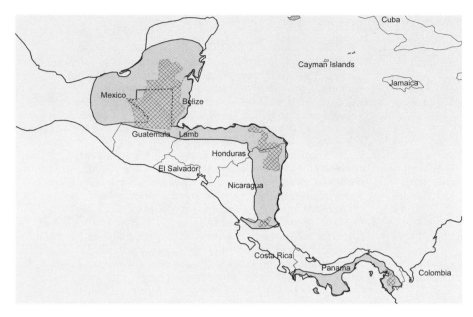

Figure 5.2. Distribution of big-leaf mahogany according to Lamb (1966) (*light gray*) compared with current distribution within Lamb's areas (*dark gray*).

information are not clear, but we have used the same taxonomic criteria. These discrepancies probably arise because the current survey encompassed farmland as well as natural and fragmented forests.

Conservation and Management

Our findings highlight the decline in this ecologically and economically important species. To combat further decline, effective conservation and management programs are essential. Several areas of natural forest already have protected status, but some of them (such as the Maya biosphere) are extraction reserves, where big-leaf mahogany is exploited. Indeed, in some reserves, it may be the only species logged. In these areas, mahogany is rare or totally absent in sections close to roads. Completely protected areas such as the Tikal National Park protect mahogany genetic resources more effectively.

In addition to *in situ* conservation, *ex situ*, or *circa situ* conservation programs should also be supported. The Lancetilla Botanic Garden in Honduras contains trees grown from seed collected along the Honduran Atlantic Coast. Our DNA studies showed this population to be extremely diverse (Gillies et al. 1999). Molecular studies have an important role in determining the effect of logging on genetic diversity and in targeting priority areas for conservation. In our study, populations with a long history of logging were less diverse than protected populations. An analysis using random polymorphic DNAa (RAPDs) showed the population from Quintin (Panama) was one of the most diverse, but populations in México maintained low diversity (Gillies et al. 1999). Quintin is a virgin forest fragment surrounded by areas of destroyed forest. Given Quintin's high diversity and vulnerability, effective conservation moves are urgently required. In conserving a species like this one with a wide geographic range, promoting collaboration between countries for collecting, characterizing, and conserving genetic resources becomes increasingly important.

Forest management can play an important role in regenerating this species, which is light demanding and has seedlings that require high irradiance to become established (Mayhew and Newton 1998). To ensure successful regeneration, gaps should be opened in the forest. Selectively logging other species first will provide gaps for regenerating mahogany. Gaps must be opened at least 2 years before logging big-leaf mahogany to stimulate the mature trees to produce more fruit and to provide open ground for seed germination.

Three provenance trials have been planted in Costa Rica, incorporating material from this latest collection. Assessments of tree growth and resistance to the shoot borer *Hypsipyla grandella* are in progress.

Acknowledgments. This project was partly funded by the European Commission, under contract number TS3*-CT94-0316. We gratefully acknowledge the information, guidance, and assistance in collecting provided by numerous individuals and organizations, including, in Belize, the Forest Department, Las Cuevas Experimental Station, and Río Bravo Conservation Area; in Guatemala, the Project Bosques Naturales (CATIE), Centro Maya, the Consejo Nacional de Areas Protegidas, and Tikal National Park; in Honduras, La Escuela Nacional de Ciencias Forestales and the Proyecto Desarrollo de Bosques Latifoliados; in Nicaragua, El Ministerio del Ambiente y Recursos Naturales and the Centro de Mejoramiento Genetico y Banco de Semillas; in Costa Rica, Area de Conservacion Guanacaste and private landowners including Rigoberto Abarca and Jose E. Rodriguez; in Panama, Asociación para la Conservacion de la Naturaleza and Instituto de Recursos Naturales Renovables; in Mexico, Instituto Nacional de Investigaciones Forestales y Agropecuarias and the International Centre for Research in Agroforestry. We also gratefully acknowledge the tremendous efforts of the tree climbers: Leonel Coto, Victor Alvarado, Carlos Chi, Jorge Sosa, and all those others who provided assistance, especially Patricia Negreros and Jonathan Cornelius.

Literature Cited

Boone, R.S., and Chudnoff, M. 1970. Variations in wood density of the mahoganies of Mexico and Central America. *Turrialba* **20**(3):369–371.

Chase, M.W., and Hills, H.H. 1991. Silica gel: an ideal material for field preservation of leaf samples for DNA studies. *Taxon* **40**:215–220.

Frankie, G.W., Opler, P.A., and Bawa, K. 1976. Foraging behavior of solitary bees: implications for outcrossing of a Neotropical forest tree species. *Journal of Ecology* **64**:1049–1057.

Gillies, A.C.M., Cornelius, J.P., Newton, A.C., Navarro, C., Hernández, M., and Wilson, J. 1997. Genetic variation in Costa Rican populations of the tropical timber species *Cedrela odorata* L., assessed using RAPDs. *Molecular Ecology* **6**:1133–1146.

Gillies, A.C.M., Navarro, C., Lowe, A.J., Newton, A.C., Hernandez, M., Wilson, J., and Cornelius, J.P. 1999. Genetic diversity in Mesoamerican populations of mahogany (*Swietenia macrophylla*), assessed using RAPDs. *Heredity* **83**:722–732.

Gullison, R.E., Panfil, S.N., Strouse, J.J., and Hubbell, S.G. 1996. Ecology and management of mahogany (*Swietenia macrophylla* King) in the Chimanes forest, Beni, Bolivia. *Botanical Review of the Linnean Society* **122**(1):9–34.

Janzen, D.H. 1971. Euglossine bees as long-distance pollinators of tropical plants. *Science* **171**:203–205.

Lamb, F.B. 1966. *Mahogany of Tropical America: Its Ecology and Management.* University of Michigan Press, Ann Arbor.

Matamoros, Y., and Seal, U.S., eds. 1996. *Report of Threatened Plants of Costa Rica.* Workshop, 4–6 October. IUCN/SSC Conservation Breeding Specialist Group, Apple Valley, MN.

Mayhew, J.E., and Newton, A.C. 1998. *The Silviculture of Mahogany.* CABI Publishing, Wallingford.

Nason, J.D., Herre, E.A., and Hamrick, J.L. 1996. Paternity analysis of the breeding structure of strangler fig populations: evidence for substantial long-distance wasp dispersal. *Journal of Biogeography* **23**:501–512.

Patiño, F. 1997. Genetic resources of *Swietenia* and *Cedrela* in the Neotropics: proposals for coordinated action. Based on contractual work for FAO by P.Y. Kageyama, C. Linares B., C. Navarro P., and F. Patiño V. Forest Resources Division, Forestry Department, FAO, Rome.

Pennington, T.D., and Styles, B.T. 1975. A generic monograph of the Meliaceae. *Blumea* **22**(3):419–540.

Styles, B.T., and Khosla, P.K. 1976. Cytology and reproductive biology of Meliaceae. In *Tropical Trees: Variation, Breeding and Conservation*, eds. J. Burley and B.T. Styles, pp. 61–67. Linnean Society, London.

Wilson, J., Gillies, A.C.M., Newton, A.C., Cornelius, J.P., Navarro, C., Hernandez, M., Kremer, A., Labbe, P., and Caron, H. 1995. *Assessment of Genetic Diversity of Economically and Ecologically Important Tropical Tree Species of Central America and the Caribbean: Implications for Conservation, Sustainable Utilization and Management.* First Annual Scientific Report to the European Commission, 1 November 1994–31 October 1995. Contract TS3*-CT94-0316.

Wilson, J., Gillies, A.C.M., Navarro, C., Cornelius, J.P., Hernandez, M., Kremer, A., Labbe, P., and Caron, H. 1997. *Assessment of Genetic Diversity of Economically and Ecologically Important Tropical Tree Species of Central America and the Caribbean: Implications for Conservation, Sustainable Utilization and Management.* Final Scientific Report to the European Commission, 1 November 1994–31 October 1997. Contract TS3*-CT94-0316. Brussels, INCO (International Cooperation for Developing Countries).

2. Ecophysiology and Regeneration

6. Photosynthetic Response of Hybrid Mahogany Grown Under Contrasting Light Regimes

Ned Fetcher, Shiyun Wen, Adiscl Montaña,
and Francisco de Castro

Abstract. In an experimental trial in Río Piedras, Puerto Rico, saplings of mahogany hybrids (small-leaf × big-leaf mahogany) were transplanted into environments with low (6% of full sun), medium (33% of full sun), and high (80% of full sun) light. The saplings showed wide variation in height growth in all environments. We measured photosynthetic response to light and CO_2 as well as chlorophyll fluorescence characteristics to determine if physiological traits showed significant variation that might explain the differences in height growth under high light. Mean light-saturated photosynthetic rates were $5.2 \pm 0.1 \mu mol\,m^{-2}s^{-1}$ for plants grown in low light, 5.6 ± 0.2 for medium light, and 6.6 ± 0.2 for high light. Mean light compensation point was $8.5 \mu mol\,m^{-2}s^{-1}$ in low light, indicating that these hybrids are capable of positive photosynthesis under heavy shade. The shade treatments significantly affected specific leaf mass, N per unit mass, and A_{max} per unit mass, but did not affect A_{max} per unit N. The variation in height growth was apparently explained by the capacity of the plants to acquire nitrogen because shorter plants had lower concentrations of foliar N than did taller ones. Large plants had larger leaf-specific mass than did shorter ones. No strong evidence for chronic photoinhibition in high light was found, although considerable energy was dissipated through photoprotective mechanisms, as shown by reductions in intrinsic Photosystems II efficiency.

Keywords: Mahogany, *Swietenia*, Photosynthesis, Photoinhibition, PFD, Light compensation point, Nitrogen, Growth, Internal CO_2

Introduction

Investigators and field workers who have tried to establish plantations of tropical trees have occasionally found that young saplings fail to grow in unshaded conditions. The saplings can remain for several months without a significant increment in height, which allows them to be overtopped by herbaceous vegetation. Hence, weeding is nearly always required for successful establishment of plantations (Wadsworth 1997: 245). The experimental array of saplings discussed by Medina et al. (Chapter 8, this volume) varied greatly at about 6 months after planting. The shortest saplings were less than half as tall as the tallest ones (H. Wang, Taiwan Forestry Research Institute, Taipei, Taiwan, personal communication). The objective of this study was to determine which physiological factors might have been responsible for the apparent lack of height growth.

Although the physiology of tropical trees has often been studied (Kitajima 1996; Strauss-Debenedetti and Bazzaz 1996; Swaine 1996), mahogany has so far escaped attention despite its commercial importance. Consequently, little information is available about the responses of mahogany seedlings growing under natural conditions to changes in light availability and CO_2 concentration. Here, we present this information in the context of trying to understand what limits the growth of mahogany tree seedlings under high light.

Methods

Seedlings of the hybrid between small-leaf and big-leaf mahogany were planted at the International Institute of Tropical Forestry (The Institute) in December 1996. The seedlings, in soil in raised concrete beds (Medina et al., Chapter 8, this volume), received three shade treatments: low, medium, and high. The low and medium treatments were provided by shadecloth. To determine light availability for each of the three treatments, we used paired quantum sensors (LiCor, Lincoln, Nebraska). One instrument was set up on the roof of a nearby building, and the other was used to take measurements next to each sapling.

Photosynthetic photon flux density (PFD) averaged 6.2% (SE = 0.4%, $n = 24$) of the values on the roof in low light; 32.9% (SE = 0.6%, $n = 24$) in medium light; and 79.8% (SE = 1.1%, $n = 17$) in high light. Thus, the plants in medium and low light received 41% and 8%, respectively, of the PFD received by the plants in high light.

Photosynthetic response to light and CO_2 was measured with a portable photosynthesis system (LI-6400; LiCor). We sampled 6 plants in the low and medium treatments and 10 plants under high light. Half the plants in each environment were short and half were tall. On August 1, 1997, the mean height of the short plants in low light was 38 ± 0.4 (SE) cm; mean height of the tall plants was 71.3 ± 0.8 cm (H. Wang, personal communication). In medium light, the short plants averaged 47 ± 1.3 cm and the tall plants, 130 ± 5.3 cm. In high light, the short plants averaged 34.1 ± 1.6 cm and the tall plants, 118 ± 10.3 cm.

Photosynthetic response to light and CO_2 was determined at two temperatures, 27°C and 33°C. For the light curves, CO_2 concentration was maintained at $360 \mu mol\, mol^{-1}$. For the CO_2 response curves, PFD levels were maintained at $1000 \mu mol\, m^{-2} s^{-1}$. We followed the manufacturer's recommendations for calibrating with CO_2 calibration gases from Scott Specialty Gases (Plumsteadville, PA) and a water vapor generator (LiCor). For the leaves used for the light curves, we measured the N concentration and dry mass. Nitrogen was analyzed by the micro-Kjeldahl technique at the Institute.

To determine whether the photosynthetic apparatus of the short plants responded differently to ambient light conditions than did tall plants, we measured fluorescence characteristics. We used a portable fluorometer (Optisciences OS-500; Haverhill, MA) to assay for chronic photoinhibition by covering the leaves overnight and measuring the ratio of variable to maximum fluorescence (F_v/F_m) in the morning before exposure to light (Castro et al. 1995). For an assay of quantum yield under ambient light, we followed the protocol of Demmig-Adams and Adams (1996) to estimate the ratio of variable to maximum fluorescence under ambient light (F_v'/F_m'), also called intrinsic Photosystems II efficiency (Demmig-Adams and Adams 1996).

Parameters were extracted from the light and CO_2 responses and tested by analysis of variance (ANOVA). From the light curves, we extracted the maximum rate of photosynthesis, the initial slope, the rate of dark respiration, the light compensation point, and the degree of curvature of a line fitting the data. This degree of curvature is represented by the parameter, b, in the equation from Küppers and Schulze (1985):

$$A = A_{max}\left[1 - e^{-b(I - I_c)}\right]$$

which was fit to our data using nonlinear least squares; A is the rate of
assimilation at PFD level I, A_{max} is the light-saturated rate of photosynthesis, and I_c is the light compensation point. From the CO_2 response curves,
we extracted the maximum photosynthetic rate under high ambient CO_2
concentrations ($1000\,\mu mol\,mol^{-1}$), the initial slope of the curve, the CO_2
compensation point, and the respiration rate under high light and with no
CO_2.

Factors in the analysis of variance were shade treatment, size of the plant
(short, tall), and temperature ($27°C$, $33°C$). Because the same plants were
measured at both temperatures, this factor was treated as a repeated
measure in ANOVA for repeated measures. One-way ANOVA was used
for F_v/F_m and F_v'/F_m' data, except when the data were not normally distributed; then, we used the nonparametric Kruskal–Wallis test.

Results

Photosynthetic Responses to Light and CO_2

Light-saturated photosynthesis (A_{max}) increased significantly as shade
decreased (Fig. 6.1; Tables 6.1, 6.2). The mean values were $5.2 \pm 0.1\,\mu mol$
$m^{-2}s^{-1}$ for low light, $5.6 \pm 0.2\,\mu mol\,m^{-2}s^{-1}$ for medium light, and $6.6 \pm$
$0.2\,\mu mol\,m^{-2}s^{-1}$ for high light. The curvature parameter, b, had higher mean
values in low (0.01058) and medium (0.00957) than in high (0.00746) light.
This result indicated that the bend in the light-response curve had a smaller
radius under shade. Mean light compensation point increased significantly
from $8.5\,\mu mol\,m^{-2}s^{-1}$ in low to $14.2\,\mu mol\,m^{-2}s^{-1}$ in medium and $18.6\,\mu mol$
$m^{-2}s^{-1}$ in high light. Quantum yield (overall mean = 0.044 ± 0.0006) was
unaffected by any of the treatments. Dark respiration also increased as
shade decreased. The A_{max} was slightly higher for tall plants than for short
ones (6.4 vs. $5.5\,\mu mol\,m^{-2}s^{-1}$), but none of the other parameters of the light
curves was affected by size. Temperature had no effect on A_{max}, but respiration increased and light compensation point decreased with increasing
temperature (Tables 6.1, 6.2).

Photosynthetic rates at high light and CO_2 concentrations increased significantly from low to medium to high light (Fig. 6.2; Tables 6.1, 6.2). Their
behavior was similar to that of photosynthesis at ambient CO_2. The initial
slope of the CO_2 response curve was higher in high (0.0276) than in medium
(0.0228) and low (0.0209) light. Respiration at high light and 0 CO_2
increased with increasing light availability and temperature (Tables 6.1,
6.2). Presumably, because photorespiration was reduced at high CO_2, temperature had a significant effect on photosynthesis, which was somewhat
higher at $33°C$ ($11.7\,\mu mol\,m^{-2}s^{-1}$) than at $27°C$ ($10.5\,\mu mol\,m^{-2}s^{-1}$). The initial
slope also increased with temperature (Table 6.1). The CO_2 compensation

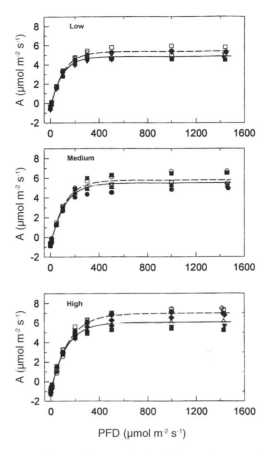

Figure 6.1. Response of net photosynthesis (A) to photosynthetic photon flux density (PFD) at 33°C for mahogany hybrids grown under low, medium, and high levels of light. Short plants, *solid symbols, solid line*; tall plants, *open symbols, dashed line*. Each symbol represents an individual plant. Curves drawn using equation from Küppers and Schulze (1985).

point was marginally lower in low than in high and medium light (Table 6.1). It was unaffected by temperature because the increase in initial slope was balanced by the increase in respiration (Fig. 6.2). The A_{max} was higher for tall than for short plants (see Tables 6.1, 6.2), but none of the other variables was affected by plant height.

The shade treatment had a significant effect on specific leaf mass, N per unit mass, and A_{max} per unit mass, but did not affect A_{max} per unit N (Tables 6.1, 6.3). Specific leaf mass was greater for the larger plants. Thus, A_{max} per unit N remained relatively constant under the different shade treatments (Table 6.3). Size also had a significant effect on N per unit area, which was smaller for the smaller plants. The percentage of N per unit mass was larger

Table 6.1. Repeated Measures ANOVA for Effect of Shade, Size, and Temperature

Parameters for photosynthetic response to light

Source	Photosynthetic Rate, A_{max} Prob > F[a]	Response Curve Prob > F[b]	Light comp Prob > F[c]	QY Prob > F[d]	R_d Prob > F[e]
Size	0.011	0.100	0.844	0.027	0.214
Shade	0.001	<0.0001	<0.0001	0.895	<0.0001
Size* shade	0.563	0.741	0.846	0.399	0.946
Tree [size, shade]	0.025	0.373	0.352	0.096	0.012
Temperature	0.520	0.492	0.002	0.073	0.001

Parameters for photosynthetic response to internal CO_2

Source	$A_{max}(HC)$[f] Prob > F	IS[g] Prob > F	CO_2 comp[h] Prob > F	Resp[i] Prob > F
Size	0.004	0.037	0.062	0.463
Shade	<0.0001	<0.0001	0.004	<0.0001
Size* shade	0.451	0.172	0.953	0.159
Tree [size, shade]	0.011	0.045	0.398	0.497
Temperature	<0.0001	<0.0001	0.078	<0.0001

Specific leaf mass, foliar N, and photosynthesis per unit mass and per unit N

Source	SLM[j] Prob > F	N_{mass}[k] Prob > F	N_{area}[l] Prob > F	A_{max_mass}[m] Prob > F	A_{max_N}[n] Prob > F
Size	<0.0001	0.0002	0.014	0.0101	0.3718
Shade	<0.0001	<0.0001	<0.0001	<0.0001	0.094
Size* shade	0.0202	0.5289	0.0434	0.4859	0.4824
Tree (size, shade)	0.0145	0.1101	0.0456	0.1468	0.009
Temperature	0.0001	0.2002	0.0019	0.0037	0.0005

[a] A_{max}, light-saturated photosynthetic rate per unit area.
[b] Response curve, curvature of photosynthetic response curve.
[c] Light comp, light compensation point.
[d] QY, quantum efficiency.
[e] R_d, dark respiration.
[f] $A_{max}(HC)$, light-saturated photosynthetic rate at high CO_2.
[g] IS, initial slope of CO_2 response curve.
[h] CO_2 comp, CO_2 compensation point.
[i] Resp, respiration under saturating light and zero CO_2.
[j] SLM, specific leaf mass.
[k] N_{mass}, N concentration per unit mass.
[l] N_{area}, N concentration per unit area.
[m] A_{max_mass}, light-saturated photosynthetic rate per unit mass.
[n] A_{max_N}, light-saturated photosynthetic rate per unit N.

in small plants (Tables 6.1, 6.3). The leaves sampled for measuring at 27°C apparently had higher N per unit area than leaves sampled for measuring at 33°C (Tables 6.1, 6.3). This difference is hard to understand unless the plants experienced some developmental change during the 3 weeks that separated sampling periods. Temperature had a significant effect on photosynthetic rate per unit of N; it was somewhat higher at the higher temperature (Tables 6.1, 6.3).

Table 6.2. Mean Values (SE) for Height on August 1 and October 15, 1998, of Plants at Low (LL), Medium (ML), and High (HL) Light

Light	Temperature (°C)	Size	Ht Aug. 1 (cm)	Ht Oct. 15 (cm)	A_{MAX}[a] (μmol m^{-2} s^{-1})	b[b] $\times10^2$	Light comp[c] (μmol m^{-2} s^{-1})	QY[d] $\times10^2$	Rd[e] (μmol m^{-2} s^{-1})	A_{MAX}(HC)[f] (μmol m^{-2} s^{-1})	IS[g] $\times10^2$	Resp[i] (μmol m^{-2} s^{-1})	CO_2 comp[h] (μmol/mol)
LL	27	Short	38 (0.60)	43.8 (2.10)	5.06 (0.26)	1.06 (0.12)	7.6 (0.30)	4.5 (0.10)	-0.38 (0.04)	7.45 (0.48)	1.95 (0.15)	-0.8 (0.11)	40.6 (2.30)
LL	27	Tall	71.3 (1.20)	85.7 (3.60)	5.55 (0.05)	1.05 (0.04)	7.4 (0.71)	4.7 (0.10)	-0.38 (0.06)	7.92 (0.23)	2.03 (0.09)	-0.85 (0.05)	41.6 (1.30)
LL	33	Short	38 (0.60)	43.8 (2.10)	4.84 (0.24)	1.09 (0.03)	8.8 (1.44)	4.3 (0.10)	-0.44 (0.08)	8.02 (0.65)	2.03 (0.13)	-1.00 (0.01)	49.7 (2.6)
LL	33	Tall	71.3 (1.20)	85.7 (3.60)	5.36 (0.27)	1.03 (0.07)	10.1 (0.54)	4.2 (0.10)	-0.57 (0.04)	10.59 (0.56)	2.3 (0.06)	-1.01 (0.02)	44 (1.10)
ML	27	Short	47 (2.00)	61.2 (3.60)	4.94 (0.52)	1.1 (0.14)	12.9 (0.63)	4.2 (0.10)	-0.66 (0.02)	9.47 (0.42)	2.1 (0.15)	-1.07 (0.01)	51.3 (3.30)
ML	27	Tall	130 (8.30)	160.3 (3.50)	6.11 (0.32)	0.94 (0.09)	12.6 (1.43)	4.8 (0.30)	-0.69 (0.03)	10 (0.29)	2.13 (0.12)	-1.13 (0.01)	53.1 (2.80)
ML	33	Short	47 (2.00)	61.2 (3.60)	5.52 (0.52)	0.88 (0.06)	16 (1.68)	4.1 (0.20)	-0.75 (0.10)	10.37 (0.83)	2.47 (0.15)	-1.43 (0.16)	58 (5.90)
ML	33	Tall	130 (8.30)	160.3 (3.50)	5.81 (0.38)	0.9 (0.08)	15.4 (1.08)	4.4 (0.00)	-0.81 (0.05)	11.6 (0.76)	2.43 (0.07)	-1.23 (0.06)	50.3 (1.40)
HL	27	Short	34.1 (2.40)	38.1 (2.50)	6 (0.60)	0.8 (0.07)	18 (0.96)	4.2 (0.30)	-0.88 (0.09)	11.05 (0.80)	2.34 (0.18)	-1.26 (0.07)	54.5 (2.40)
HL	27	Tall	118.1 (15.40)	158.1 (10.60)	7.46 (0.32)	0.67 (0.02)	17.3 (0.48)	4.7 (0.10)	-0.91 (0.05)	13.46 (0.50)	2.8 (0.03)	-1.32 (0.07)	47.4 (3.00)
HL	33	Short	34.1 (2.40)	38.1 (2.50)	6.06 (0.34)	0.8 (0.05)	18.9 (1.57)	4.2 (0.10)	-0.93 (0.10)	12.54 (0.95)	2.72 (0.22)	-1.47 (0.10)	54.6 (2.60)
HL	33	Tall	118.1 (15.40)	158.1 (10.60)	6.99 (0.25)	0.71 (0.04)	20.2 (1.14)	4.7 (0.00)	-1.07 (0.03)	14.76 (0.09)	3.18 (0.12)	-1.68 (0.03)	53.3 (2.30)

[a] A_{max}, light-saturated photosynthetic rate per unit area.
[b] b, curvature of photosynthetic response curve.
[c] Light comp, light compensation point.
[d] QY, quantum efficiency.
[e] R_d, dark respiration.
[f] Amax(HC), light-saturated photosynthetic rate at high CO_2.
[g] IS, initial slope of CO_2 response curve.
[h] CO_2 comp, CO_2 compensation point.
[i] Resp, respiration under saturating light and zero CO_2.

123

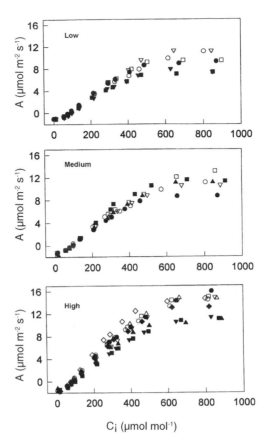

Figure 6.2. Response of net photosynthesis (A) to internal CO_2 concentration (C_i) at 33°C for mahogany hybrids grown under low, medium, and high levels of light. Short plants, *solid symbols*; tall plants, *open symbols*. Each symbol represents an individual plant.

Chlorophyll Fluorescence

The plants in high light showed some indication of being chronically photoinhibited. The ratio of variable to maximum fluorescence (F_v/F_m) had a median value of 0.687 ($n = 23$), significantly lower than the value of 0.764 ($n = 5$) observed for the plants in medium light (Mann–Whitney test, $p < 0.041$). Because the difference in F_v/F_m between the two shade treatments was not great, the chronic photoinhibition apparently did not greatly affect the performance of the plants in high light.

The intrinsic PSII efficiency as measured by F_v'/F_m' was significantly lower in high (median = 0.513, $n = 23$) than in medium (median = 0.717, $n = 5$) or low (median = 0.731, $n = 5$) light (Kruskal–Wallis test, $p < 0.0001$). Much more incoming light was being dissipated through photoprotective processes for the plants in high than in low light.

Table 6.3. Mean Values (SE) Under Low (LL), Medium (ML), and High (HL) Light

Light	Temperature (°C)	Size	SLM[a] (g/m²)	N_{mass}[b] (%)	N_{area}[c] (g/m²)	A_{max_mass}[d] (μmol g^{-1}s^{-1})	A_{max}N[e] (μmol gN^{-1}s^{-1})
LL	27	Short	34.9 (0.96)	1.92 (0.159)	0.67 (0.049)	0.14 (0.009)	7.55 (0.32)
LL	27	Tall	48.2 (3.41)	1.46 (0.117)	0.70 (0.024)	0.11 (0.008)	7.79 (0.15)
LL	33	Short	33.1 (0.48)	1.81 (0.074)	0.60 (0.018)	0.15 (0.011)	8.10 (0.50)
LL	33	Tall	40.0 (3.34)	1.67 (0.156)	0.66 (0.018)	0.14 (0.015)	8.41 (0.19)
ML	27	Short	61.0 (3.06)	1.28 (0.031)	0.78 (0.022)	0.09 (0.010)	6.68 (0.68)
ML	27	Tall	71.5 (6.35)	1.18 (0.038)	0.84 (0.060)	0.09 (0.010)	7.28 (0.62)
ML	33	Short	57.6 (1.88)	1.37 (0.116)	0.79 (0.061)	0.10 (0.011)	7.13 (0.49)
ML	33	Tall	67.7 (6.27)	1.13 (0.070)	0.76 (0.052)	0.09 (0.006)	7.72 (0.20)
HL	27	Short	69.6 (2.83)	1.25 (0.043)	0.86 (0.024)	0.09 (0.005)	6.81 (0.31)
HL	27	Tall	95.3 (5.24)	1.09 (0.046)	1.03 (0.050)	0.08 (0.004)	7.13 (0.45)
HL	33	Short	58.0 (2.60)	1.30 (0.015)	0.75 (0.030)	0.10 (0.004)	7.93 (0.20)
HL	33	Tall	85.2 (2.38)	1.14 (0.039)	0.97 (0.028)	0.08 (0.001)	7.3 (0.15)

[a] SLM, specific leaf mass.
[b] N_{mass}, N concentration.
[c] N_{area}, N per unit area.
[d] A_{max_mass}, light-saturated photosynthetic rate per unit mass.
[e] A_{max}N, light-saturated photosynthetic rate per unit N.

Discussion

Based on the information we present here, mahogany seems to be able to acclimatize to a variety of forest habitats. The mean maximum photosynthetic rate in high light was rather low, 6.6 μmol m^{-2}s^{-1}, and the maximum photosynthetic rates in low and medium light were moderately reduced (79% and 85% of the rate in high light, respectively). This pattern is similar to that of late-successional species such as *Dipteryx panamensis* and *Carapa guianensis* (Fetcher et al. 1987), as well as many others reviewed in Strauss-Debenedetti and Bazzaz (1996). The light compensation point in low light also falls within the range of values reported for late-successional species by Fetcher et al. (1987). From the standpoint of regenerating forest, it is important to recognize that mahogany can maintain a positive carbon balance when light is as low as 6% of the values in an open environment.

We have good reasons to believe that mahogany seedlings can survive with even lower light. The light compensation point for plants grown in low light was 8.5 μmol m^{-2}s^{-1}, which would be 0.367 mol m^{-2}d^{-1}, assuming a 12-h day. Allowing for losses to leaf respiration at night, as well as respiration by stems and roots, we estimated the whole-plant light compensation point to be about $20\,mol\,m^{-2}\,s^{-1}$, equivalent to $0.864\,\mu mol\,m^{-2}\,d^{-1}$, assuming a 12-h day. Fernández and Fetcher (1991) found that mean daily PFD dropped to $0.8\,mol\,m^{-2}\,d^{-1}$ 15 months after Hurricane Hugo passed over the forest in September 1989. Myster and Fernández (1995) reported daily PFD values of 0.8 for forest understory and $3.9\,mol\,m^{-2}\,d^{-1}$ for the border of a landslide. Based on these data, undisturbed areas of the Luquillo Experimental Forest would apparently have too little light for mahogany hybrids to survive, but sites such as treefalls and the edges of landslides would provide adequate light for survival and growth.

The variation in size of the plants after 9 months of growth probably resulted from their hybrid parentage. Nevertheless, examining some of the physiological differences between the tall and short saplings should help to determine if variation in physiology could have resulted in variation in growth. The taller plants had higher light-saturated photosynthetic rates than did the shorter ones. Over several months, this difference could be magnified into the observed dramatic difference in size. The difference in photosynthetic rate appeared to result from a difference in specific leaf mass, which was lower for the smaller plants, and in foliar N concentrations because foliar N concentrations were lower for the smaller plants. Photosynthetic rate per unit N was not different between the two sizes of plants, suggesting that the smaller plants may have been less successful in acquiring N, which, in turn, would result in a reduced photosynthetic rate per unit area.

Although the plants in high light had slightly lower F_v/F_m, these values were not low enough to suggest chronic photoinhibition, as was observed in *Dipteryx panamensis* by Castro et al. (1995). The lower values for intrinsic PSII efficiency (F_v'/F_m') for the plants in high light indicate that photoprotective mechanisms were activated, most probably by the conversion of the pigment violaxanthin to zeaxanthin and antheraxanthin (Demmig-Adams and Adams 1992). Thus, the saplings seemed able to withstand continuous exposure to direct sun without damage to the photosynthetic apparatus.

Another mechanism besides photoinhibition needs to be invoked to explain the slow growth in some of the saplings. Conductances in all shade treatments were quite high (data not shown), and the ratios of internal to external CO_2 concentration ranged from 0.8 to 0.95, considerably higher than the expected optimum of 0.7 for C_3 plants. Energy not used for photochemistry is dissipated as heat. If leaf temperatures become too high (more than 35°C), the likelihood of photodamage increases (Demmig-Adams and Adams 1992; Gamon and Pearcy 1989). The high rates of transpiration may have been necessary to maintain leaf temperatures sufficiently low to prevent damage to the photosynthetic apparatus by a combination of high temperatures and high light. Thus, the limitation to growth in high light environments may be a result of the indirect costs of avoiding photodamage and temperature stress, as well as acquiring nutrients. Because of the requirement for transpirational cooling in high light, plants whose normal environment is shaded may have to allocate so much carbon belowground that they are unable to grow. If the stomata close, then the leaf becomes warmer and more susceptible to photodamage. This response may be exacerbated by low availability of nutrients. Low nutrient availability promotes increased allocation to roots (Mooney and Winner 1991) and may make leaves more susceptible to photodamage (Castro et al. 1995). Further investigation will be required to clarify these relations for saplings growing in high-light environments.

Acknowledgments. We thank Hsiang-hua Wang for collecting the height data, the International Institute of Tropical Forestry for the analyses of foliar N, and Ernesto Medina for helpful comments. We acknowledge grant BSR-881190 from the National Science Foundation to the Terrestrial Ecology Division, University of Puerto Rico, and the International Institute of Tropical Forestry, as part of the Long-Term Ecological Research Program in the Luquillo Experimental Forest. Additional support was provided by the USDA Forest Service, the University of Puerto Rico, a NASA Institutional Research Award to the University of Puerto Rico, and the Center for Tropical Atmospheric Sciences at the University of Puerto Rico.

Literature Cited

Castro, Y., Fetcher, N., and Fernández, D.S. 1995. Chronic photoinhibition in seedlings of tropical trees. *Physiologia Plantarum* **94**:560–565.

Demmig-Adams, B., and Adams, W.W. III. 1992. Photoprotection and other responses of plants to high light stress. *Annual Review of Plant Physiology and Plant Molecular Biology* **43**:599–626.

Demmig-Adams, B., and Adams, W.W. III. 1996. Xanthophyll cycle and light stress in nature: uniform response to excess direct sunlight among higher plant species. *Planta* **198**:460–470.

Fernández, D.S., and Fetcher, N. 1991. Changes in light availability following Hurricane Hugo in a subtropical montane forest in Puerto Rico. *Biotropica* **23**(4a):393–399.

Fetcher, N., Oberbauer, S.F., Rojas, G., and Strain, B.R. 1987. Efectos del regimen de luz sobre la fotosintesis y el crecimiento en plántulas de árboles de un bosque lluvioso tropical de Costa Rica. *Revista de Biología Tropical* **35**(suppl. 1):97–110.

Gamon, J.A., and Pearcy, R.W. 1989. Leaf movement, stress avoidance and photosynthesis in *Vitis californica*. *Oecologia* (Berlin) **79**:475–481.

Kitajima, K. 1996. Ecophysiology of tropical tree seedlings. In *Tropical Forest Plant Ecophysiology*, eds. S.S. Mulkey, R.L. Chazdon, and A.P. Smith, pp. 559–596. Chapman & Hall, New York.

Küppers, M., and Schulze, E.D. 1985. An empirical model of net photosynthesis and leaf conductance for the simulation of diurnal courses of CO_2 and H_2O exchange. *Australian Journal of Plant Physiology* **12**:513–526.

Mooney, H.A., and Winner, W.E. 1991. Partitioning response of plants to stress. In *Response of Plants to Multiple Stress*, eds. H.A. Mooney, W.E. Winner, and E.J. Pell, pp. 129–141. Academic Press, New York.

Myster, R.W., and Fernández, D.S. 1995. Spatial gradients and patch structure on two Puerto Rican landslides. *Biotropica* **27**:149–159.

Strauss-Debenedetti, S., and Bazzaz, F.A. 1996. Photosynthetic characteristics of tropical trees along successional gradients. In *Tropical Forest Plant Ecophysiology*, eds. S.S. Mulkey, R.L. Chazdon, and A.P. Smith, pp. 162–186. Chapman & Hall, New York.

Swaine, M.D., ed. 1996. *The Ecology of Tropical Tree Seedlings*. UNESCO, Paris.

Wadsworth, F.H. 1997. *Forest Production for Tropical America. Agriculture Handbook 710*. U.S. Department of Agriculture, Forest Service, Washington, DC.

7. Comparative Analysis of the Nutritional Status of Mahogany Plantations in Puerto Rico

Ernesto Medina and Elvira Cuevas

Abstract. Mahogany species from the Caribbean, Central America, and South America and the hybrid of small-leaf and big-leaf mahogany were planted on contrasting soils under different climates in Puerto Rico. Species performance differed markedly according to water availability. Big-leaf mahogany and the hybrid survived and grew at higher rates in the wetter northeastern sites, and small-leaf mahogany was more successful in the drier southwestern sites. We tested the hypothesis that this pattern results from differences in nutrient acquisition, as determined by soil acidity. We postulated that small-leaf mahogany is drought tolerant and sensitive to low pH and high Al mobility; big-leaf mahogany is drought sensitive and less tolerant to high concentrations of soluble soil Ca. The acid soils of wetter sites have a higher Al mobility that may interfere with P uptake in species sensitive to low soil pH. Calcareous soils of drier sites with low Fe availability and lower water-retention capacity may impair growth of drought-sensitive species. Concentrations of minerals (N, P, K, Ca, Mg, and occasionally Al and Mn) in the leaves showed significant differences among species and sites. In wetter sites, N, K, Mg, and Al tended to be higher; Ca was higher in sites with calcareous soils. The N:P molar ratios were normally higher in wetter sites (>45); the Ca:Mg molar ratios reflected the Ca availability and tended to be higher in the drier sites (>4). We emphasize the need to conduct experiments at the seedling and sapling stages to measure the pH-modifying character of the different mahogany species and their hybrids, as a predictor of nutrient uptake capacity under natural conditions.

Keywords: *Swietenia mahagoni, Swietenia macrophylla, Swietenia humilis*, Plantations, Growth, Nutrients relations, Mahogany

Introduction

The family Meliaceae contains several of the most valuable tropical timber trees (Little and Wadsworth 1964). In the Neotropics, species of the genus *Swietenia* produce a highly valued wood that has been exported to Europe since colonial times (Rizzini 1971). Three species are recognized: small-leaf mahogany, native to the Caribbean Islands; Pacific coast mahogany, distributed along the Pacific side of México and Central America; and big-leaf mahogany, the species with the largest range, from southern México to central-western Amazonia in Brazil, Ecuador, Perú, and Bolivia (Little and Wadsworth 1964; Lamb 1966).

Big-leaf mahogany was introduced and widely planted in Puerto Rico during colonial times (Little and Wadsworth 1964). More recently, this species and a hybrid between the big- and small-leaf mahoganies have been planted in selected areas under forest-cover rehabilitation programs in the Luquillo mountains (Bauer 1987).

In 1963, the Institute of Tropical Forestry of the USDA Forest Service began a comprehensive program to establish plantations of the three mahogany species in different places in Puerto Rico (Barres 1963). The goals were to develop a nursery with known provenances of mahogany from Central and South America and to measure their performance under field conditions. Growth rate and resistance to environmental stress and insect plagues were assessed. The species × provenance experiments (S-P experiment) were successfully established in sites of contrasting soil and climatic conditions (Geary et al. 1973).

Table 7.1. Survival (%) of Mahogany from Seed Sources in Puerto Rico as Influenced by the Moisture Regime of the Planting Location[a]

Species	Origin	Wetter Sites	Drier Sites
Big-leaf mahogany	México Guatemala	76	56
Pacific coast mahogany	México	27	69
Small-leaf mahogany	St. Croix, VI	23	68

[a] From Geary et al. (1973).

Measurement of growth and survivorship in several mahogany plantations throughout Puerto Rico and results of the species × provenance experiments yielded useful information on a variety of aspects: the apparent higher drought tolerance in small-leaf and Pacific coast mahogany than in big-leaf mahogany (Marrero 1950; Nobles and Briscoe 1966); the poor performance of small-leaf mahogany compared to big-leaf mahogany in wet sites of the Luquillo Mountains (Weaver and Bauer 1986); and the suitability of the hybrid mahogany for a wider range of environments than its parents (Geary et al. 1972).

A most indicative set of data on differential drought tolerance of the mahogany species cultivated in Puerto Rico was published by Geary et al. (1973). Survival of big-leaf mahogany was higher in wetter sites, where the other two species performed poorly. These two species showed the highest survival in drier sites (Table 7.1). The difference in survivorship between drier and wetter sites was only +20% in big-leaf mahogany, but it reached −42% in Pacific coast and −45% in small-leaf mahogany.

These observations have been confirmed by more recent measurements of the growth of the three species planted in that study in Coamo, El Verde, and Guavate (Fig. 7.1). Big-leaf mahogany performed best in Guavate followed by El Verde, and small-leaf mahogany performed best in Guavate, followed by Coamo. Growth of this species in El Verde was particularly poor. Pacific coast mahogany attained the largest diameters in Coamo, but its performance in any of the sites was well below that of the other two species. Survivorship of both Pacific coast and small-leaf mahogany was lower in the wet sites, but big-leaf mahogany was similar in the three sites examined (data not shown).

These results inspired us to evaluate the nutritional status of mahogany plantations in Puerto Rico. Our goal was to investigate the relations between nutrient status (as evaluated through the nutrient concentration of mature leaves) and soil types and site humidity, factors that could be used to explain the performance of mahogany plantations in different sites. So far as we know, the nutritional status of mahogany, in either natural forests or plantations, has seldom been assessed. We used some results obtained in two species of the genus *Khaya* (Meliaceae), the African

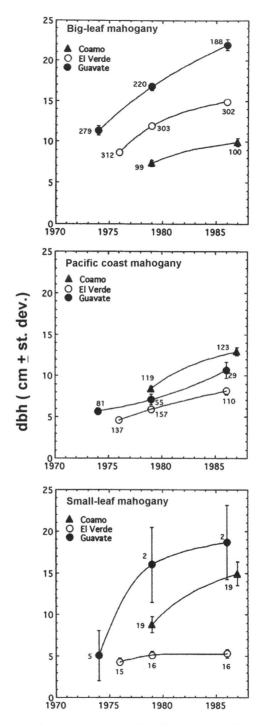

Figure 7.1. Growth performance measured as diameter at 1.5-m height (dbh in cm) in the three species of mahogany grown in sites of contrasting climate and soil type (site description in Table 7.2). Number of trees measured indicated at each point. Trees were planted in 1965.

mahogany that resembles *Swietenia* spp. in its nutritional requirements (Lugo et al. 1990; Cuevas and Lugo 1998). Our goal was to test these hypotheses: the total concentration of cations in the leaves reflects their availability in the soil; total growth rate and survival are associated with the availability of P in the soil; and differences in survivorship and growth among the species analyzed are correlated with drought tolerance, sensitivity to the mobility of potentially toxic ions such as Al and Mn in wet, acid soils, or both.

Materials and Methods

The sites sampled included several described in the species × provenance experiment (Geary et al. 1973) and other sites where big-leaf mahogany and the hybrid mahogany were planted. Most of the plantations were established in 1963 (Barres 1963; Weaver and Bauer 1986).

In most sites, we collected leaf samples from two to five trees, carefully selecting undamaged, healthy leaves. We stratified the leaf sampling when clear differences in leaf age were observed in the field. We differentiated qualitatively among young, mature, and old leaves according to texture, branch position, and color. Considering the deciduous character of mahogany species, differences in age among these qualitatively differentiated groups are about 2 to 3 months for young leaves, 4 to 6 months for adult leaves, and more than 7 months for old leaves. This stratification gave only a qualitative separation that was tested later against the results of nutrient analyses. The sequence young–mature–old should result in decreasing concentrations of P, N, and K and increasing concentrations of Ca and Mg (Medina 1984). In several sites, we also collected samples of bark from the main stem, at around 1.5 m in height. In these sets, we included samples obtained from big-leaf mahogany trees growing in the Institute's Arboretum in the Luquillo Experimental Forest. Bark samples were obtained with a leather punch 2 cm in diameter. Bark samples included the secondary phloem, the primary tissues that might still be present, and the periderm (Esau 1977). The most external dead tissues were removed to avoid differences between trees of different age.

Samples were collected in paper bags, transported to the laboratory within 6 h after collection, and oven-dried at 60°C to constant weight. Milled subsamples (100–500 mg dry weight) were acid digested. The resulting solution was analyzed in a plasma spectrometer for P, K, Ca, Mg, and, sometimes, for Al and Mn. Nitrogen was analyzed by a micro-Kjeldahl procedure. Element concentrations are given on an atomic basis, mmol kg^{-1} dry weight. This notation allows a direct comparison of the number of atoms of each nutrient incorporated into the tissues analyzed (Medina et al. 1994). Values are given as averages and the standard deviation of the mean.

Results and Discussion

General Soil and Climate of the Sampling Sites

The mahogany sampling sites covered most of the range of climates and soil types in Puerto Rico (Table 7.2). Holdridge's life zone system was used to classify the sites as dry, moist, and wet (Ewel and Whitmore 1973), and soil fertility was described by the U.S. Soil Survey classification (Soil Survey of Puerto Rico 1975–1982). The drier areas in the south included Guánica, Susúa, and Coamo. The soils have low to medium fertility and a pH neutral to slightly alkaline in Guánica (calcium carbonate substrate) and Susúa (magnesium-rich serpentine). Sabana Hoyos, Guajataca, and Río Abajo have a moist climate, although at higher altitude. Río Abajo has a slightly more humid soil than the other two sites. Fertility in these three sites is medium to high, calcium availability is high, and pH varies little around neutral. Finally, the sites in the mountain complex northeast of Puerto Rico have a wet climate, with relatively acid soils and medium soil fertility. Guavate is the site with higher effective humidity because of its altitude.

Variations in Nutrient Concentration as Determined by Leaf and Plant Age

Evaluating soil fertility of a site by analyzing plants required several precautions to reduce variability of the nutrient indices of the tissues analyzed. The analysis of total nutrient concentration posed some problems for interpreting the physiologically active fractions of the elements analyzed. Nitrogen and phosphorus are constituents of physiologically active compounds seldom accumulated in metabolically inactive pools; thus, N and P concentrations may be related to the physiological capacity of the plant (Medina 1984). Potassium, which does not constitute any biological structure, is found in the hydration layers of proteins in the cytosol and as a component of vacuolar sap (Marschner 1995); thus, its concentration in leaves is equal, or nearly so, to the amount of soluble, physiologically active compound. Potassium is related to enzymatic activity and intervenes in forming cell turgor. Calcium and Mg, however, tend to be accumulated in plant tissues in amounts related to their availability in the soil. In leaves, cation accumulation is related to the rate of transpiration. The concentrations of P, N, and K decrease with leaf age, in part because of retranslocation to growing tissues, and also because of the dilution effect represented by the increase in cell wall thickness with leaf age. Calcium concentration increases with leaf age because this element is practically immobile once it has been transported to the leaves. Magnesium presents different patterns according to soil availability and physiological demands.

The patterns of nutrient concentrations in leaves we described were detected in the mahogany samples when young, mature, and old leaves of

Table 7.2. General Soil and Climatic Characteristics of the Sites Sampled in Puerto Rico[a]

Sites	Elevation (m)	Holdridge's Subtropical Life Zone[b]	Soil Classification	Series	pH
Guánica	30	Dry	Udic Pellustert	Guánica	Neutral–alkaline
Susúa	100	Dry	Serpentine outcrops		Neutral–alkaline
Coamo	250	Dry	Typic Ustropepts	Callabo	Neutral–acid
Sabana Hoyos	50	Moist	Tropeptic Eutrorthox	Matanzas	Neutral–acid
Guajataca	230	Moist	Eutropeptic Rendolls	Soller-Colinas	Neutral–alkaline
Río Abajo	330	Moist-Subhumid	Lithic Ustorthents	Soller	Neutral–alkaline
El Verde-Harvey	100	Wet	Typic Tropohumult	Humatas	Acid
Sabana-Río Chiquito	100	Wet	Lithic Tropaquepts	Guayabota-Los Guineos	Acid
Guavate	600	Wet	Epiaquic Tropohumult	Los Guineos	Acid

[a] Sources: Basic information on the species x provenance experiments of the International Institute of Tropical Forestry from Geary et al. (1973). Other information and actualization from the Soil Survey of Puerto Rico (1975, 1977, 1979, 1982); Lugo-López and Rivera (1976, 1977).
[b] Life zone designation according to Ewel and Whitmore (1973).

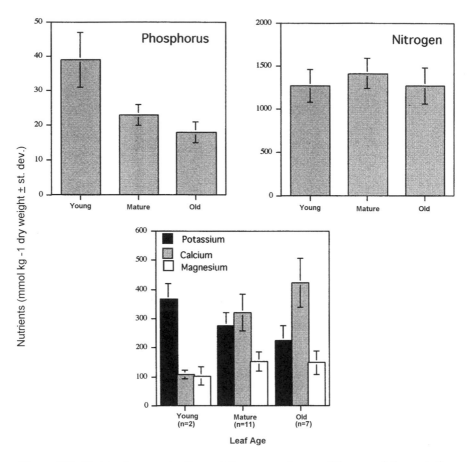

Figure 7.2. Nutrient concentrations of three age-classes of leaves of the putative mahogany hybrid between big-leaf and small-leaf mahogany, sampled in the Sabana site, Luquillo Mountains (number of samples in *parentheses*).

adult trees were analyzed separately. In a plantation of the mahogany hybrid in Sabana, N concentration remained almost constant in the three age categories, but P and K decreased markedly (Fig. 7.2). On the other hand, Ca concentration increased as a mirror image of P, increasing almost four times within the age categories. Magnesium increased only slightly with leaf age. This pattern of nutrient concentration may indicate that availability of N in this site was higher than that of P and K. During the life of a tree, changes in nutrient concentrations of fully developed leaves can also be detected, resulting from the changing nutritional conditions during tree growth. An age-series constituted by leaves of seedlings, regeneration, and canopy trees in small-leaf mahogany plantations were measured in Guánica. With plant age, N, P, and K clearly decreased, but Ca and Mg remained similar, showing the change in nutrient availability during plant growth (Fig. 7.3).

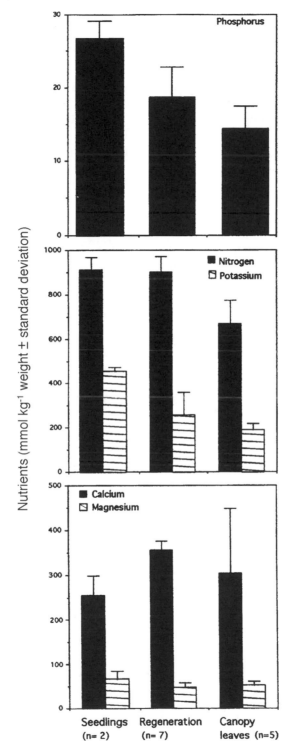

Figure 7.3. Mean nutrient concentration of fully expanded leaves of small-leaf mahogany seedlings, regeneration, and adult trees in the Guánica site. Differences result from plant age and partial shading of regeneration and seedlings (number of samples in *parentheses*).

Interspecific Differences in Leaf Nutrient Concentrations

Leaf analyses of the three species grown in Coamo showed no significant differences among the concentrations of P, N, and K (Fig. 7.4). Small-leaf mahogany, however, had significantly lower concentrations of Ca and Mg, and big-leaf mahogany showed lower Al concentrations. Manganese was barely detectable, and differences were erratic. These results were potentially significant in explaining the differences in performance of the mahogany species. On calcareous or serpentine substrates with high concentrations of available Ca or Mg, small-leaf mahogany would be able to restrict the uptake of these cations, thereby avoiding excessive accumulations. In humid and acid substrates, where Al mobility increases, the big-leaf species would more effectively restrict Al uptake than would the other two species. The Pacific coast mahogany was apparently less efficient in restricting the uptake of these cations.

Variations in Nutrient Concentrations on Different Sites

Comparison of the nutrient concentrations of all leaf samples collected showed distinct patterns when the sites were arranged according to water availability. Following the approach of Geary et al. (1973), we separated wetter from drier sites. In the graphic representation of nutrient values, we included the average for the whole set to facilitate interpreting the data (Fig. 7.5).

Phosphorus concentration of mature leaf samples was similar for most sites. Only young leaves from Sabana and Vieques had significantly above-average concentrations. The samples from Guánica showed significantly below-average concentrations of P, indicating a potential limitation for plant growth as suggested by Lugo and Murphy (1986). Nitrogen concentrations were clearly higher in most of the wetter sites, followed by the moist and seasonally dry sites. Guánica samples were significantly below the average for the whole set. Potassium had a pattern similar to N, although variability in the wetter sites was more pronounced. The pattern of distribution of Ca was erratic. Humid calcareous substrates, such as those from sites 5 to 8, appear to have large Ca availability for plants resulting from the solubility of calcium carbonate. As expected, the young leaves collected in sites 4 and 11 showed Ca concentrations as low as the one recorded for Susúa, the site on serpentine. Magnesium showed a tendency to be higher than average and with more variable concentrations in the wetter sites. From the drier sites, Vieques and Guánica, samples had consistently below-average concentrations, but the Susúa samples, as expected from the serpentine substrate, had above-average Mg concentrations. Concentrations of Al did not follow a clear pattern, but again the samples from wetter sites apparently tended to show higher Al concentrations than the samples from drier sites.

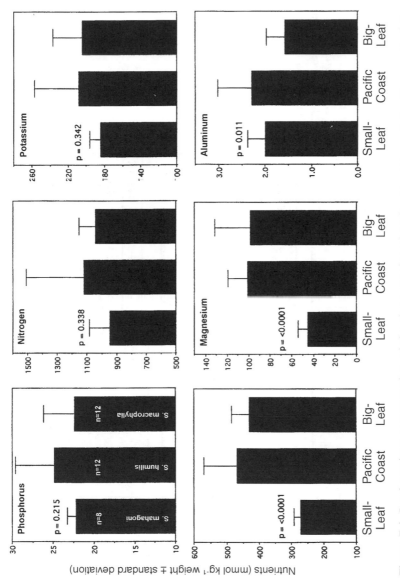

Figure 7.4. Leaf nutrient concentration of the three species of mahogany grown at the Coamo site (site description in Table 7.2). The significance level of the one-way analysis of variance is shown in the *first column* of each nutrient. Number of samples is included in the upper *left graph*.

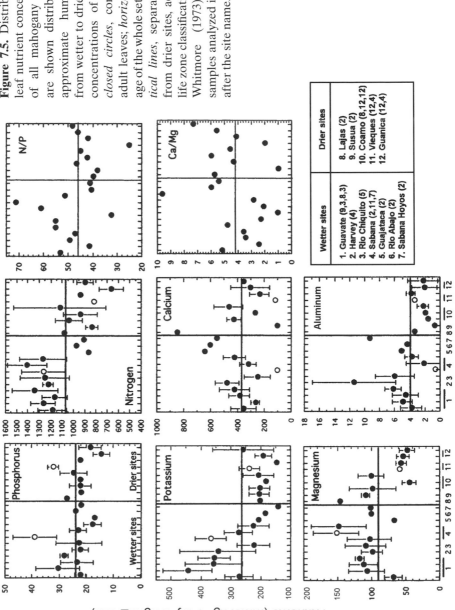

Figure 7.5. Distribution of mean leaf nutrient concentration (±SD) of all mahogany samples. Sites are shown distributed along an approximate humidity gradient from wetter to drier. *Open circles*, concentrations of young leaves; *closed circles*, concentrations of adult leaves; *horizontal line*, average of the whole set of samples; *vertical lines*, separation of wetter from drier sites, according to the life zone classification of Ewel and Whitmore (1973). Number of samples analyzed is in *parentheses* after the site name.

Nutrient Concentration in the Bark

The bark grows continuously through the life of a tree, and the dead tissues of the outer peridermis are regularly shed as the tree grows (Esau 1977). The concentration of nutrients in the bark may be used to assess the nutrient status of trees in natural forests or plantations. Our data set was too small to attempt a thorough assessment, but it provided a pattern for the nutrient economy of the plant. The concentration of N and P in the bark was well below the average concentration of these elements in the functional leaves ($24\,\mathrm{mmol\,kg^{-1}}$ for P and $1078\,\mathrm{mmol\,kg^{-1}}$ for N; Fig. 7.6). The proportion of P in bark compared to leaf tissues was much larger than that

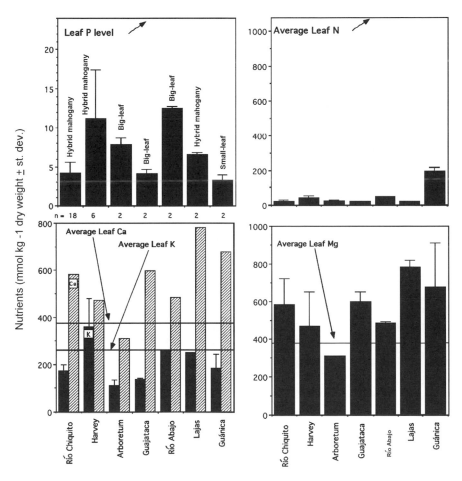

Figure 7.6. Nutrient concentration of bark samples from different species of mahogany trees growing in different sites in Puerto Rico. *Horizontal* lines indicate the average living leaf concentration of the corresponding nutrient obtained from Figure 7.5.

of N. The concentration of K in bark tissues was similar or slightly lower than in living leaves. The concentration of Ca and Mg, however, was clearly larger than in living leaves of the same tree. This accumulation of Ca and Mg in the bark represented a net loss of these cations from living plant tissues, but it contributed to their enrichment of the upper soil layers of the plantation as the bark was periodically shed from the older stems.

Range for Leaf Nutrients of Mahogany Planted in Puerto Rico

We are not aware of any published set of data on the nutrient status of mahogany plantations. The concentration of nutrients of the mahogany sample set in Puerto Rico was highly variable, but, for most sites, nutritional conditions were satisfactory. The results reported here were compared with averages for fertilized plantations of *Khaya senegalensis* in Africa (Rance et al. 1983) or for several common temperate nonlegume trees cultivated in the field (Mills and Jones 1996) (Table 7.3).

Comparison of the averages of *K. senegalensis* with our values indicated that P concentrations of mahogany planted in Puerto Rico were lower than expected for optimum performance. Our results on leaf P were more similar to those reported by Rance et al. (1983) than to the values reported by Nwoboshi (1982). The latter values for P are too high, possibly resulting from an excessive supply of P at the seedling stage. The P concentrations for mahogany are also well below the average calculated for temperate non-legume dicot trees. The range of P concentrations measured in mahogany trees in Puerto Rico resembled the values for the *K. senegalensis* cultivated without P fertilization ($17-28\,mmol\,kg^{-1}$) (Rance et al. 1983).

Average N concentration was slightly below the optimum, and the deficiencies were more frequent in the drier sites. Both P and Mg were within the expected range for optimum performance, but Ca was much higher on

Table 7.3. Range of Leaf Nutrient Concentrations for Three Mahogany Species[a] and the Hybrid of Small- and Big-Leaf Mahogany Under Natural Conditions in Puerto Rico

Nutrients mmol/kg	Mahogany spp		*Khaya senegalensis*		Temperate Trees	
	Average	Range	1[b]	2[c]	Average	Range
P	24	19–29	103	32–39	66	36–103
N	1078	882–1276	1286	1071–1429	1509	847–2207
Ka	267	176–358	292	128–179	270	145–509
Cag	376	201–551	125		384	15–1109
Mg	93	61–124	95		140	42–290

Data were compared with fertilization experiments on *Khaya senegalensis* (Nwoboshi 1982; Rance et al. 1983) and the average and range of 27 temperate trees. Adult leaves were selected from trees planted in the field; data were selected from Mills and Jones (1996).
[a] Big-leaf, small-leaf, Pacific coast mahogany.

average in Puerto Rico compared to the optimum values for *K. senegalensis*. These higher concentrations were surely associated with the widespread calcium carbonate-rich soils in Puerto Rico. Most of this Ca, however, was certainly insoluble and associated with cell walls (Medina 1984). Compared to the temperate tree average, mahogany samples appeared to have lower concentrations of P, N, and Mg, and similar concentrations of K and Ca.

Conclusions

Big-leaf, small-leaf, and hybrid mahogany grow satisfactorily in a wide variety of soils and climates in Puerto Rico. Differences in growth rates and nutrient requirements among these taxa accounted in part for the better performance of big-leaf mahogany in wetter, acid sites. The poor performance of small-leaf mahogany in wet sites may be associated with its intolerance to acid soils with higher Al mobility. The higher survival of small-leaf mahogany in drier climates was probably not only the consequence of higher drought tolerance, but also of the requirement for better soil aeration and tolerance to high concentrations of soluble Ca. Calcareous substrates, such as those predominating in the Guánica site, were characterized by a high degree of soil heterogeneity, with numerous sinkholes where water accumulates during the rainy season. This water is available for tree growth and allows the production of new foliage before the onset of the rainy season (Medina and Cuevas 1990). The behavior of Pacific coast mahogany was more difficult to interpret because of the small number of samples from this species, but it seemed to require well-aerated soils, medium to high rainfall, and neutral to slightly acid soils. Thus, big-leaf mahogany appeared to be the plant with wider environmental tolerances, and Pacific coast mahogany was more habitat specific.

The nutritional status of mahogany in Puerto Rico appeared to be satisfactory, with only a few extreme soil types or climates being inadequate for plantation development. Therefore, our results supported and expanded the conclusions of the Institute in its perspectives for developing mahogany plantations in the Caribbean (Geary et al. 1973; Weaver and Bauer 1986; Bauer 1987).

Acknowledgments. We acknowledge the financial support of the Guggenheim Foundation (E.M.), the Venezuelan Institute for Scientific Research, and the International Institute of Tropical Forestry (E.M. and E.C.). The technical support of Mary Jeane Sánchez and Edwin López of the International Institute of Tropical Forestry laboratory is gratefully recognized.

Literature Cited

Barres, H. 1963. Mahogany provenance study plan. 1.4-1 2421. U.S. Department of Agriculture, Forest Service, Institute of Tropical Forestry, Río Piedras, PR.

Bauer, G.P. 1987. Reforestation with mahogany (*Swietenia* spp.) in the Caribbean National Forest, Puerto Rico. Seminario-Taller de Cooperación y Manejo de Bosques Tropicales. Lima, Selva Central, Pucallpa, Perú.

Cuevas, E., and Lugo, A.E. 1998. Dynamics of organic matter and nutrient return from litterfall stands of ten tropical tree plantation species. *Forest Ecology and Management* **112**:263–279.

Esau, K. 1977. *Anatomy of Seed Plants*, second edition. Wiley, New York.

Ewel, J., and Whitmore, L. 1973. *The Ecological Life Zones of Puerto Rico and the U.S. Virgin Islands*. Research Paper ITF-18. U.S. Department of Agriculture, Forest Service, Institute of Tropical Forestry, Río Piedras, PR.

Geary, T.F., Barres, H., and Briscoe, C.B. 1972. *Hybrid Mahogany Recommended for Planting in the Virgin Islands*. Research Paper ITF-15. U.S. Department of Agriculture, Forest Service, Institute of Tropical Forestry, Río Piedras, PR.

Geary, T.F., Barres, H., and Ybarra-Coronado, R. 1973. *Seed Source Variation in Puerto Rico and Virgin Islands Grown Mahoganies*. Research Paper ITF-17. U.S. Department of Agriculture, Forest Service, Institute of Tropical Forestry, Río Piedras, PR.

Lamb, F.B. 1966. *Mahogany of Tropical America: Its Ecology and Management*. University of Michigan Press, Ann Arbor, MI.

Little, E.L., Jr., and Wadsworth, F.H. 1964. *Common Trees of Puerto Rico and the Virgin Islands, Vol. 1. Agriculture Handbook 249*. U.S. Department of Agriculture, Forest Service, Washington, DC.

Lugo, A.E., and Murphy, P.G. 1986. Nutrient dynamics of as Puerto Rican subtropical dry forest. *Journal of Tropical Ecology* **2**:55–72.

Lugo, A.E., Cuevas, E., and Sánchez, M.J. 1990. Nutrients and mass in litter and top soil of ten tropical tree plantations. *Plant and Soil* **125**:263–280.

Lugo-López, M.A., and Rivera, L.H. 1976. *Taxonomic Classification of the Soils of Puerto Rico, 1975*. Bulletin 245. Agricultural Experiment Station, College of Agricultural Sciences, University of Puerto Rico, Río Piedras, PR.

Lugo-López, M.A., and Rivera, L.H. 1977. *Updated Taxonomic Classification of the Soils of Puerto Rico*. Bulletin 258. Agricultural Experiment Station, College of Agricultural Sciences, University of Puerto Rico, Río Piedras, PR.

Marrero, J. 1950. Results of forest planting in the insular forests of Puerto Rico. *Caribbean Forester* **11**:107–147.

Marschner, H. 1995. *Mineral Nutrition of Higher Plants*. Academic Press, London.

Medina, E. 1984. Nutrient balance and physiological processes at the leaf level. In *Physiological Ecology of Plants of the Wet Tropics*, eds. E. Medina, H.A. Mooney, and C. Vázquez Yanes, pp. 139–154. Junk, The Hague.

Medina, E., and Cuevas, E. 1990. Propiedades fotosintéticas y eficiencia de uso de agua de plantas leñosas del bosque decíduo de Guánica: consideraciones generales y resultados preliminares. *Acta Científica, Puerto Rico* **4**(1–3):25–36.

Medina, E., Cuevas, E., Figueroa, J., and Lugo, A.E. 1994. Mineral content of leaves from trees growing on serpentine soils under contrasting rainfall regimes in Puerto Rico. *Plant and Soil* **158**:13–21.

Mills, H.A., and Jones, J.B., Jr. 1996. *Plant Analysis Handbook II*. Revised edition. MicroMacro Publishing, Athens, GA.

Nobles, R.W., and Briscoe, C.B. 1966. *Height Growth of Mahogany Seedlings, St. Croix, Virgin Islands*. Research Note ITF-10. U.S. Department of Agriculture, Forest Service, Institute of Tropical Forestry, Río Piedras, PR.

Nwoboshi, C.L. 1982. Indices of macronutrient deficiencies in *Khaya senegalensis* Juss seedlings. *Communications in Soil Science and Plant Analysis* **13**:667–682.

Rance, S.J., Cameron, D.M., and Williams, E.R. 1983. Nutritional requirements and interactions of *Khaya senegalensis* on tropical red and yellow earths. *Communications in Soil Science and Plant Analysis* **14**:167–183.

Rizzini, C.T. 1971. *Arvores e Madeiras Úteis do Brasil. Manual de Dendrologia Brasileira*. Editôra Edgard Blücher, Editora da Universidade de São Paulo, Brasil.
Soil Survey of Puerto Rico. 1975–1982. Soil Survey of Puerto Rico (1975, 1977, 1979, 1982). U.S. Department of Agriculture, Soil Conservation Service and University of Puerto Rico, College of Agricultural Sciences, Río Piedras, PR.
Weaver, P.L., and Bauer, G.P. 1986. Growth, survival, and shoot borer damage in mahogany plantings in the Luquillo Forest in Puerto Rico. *Turrialba* **36**:509–522.

8. Growth-, Water-, and Nutrient-Related Plasticity in Hybrid Mahogany Leaf Development Under Contrasting Light Regimes

Ernesto Medina, Hsiang-Hua Wang,
Ariel E. Lugo, and Nathaniel Popper

Abstract. Growth responses and changes in leaf chemical composition of hybrid mahogany (big-leaf × small-leaf) seedlings grown under contrasting light regimes were measured over 417 days. The seedlings showed strong reductions in relative growth rates under low light, but maintained a positive carbon gain, indicated by continuous biomass accumulation under these conditions. Light intensity seemed to affect more markedly the growth of individuals with the largest intrinsic growth capacity. Growth under full light resulted in pronounced water stress, particularly in seedlings without a large root system. The seedlings showed high plasticity in leaf area but not in leaf dry weight, which explains the larger area:weight ratios of seedlings developing under low light. Nitrogen concentration was significantly higher in low-light seedlings, but because of the larger area:weight ratios of low-light leaves, the N concentration per unit area was higher under the intermediate- and high-light ($\sim$70 mmol m^{-2}) than in the low-light treatment (55 mmol m^{-2}).

146

Keywords: Mahogany, Shade tolerance, Nutrients, Water stress, Phenolic compounds, *Swietenia*

Introduction

Tree species in the family Meliaceae produce highly valued woods for the international market (Lamb 1966). Big-leaf mahogany is extensively logged from natural forests in Central and South America to supply the international trade. A discussion has developed around harvest of this species from natural forests that essentially considers the capacity of big-leaf mahogany to regenerate under natural conditions (Veríssimo et al. 1995; Snook 1996).

Lamb (1966), Gullison et al. (1996), and Snook (1996) have summarized the known characteristics of big-leaf mahogany regeneration; the species appears to require large-scale disturbances, resulting in large clearings, to regenerate successfully. The seed source for occupying those clearings is surviving adult trees. After establishing in large clearings or pastures, big-leaf mahogany trees develop for centuries in almost even-aged populations. Big-leaf mahogany seedlings do not survive for prolonged periods in the forest understory, however. Some authors (Lamb 1966; Gullison et al. 1996; Snook 1996) consider big-leaf mahogany a late-successional species, implying that it is at least partially shade tolerant (Ramos and Grace 1990). Gerhardt (1994) observed natural and enrichment-planting regeneration of big-leaf mahogany in secondary, seasonally dry forests of Guanacaste, Costa Rica.

Big-leaf mahogany grows in a range of climatic conditions, from yearlong wet to seasonally dry lowlands in Central and South America (Lamb 1966), which may at least partially explain the contradictory reports on the regeneration ecology of this species. A literature search showed that very little work has been done on the ecophysiology of *Swietenia* species and that not much has been done in this direction since Lamb's review.

We studied growth and nutrient relations of hybrid mahogany seedlings cultivated under contrasting light conditions in Puerto Rico. Our goal was to measure the tolerance to shade, as expressed by biomass accumulation, nutrient absorption, and modifications in leaf structure and osmotic properties of the leaf sap. We studied a hybrid of big-leaf and small-leaf mahogany developed in Puerto Rico (Whitmore and Hinojosa 1977) that has been recommended as a plantation tree in the Caribbean for its performance under seasonal climates (Marrero 1950; Nobles and Briscoe 1966; Geary et al. 1972). Both big- and small-leaf mahogany grow naturally in contrasting climates, depending on water availability. Big-leaf mahogany inhabits lowland evergreen to seasonally humid forests in an area extending from southern México to Bolivia (Lamb 1966; Newton et al. 1993).

Small-leaf mahogany grows naturally in seasonally dry forests in the western Caribbean islands and southern Florida. Both species are canopy trees that lose their leaves in markedly seasonal climates, but big-leaf mahogany appears to be more tolerant of acid than small-leaf mahogany, which appears to require nearly neutral soils in drier climates in the Caribbean (Medina and Cuevas, Chapter 7, this volume). Although the physiological behavior of seedlings obtained from hybrid plants may be more variable, we believe that the pattern of response to light would be comparable to the light requirements of either parent species.

The experiment was expected to provide information to test the following hypotheses:

- Under low light, shade-tolerant plants maintain lower relative rates of growth in height and volume than do shade-intolerant plants.
- Tolerance to reduced daylight is associated with the capacity of the plant to maintain a positive carbon balance, expressed in biomass accumulation.
- Regardless of shade tolerance, water stress on plants grown under full daylight is more pronounced than in plants grown under reduced daylight.
- Shade tolerance is associated with the plasticity of leaf structural properties (leaf area, specific leaf area, nutrient uptake and allocation, and concentration of osmotically active solutes in leaf sap, or leaf sap osmolality). Shade-tolerant plants are less plastic than shade-intolerant plants.

Materials and Methods

Seedling Treatment and Planting

We obtained 3-month-old seedlings of the hybrid mahogany from the Commonwealth Department of Natural Resources and the Environment nursery at Cambalache, Puerto Rico, in December 1996. Seedlings grown from seed were transplanted from plastic bags to six nursery beds filled with garden soil to 1-m depth at the International Institute of Tropical Forestry. They were planted at 15- by 15-cm spacing, 58 seedlings in beds one to four, 46 seedlings in bed five, and 32 seedlings in bed six. Initially, all seedlings were exposed to full sunlight for 2 weeks to allow recovery from transplantation. Then the seedlings were exposed to the following light treatments (Fetcher et al., Chapter 6, this volume), with two nursery beds per light treatment: 6.2% full daylight (low light), 32.9% full daylight (intermediate light); and 79.8% full daylight (high light). These light intensities were obtained by using green shading screens (saran net) to cover the nursery beds. We did not measure red:far-red (R:FR) ratios in our experiment, but they were assumed not to differ among treatments. Under natural forest canopies, the R:FR ratio of daylight is reduced about five times in passing through a closed forest canopy (Tinoco-Ojanguren and Pearcy 1995; Lee et al. 1996). Large reductions in R:FR ratios affect morphogenesis and carbon allocation to roots and shoots, but we were interested in how availability of photosynthetically active radiation affected photosynthesis, total biomass accumulation, and leaf composition. In the discussion, we address the possible influence of R:FR ratios on growth patterns of saplings.

Growth Measurements

Seedling height to the tip of the stem was measured with metric tape, and stem diameter at the soil surface was measured with a caliper. For the analysis of growth, we used height and a parameter related to stem volume calculated as the product of basal area × stem height (cm^3). All seedlings were measured 30, 61, 211, 296, and 417 days after they were transplanted in December 1996. We used these dates to represent plant age in the analysis of biomass and nutrient accumulation.

Biomass of the seedlings was measured at the beginning of the experiment and at 296 and 417 days after transplanting. For biomass sampling, 10 seedlings were harvested from each light treatment and separated into two categories: stems and leaves. Because of the large variability in seedling size, we divided the population at these sampling periods into two height groups corresponding to the higher and lower 50th percentiles (tall and short populations). From each group, five seedlings were selected at random and cut for aboveground biomass measurements. The plant material was oven-dried at 65°C to constant weight and then ground for chemical analysis.

Osmotic Evaluation

Before the final harvest, leaf osmotic parameters were measured in detail. The nursery beds were irrigated twice a day for 3 days before sampling, to eliminate water potential differences that may have been caused by higher leaf and air temperatures in the intermediate- and high-light treatments. Leaves were collected before dawn to prevent leaf dehydration from transpiration after sunrise. Collections of fully expanded leaves were stratified into three classes: young (one to three leaf pairs from the top), adult (four to seven leaf pairs), and old (the lowest two leaf pairs). Six leaves were collected for each class in every treatment from which undamaged leaflets (functionally equivalent to leaves) were separated and cleaned with a soft paper towel. Seedlings sampled were selected randomly, from both nursery beds in each light treatment, but excluding the seedlings bordering each extreme of the nursery bed. This sampling allowed us to measure leaves of different ages, developed under different light regimes. The statistical verification of this visual classification of leaf types was undertaken *a posteriori*, by using the following variables: area, dry weight, water content, concentration of nutrients, osmolality of leaf sap, and concentration of flavonoid-like substances.

Sampled leaves were put in a large plastic bag to prevent dehydration and maintained in an air-conditioned room until further processing. In the laboratory, we separated the leaflet laminae from each compound leaf, and filled 10-cm^3 plastic syringes with two to three leaflets from each age-treatment group. Each syringe was sealed with its plastic plunger and frozen in a styrofoam container with dry ice. From the remaining group, we selected 20 leaflets for fresh weight and area determination with a Licor leaf-area meter. The leaflets were dried in a ventilated oven at 65°C for 72 h and weighed. Dried leaflets from each age-treatment group were separated into three subsamples (seven, seven, and six leaves) and ground for chemical analysis. Leaf water content per unit area and weight was calculated as the difference between fresh and dry weight divided by the corresponding area and weight.

Frozen leaflets in plastic syringes were allowed to thaw for 1 h at room temperature and then centrifuged at 4500 rpm for 20 min to extract the leaf sap. This procedure allowed us to obtain 1 to 3 ml of leaf sap. The sap was placed in 7-ml glass vials, sealed, and maintained in a refrigerator at 4°C until further processing. Osmolality of leaf sap was measured with a Wescor dewpoint osmometer calibrated with NaCl solutions. Values are expressed in millimoles (mmol) of osmotically active substances per kilogram of leaf sap.

Flavonoid-Like Substances

The extracted leaf sap was a pale yellow-brown that increased in intensity with leaf age and light treatment. We measured the absorption spectrum (Bausch and Lomb spectrophotometer) and found that light absorption of all saps decreased from very high at 350 nm to zero at 420 nm. This color probably results from the accumulation of flavonoids that have a strong absorption below 420 nm (Harborne 1973). We measured the light absorbance at 380 nm in diluted sap (0.2 ml → 25 ml) to ensure that all samples were measured under the same conditions.

Nutrient Analysis

Soil, leaf, and stem biomass and leaf sap were analyzed for P, K, Ca, and Mg by using plasma emission spectrometry (Spectra Span V). The total content of C, N, and S in soil and biomass samples was measured by a dry combustion method (using a LECOCNS-2000 analyzer), based on the work of Nelson and Sommers (1982) and Tabatabai and Bremmer (1991). Ash content was measured after ashing at 490°C. Exchangeable Ca and Mg in soils were determined by using the 1N KCl extraction method: K and P were determined by using Olsen-EDTA [NH$_4$F-ethylenediaminetetraacetic acid-(EDTA-)NaHCO$_3$], following the procedure of Hunter (1982).

Table 8.1. Chemical Properties of the Soil Used to Grow Mahogany Seedlings Under Different Light Regimes[a]

Light Treatment	Extractable					Total		
	P $(mm\,kg^{-1})$	K $(mm\,kg^{-1})$	Na $(mm\,kg^{-1})$	Ca $(mm\,kg^{-1})$	Mg $(mm\,kg^{-1})$	ECEC[b] $(cmm\,kg^{-1})$	C $(cmol\,kg^{-1})$	N $(cmol\,kg^{+1})$
High	0.425	4	3	561	14	15	892	57
Intermediate (medium)	0.365	4	2	671	18	18	946	50
Low	0.329	4	2	776	16	20	1121	57

[a] Average of two replicates.
[b] ECEC, effective cation-exchange capacity.

Ground plant samples used for P, K, Ca, and Mg analysis were digested by using the procedure of Huang and Schulte (1985). Samples were digested with concentrated HNO_3 and 30% H_2O_2.

Precision for most analyses was assured by running samples of known chemical composition after every 40 determinations for total C and for N, and after every 20 determinations for total S. These control samples [citrus leaves (NBS-1572), peach leaves (NIST-1547), and pine needles (NIST-1575)] were obtained from the National Institute of Standards and Technology, Gaithersburg MD, USA. The calibration standards (tobacco leaves, orchard leaves, and alfalfa) used in the total C, N, and S analyses were obtained from Leco Corp. (St. Joseph, MI, USA).

Soil used for cultivating mahogany seedlings was analyzed at the end of the experiment after 1 year under different light regimes to document its fertility. The replicate samples from the three nursery beds were very similar (Table 8.1).

Data Analysis

We compared the data obtained from the three light treatments as independent populations, and compared average growth rates and leaf chemical composition by using a factorial analysis of variance (ANOVA). Significance of differences was tested using an *a posteriori* test, Fisher's protected least significant difference (Stat View; Abacus Concepts, Berkeley, CA, 1992–1995).

Results

Seedling Structural Development

Seedling height was variable but increased steadily on average during the 417 days of growth after transplanting to the nursery bed (Fig. 8.1). Significant differences in average seedling height among treatments were detected at the time of the third harvest, and a clear tendency appeared for seedlings of the intermediate-light treatment to be taller.

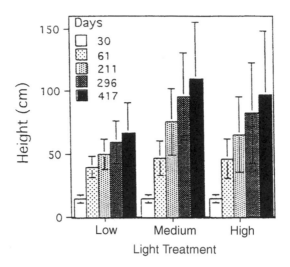

Figure 8.1. Mean (± SD) height of hybrid mahogany seedlings, over time, grown under different light regimes. Number of seedlings measured is indicated in Table 8.2.

The increase in sapling volume followed a pattern described by a power function with a significant coefficient of determination ($r^2 \geq 0.90$) in all light regimes (Fig. 8.2). A large variability in seedling development was observed throughout the experiment, probably derived from its hybrid nature. The large standard deviations of the means obtained express the variation. Cultivation under high- and intermediate-light treatments did not produce differences in seedling development, but total volume was significantly smaller under the low-light treatment. The distribution of seedling sizes overlapped throughout the experiment (Fig. 8.2). The largest seedlings of the low-light treatment were similar in size to the smaller seedlings of the intermediate- and high-light treatments. The low-light treatment, then, appeared to be limiting structural development in that fraction of the population with the largest intrinsic potential growth rate.

Relative growth rates for height and volume were calculated by using the standard formulation of Chiariello et al. (1989). Rates for the seedlings of the low-light treatment were always lower than those in the intermediate- and high-light treatments (Table 8.2). The highest rates for both height and volume were recorded during the first month after transplanting. Afterward, height growth rates remained similar until the fourth measuring period, when they decreased slightly. Volume growth rates remained similar during the last three measuring periods.

Biomass Accumulation

The accumulated biomass increased in all light treatments during the interval from 296 to 417 days after transplanting (Table 8.3). The differences

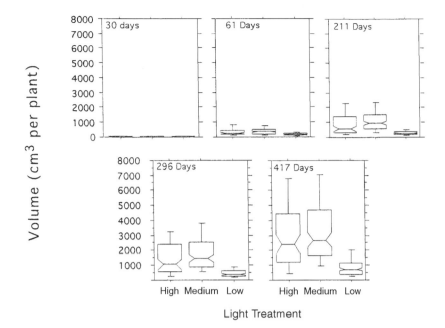

Figure 8.2. Volume development of hybrid mahogany seedlings (basal area × height) at different periods after transplanting into nursery beds and cultivated under contrasting light regimes. See Figure 8.1 for treatment designation. Values within each light treatment were fitted to a power function relating volume (y) in cm^3 to time in days: *low*, $y = 1.764\ x^{1.009}$, $r^2 = 0.91$; intermediate (*medium*), $y = 0.025\ x^{1.606}$, $r^2 = 0.92$; high, $y = 1.120\ x^{1.708}$, $r^2 = 0.91$.

Table 8.2. Mean (±SD) Relative Growth Rate (week^{-1}) in Height (cm) and Volume (cm^3) of Mahogany Seedlings Cultivated Under Different Light Regimes

Period (number of days)	Light treatment		
	High ($n = 76$)	Intermediate ($n = 114$)	Low ($n = 113$)
30–61	0.259 (0.044)[a]	0.264 (0.042)[a]	0.231 (0.014)[b]
61–211	0.015 (0.008)[a]	0.022 (0.009)[b]	0.011 (0.006)[c]
211–296	0.018 (0.009)[a]	0.019 (0.010)[a]	0.013 (0.008)[b]
30–61	523 (127)[a]	516 (95)[a]	367 (49)[b]
61–211	39 (17)[a]	45 (15)[b]	20 (10)[c]
211–296	41 (15)[a]	40 (14)[a]	33 (12)[b]
296–417	41 (15)[a]	38 (19)[b,c]	34 (24)[c]

Numbers in rows followed by the same letter are not significantly different ($p = 5\%$; Fisher's protected least significant difference).

Table 8.3. Mean (±SD) Standing Aboveground Biomass and Absolute Growth Rate per Plant Separating the 50% Lower and 50% Higher Population Percentile Under Each Treatment ($n = 5$)

Seedling Population	Treatment (g)	Leaves (g)	Stem (g)	Total (g)
Short (lower 50th percentile)	H-296	14.3 (5.7)	14.8 (4.8)	29.1 (9.4)
	M-296	21.8 (10.5)	25.6 (3.2)	47.4 (8.1)
	L-296	8.3 (5.3)	6.0 (2.7)	14.3 (7.6)
	H-417	22.5 (11.2)	43.1 (13.4)	65.6 (20.6)
	M-417	24.3 (11.7)	36.7 (13.5)	61.0 (24.5)
	L-417	8.0 (4.0)	8.5 (3.6)	16.5 (6.6)
Tall (higher 50th percentile)	H-296	68.0 (40.5)	72.8 (62.0)	140.9 (99.6)
	M-296	45.8 (11.3)	55.1 (18.8)	100.9 (27.8)
	L-296	21.1 (6.0)	12.9 (3.7)	34.0 (9.6)
	H-417	81.1 (14.1)	128.0 (40.6)	209.1 (54.1)
	M-417	119.4 (35.0)	105.1 (49.7)	224.5 (82.4)
	L-417	24.0 (10.6)	24.7 (11.1)	48.6 (21.2)
Critical difference, Fisher's PLSD, $p = 5\%$		22.6	34.6	54.7
Growth rate (mg/plant × day)				
Short (lower 50th percentile)	H	67.8	233.9	301.7
	M	20.7	92.4	112.4
	L	−2.5	20.7	18.2
Tall (higher 50th percentile)	H	108.3	456.2	563.6
	M	608.3	413.2	1021.5
	L	24.0	97.5	120.7

H, high-, M, intermediate-, and L, low-light treatment; the numbers are days of treatment. PLSD, protected least signicant difference.

among treatments and time were significant only in the tall population. For the short population, however, the low-light treatment showed a clear tendency toward smaller values; variability was high, and no statistical difference could be detected. These results showed the degree of growth suppression in the short population. Within the seedling population, a portion apparently will not grow regardless of the light regime, and light appeared to limit the growth of the portion of the population with the higher growth potential.

Calculating absolute growth rates between the two biomass harvests showed the same trends. The short population grew more slowly, but results clearly showed that seedlings grew fastest in high light and slower in low light (Table 8.3). The tall population grew faster than the short population, and the intermediate-light treatment resulted in the fastest growth rates. At the time of the last two harvests, differences in the light regime may have interacted with water stress. Seedlings in the fully exposed nursery bed had a larger atmospheric evaporative demand because of the bed's higher leaf and air temperatures.

Leaf Production

Even after 417 days of growth, the leaf scars left from fallen leaves could be distinguished, so we could count the leaves produced by each seedling by counting the leaf scars and adding the number of leaves present at harvest. The number of leaves produced during the experiment was lowest (about 25 leaves per seedling) in the low-light treatment and highest (about 50 leaves) for the large seedlings of the intermediate- and high-light treatments (Fig. 8.3). For the high- and intermediate-light treatments, the tendency was for the smaller seedlings to produce fewer leaves, although differences were significant only in the intermediate treatment.

Total Aboveground Biomass and Nutrient Accumulation

We tested whether differences in biomass accumulation caused by light conditions during growth or by the intrinsic (genetic) seedling growth rates were associated with differences in nutrient absorption. The concentration of P in leaves was not statistically different among light treatments or growth capacity (Table 8.4). Interestingly, P concentration in stems was commonly higher than that of leaves, indicating that the stems may constitute a P reservoir for this species. The N and S concentrations were always higher in leaves than in stems; with the exception of the high-light-tall group, their concentration in both tissues was significantly higher in low

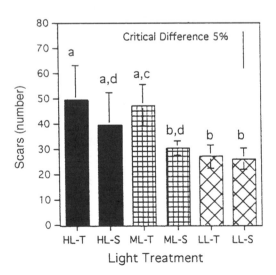

Figure 8.3. Mean (± SD) number of leaves produced by hybrid mahogany seedlings under different light regimes, as calculated from the number of leaf scars counted in seedlings sampled for biomass accumulation. The columns separate the tall (*T*) and short (*S*) seedling populations corresponding to the high (*H*) and low (*L*) 50% percentiles of the whole seedling population. See Figure 8.1 for treatment designation.

Table 8.4. Mean (+SD; $n = 5$) Leaf and Stem Biomass and Concentration of Nutrients of 1-Year-Old Hybrid Mahogany Seedlings Cultivated Under Different Light Intensities

Treatment[a]	Biomass (g)	P (mmol kg^{-1})	N (mmol kg^{-1})	S (mmol kg^{-1})	C (mol kg^{-1})	K (mmol kg^{-1})	Ca (mmol kg^{-1})	Mg (g kg^{-1})	Ash
L-T, leaves	24.0 (10.6)	47.0 (11.5)	1031 (175)	420 (81)	41.9 (0.5)	276 (10)	466 (68)	79 (13)	99.5 (12.7)
L-S, leaves	8.0 (4.0)	47.0 (11.5)	1065 (69)	420 (55)	41.4 (0.4)	286 (27)	479 (102)	94 (19)	110.4 (4.8)
L-T, stem	24.7 (11.1)	64.8 (11.6)	283 (69)	63 (17)	40.8 (0.4)	203 (27)	203 (65)	33 (5)	43.7 (7.3)
L-S, stem	8.5 (3.6)	70.4 (10.0)	294 (69)	96 (36)	40.7 (0.1)	251 (11)	174 (10)	35 (9)	46.3 (4.0)
M-T, leaves	119.4 (35.0)	36.4 (7.2)	666 (77)	280 (53)	42.3 (0.6)	352 (63)	526 (73)	71 (9)	101.1 (7.1)
M-S, leaves	24.3 (11.7)	35.0 (5.4)	740 (105)	283 (42)	41.8 (0.4)	292 (54)	434 (105)	85 (37)	98.9 (7.8)
M-T, stem	105.1 (49.7)	43.2 (8.8)	216 (41)	46 (8)	42.0 (0.4)	168 (36)	135 (58)	21 (8)	30.2 (6.3)
M-S, stem	36.7 (13.5)	44.2 (8.8)	216 (41)	46 (8)	41.6 (0.3)	167 (25)	114 (28)	25 (5)	30.3 (6.7)
H-T, leaves	81.1 (14.1)	46.6 (13.0)	647 (85)	309 (67)	43.2 (0.7)	322 (61)	497 (112)	69 (18)	100.6 (10.4)
H-S, leaves	22.5 (11.2)	36.0 (5.3)	733 (93)	377 (120)	43.3 (0.4)	251 (49)	436 (79)	74 (16)	93.0 (8.6)
H-T, stem	128.0 (40.6)	52.8 (14.0)	203 (26)	36 (4)	42.2 (0.2)	145 (20)	103 (21)	21 (3)	27.2 (6.2)
H-S, stem	43.1 (13.4)	41.2 (94)	224 (36)	50 (16)	41.7 (0.3)	155 (33)	112 (24)	19 (2)	29.5 (4.6)
LSD, leaves[b]	22.7	11.1	139	97	0.7	63	119	27	11.6
LSD, stem	36.2	14.5	63	23	0.4	34	52	7	7.8

[a] L, low-, M, intermediate-, H, high-light treatments; T, tall, and S, short, groups selected at random from low higher and lower 50% percentiles.
[b] Analysis of variance and estimation of the critical difference using Fisher's protected least significant difference (PLSD) at $p \le 5\%$.

light. Carbon concentrations were similar among treatment and tissues, but leaves and stems in low light had significantly lower concentrations than in high light. Carbon concentrations of the intermediate treatment were intermediate. Concentrations of K, Ca, and Mg were generally higher in leaves than in stems, but the concentrations of these elements were higher in the stems of the low-light treatment. The K concentration tended to be higher in the leaves of the intermediate and high treatments. Differences in the leaf concentration of Ca and Mg among treatments were not significant. Leaf ash concentrations were higher than those of the stems, possibly as a result of the preferential transportation of minerals through the transpiration stream to the leaves.

Plasticity of Leaf Structure and Composition

Area:Weight Ratios and Water Content Relative to Leaf Age

The sets of leaves differed significantly among treatments and age groups. Leaf area was largest in the low-light treatment, reduced in the intermediate-light treatment, and smallest in the high-light treatment (Fig. 8.4). The last two treatments showed a tendency for the leaf area to increase with age. Leaf weight varied similarly in the three treatments, and young leaves weighed significantly less than in the other two groups in all treatments. Water per unit leaf area increased significantly from the low- to the high-light treatment. The water per unit dry weight was always higher in young leaves, and it tended to decrease from low- to high-light treatments. Finally, the leaf area:weight ratios were larger in young leaves, and they tended to diminish as light availability increased.

Osmotic Relations and Soluble Cations

Increasing light availability during seedling growth is bound to affect water relations because of the interaction between the light environment and the demand for water transport along the soil–plant–atmosphere continuum that maintains leaf turgor. This water stress was reflected in the osmotic concentration of the leaf sap. For all treatments, young leaves showed lower osmolalities than did adult and old leaves (Fig. 8.5a). As expected, osmolality showed a clear tendency to increase from the low- to the high-light treatment.

Sap osmolality is determined by the concentration of soluble inorganic compounds, such as metallic cations, and organic compounds, such as sugars and organic acids. The concentration of soluble P in the leaf sap did not show any particular trend (Fig. 8.6), but K tended to decrease in older leaves and Ca and Mg showed large increases in concentration with the age of the leaf group. Ion concentrations tended to be lower in the low-light treatment, but the tendency was less pronounced than for total osmolality. A significant correlation was found between leaf osmolality and the sum

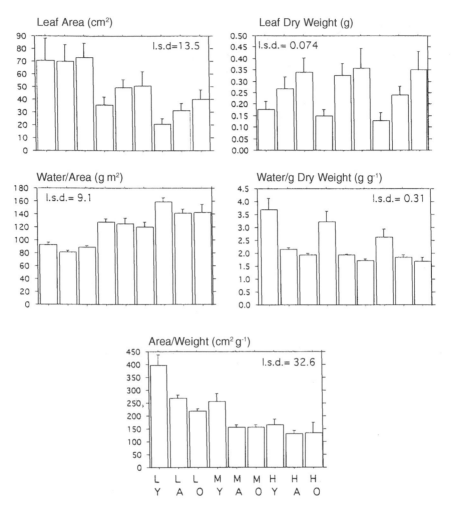

Figure 8.4. Mean (±95% CI) of leaf properties of young (*Y*), adult (*A*), and old (*O*) leaves collected from hybrid mahogany seedlings grown under contrasting light regimes (*L*, *M*, *H*, as in Fig. 8.1). *LSD*, least significant difference (95%).

of the concentrations of K, Ca, and Mg. These cations alone explained an increasing percentage of leaf sap osmolality as the leaf ages (Table 8.5). No differences were detected in the behavior of leaves from different treatments.

Accumulation of Flavonoid-Type Substances

Light absorbance at 380 nm of the leaf sap increased very strongly from the low- to the high-light treatments in all age groups (Fig. 8.5b). In the high-light treatment, sap light absorbance also increased significantly with leaf age, but it remained nearly constant in the low- and intermediate-light treatments.

Total Nutrient Concentration

Analysis of total nutrients in leaf material of seedlings grown under different light intensities is another powerful approach to evaluating plasticity as an adaptation to grow in low light. The results obtained by the biomass sampling already provide some insight into the nutrient balance of mahogany seedlings (see Table 8.4). This analysis also gives the effect of leaf age in conjunction with the light treatment. The P concentration was always higher

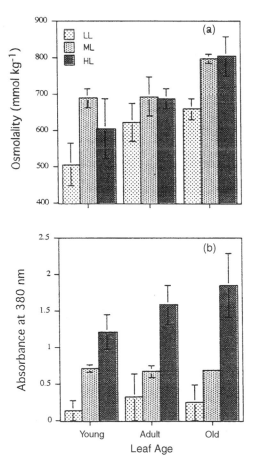

Figure 8.5a,b. Mean (±SD) osmolality (**a**) and absorbance (**b**) of leaf sap obtained from young, adult, and old leaves of hybrid mahogany seedlings grown under contrasting light regimes (*H, M, L*, as in Fig. 8.1).

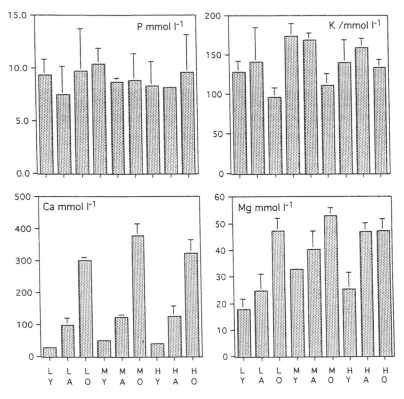

Figure 8.6. Mean (±SD) ion concentration in leaf sap of young (*Y*), adult (*A*), and old (*O*) leaves collected from hybrid mahogany seedlings grown under contrasting light regimes (*L*, *M*, *H*, as in Fig. 8.1).

Table 8.5. Mean (±SD) Osmolality and the Sum of Cations (K, Ca, and Mg) in the Sap of Leaves from Mahogany Seedlings Growing Under Different Light Intensities

Light Treatment/ Age of Leaves	Osmolality (mmol kg^{-1})	Cations (mmol l^{-1})	Percent (%) Explained
Low			
Young	505 (58)	174 (21)	34
Adult	622 (53)	264 (61)	42
Old	658 (29)	442 (13)	67
Intermediate			
Young	688 (26)	256 (12)	37
Adult	692 (54)	3332 (19)	48
Old	796 (12)	541 (42)	68
High			
Young	604 (82)	207 (40)	34
Adult	686 (28)	332 (23)	48
Old	803 (53)	504 (44)	63

in the younger leaves, but its concentration did not differ much among treatments (Fig. 8.7); N followed a similar pattern, but it had a more pronounced tendency to attain a higher concentration under low light. The K concentration decreased very clearly with leaf age, but the differences among treatments were rather small. Magnesium did not show a particular pattern; S and Ca were characteristic, in that the concentration increased with leaf age, which is a common observation for Ca but rather unexpected for S. Contrary to the results obtained by the analyses of nutrients in biomass (Table 8.4), we did not detect significant differences in S concentration in leaves among light treatments. Finally, the rather large increases of ash content with leaf age in all treatments was another indication that seedling growth did not appear to be nutrient limited under these experimental conditions.

Discussion and Conclusions

Growth Rates and Biomass Accumulation

Hybrid mahogany seedlings grew healthily in all light treatments, and no mortality was observed during 417 days. Plants grown under intermediate and high light did not differ significantly from each other, but plants in the intermediate treatment clearly tended to grow taller and have more volume (see Figs. 8.1 and 8.2). Low light reduced average height growth by 25% and volume growth by 75%, compared with high light. The low-light treatment also reduced the relative growth rate in height and volume (see Table 8.2) but to a much lesser degree. Observed differences tended to disappear with time, apparently a consequence of limitations of space for root growth in the nursery beds.

Seedling populations in all treatments showed high variability, possibly as a result of their hybrid nature. Big-leaf mahogany has a larger intrinsic growth potential than does small-leaf mahogany (Ward and Lugo, Chapter 3, this volume); therefore, seedlings expressed the full range of variation of growth potential among the parents. Biomass accumulation in the high- and intermediate-light treatments were similar, although the plants with intermediate- and low-light treatment tended to produce heavier stems and maintain a larger leaf mass at both biomass sampling dates (see Table 8.3). Differences among treatments were more pronounced when the absolute growth rate between 296 and 417 days was considered. During this period, the plants of the intermediate treatment accumulated almost 2 and 10 times as much organic matter as the high- and low-light plants, respectively. Differences in growth and biomass accumulation were not related to nutrient concentration because the plants appeared not to be limited by nutrient supply or transport (Table 8.4).

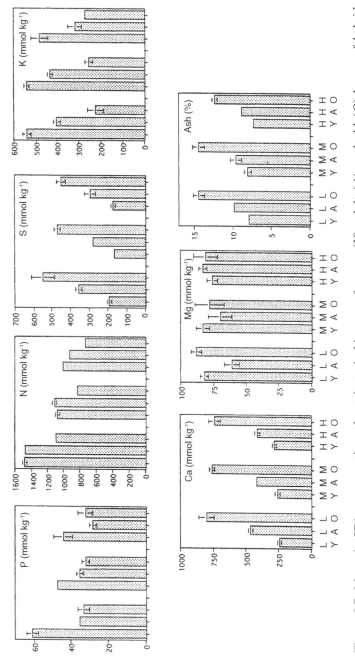

Figure 8.7. Mean (± SD) concentration of nutrients in biomass of young (*Y*), adult (*A*), and old (*O*) leaves of hybrid mahogany seedlings grown under contrasting light regimes (*L*, *M*, *H*, as in Fig. 8.1).

We concluded from these results that the hybrid mahogany could grow vigorously under light regimes of about 40% of full daylight. We believe that the reason intermediate-light plants performed as well or even better than high-light plants was likely because of water stress under the high-light treatment. We did not measure water stress continuously during the experiment, but we believe that during clear, rainless days the water potential of the high-light plants can safely be assumed lower than for the intermediate- and low-light plants. Plants in the low-light treatment were clearly limited by the availability of light energy, but the growth potential of the seedlings was so variable that plants overlapped in size and weight in all treatments throughout the experiment. The tall population of the low-light treatment was similar in height, volume, and biomass accumulation to the short population of the intermediate- and high-light treatments. The low-light treatment seems to affect the seedlings with higher growth potential more strongly than those with less.

Plants intolerant to shade tend to grow faster in height than do shade-tolerant plants, as shown in *Heliocarpus appendiculatus* (Fetcher et al. 1983) and *Cedrela odorata* (Ramos and Grace 1990). By this criterion, the hybrid mahogany appears to have higher shade tolerance than pioneer and early-successional species. Our results with relative growth rate in volume co-incided with the measurements of Ramos and Grace (1990), which showed that big-leaf mahogany can produce biomass actively under low light, outgrowing the light-demanding *C. odorata* under the same illumination regime. In a similar experiment comparing a pioneer (*Trema micrantha*) with a long-lived emergent tree (*Militia excelsa*), Lehto and Grace (1994) found similar reductions in biomass accumulation in both species induced by the low-light treatment. The pioneer leaves were much more plastic in leaf area:weight ratios, as observed also by Fetcher et al. (1983). Similarly, Strauss-Debenedetti and Bazzaz (1991), comparing five co-occurring Moraceae species, showed that early-successional species, possibly less shade tolerant, are much more plastic in their photosynthetic responses than are the shade-tolerant, late-successional species. From her study of 13 species differing widely in shade tolerance, Kitajima (1994) concluded that survivorship in shade apparently goes in the opposite direction to plastic response to shade. Shade-tolerant species are less plastic, both morphologically and physiologically, than are shade-intolerant species. These findings were confirmed with several species belonging to different families in northern Queensland (Osunkowya et al. 1994).

Plasticity in Leaf Characteristics

Shade treatment induced development of larger, thinner, less succulent leaves in our study, as has been observed in most experiments dealing with acclimation of seedlings to low light intensity (Kitajima 1996). Osmolality was significantly lower in the low-light treatment, but the fraction of osmolality explained by the accumulation of cations in leaf sap was nearly identical to that of other light treatments (see Table 8.5). This finding suggested that, although plants grown under higher light intensities are probably subject to water stress during rainless, sunny days, the plants in all treatments were able to obtain a constant fraction of ions required for building osmotic potential in the vacuole. Thus, the accumulation of ions in the leaf sap was very similar for leaves of increasing age in each treatment (see Fig. 8.6).

Accumulation of phenolic substances in foliage of rain forest plants has been shown to be associated with light (Mole et al. 1988). Our results indicated that the concentration of these substances in leaf sap was strongly determined by the light regime. Although growth characteristics and osmolality of the intermediate and light treatments were not significantly different, absorption at 380 nm was always much higher in the high-light treatment (Fig. 8.5b). Considering the potential significance of these compounds as photoprotectants (Waterman and Mole 1998), we propose to conduct a systematic analysis using the capacity to accumulate these compounds as an index for light sensitivity of mahogany leaves and those of other tropical tree species.

Nutrient accumulation in leaves showed a similar pattern across age and treatment (see Fig. 8.7). The low-light treatment produced leaves richer in N and S, but concentrations of P, K, Ca, and Mg were similar among treatments. As expected, the most metabolically important elements—P, N, and K—were more concentrated in young leaves. We expected to observe the same pattern for S, but this element behaved similarly to Ca, an element accumulated in leaves and not retranslocated in phloem (Marschner 1995). Our measurements of total S includes both organic and inorganic fractions. The accumulation of Ca could be accompanied by accumulation of sulfate, a question that remains to be answered by future research.

The Significance of Red:Far-Red Ratios for Seedling Development in Shade

Large reductions in the ratio of red:far-red light (R:FR ratios), particularly at a low intensity of photosynthetically active radiation, affect morphogenesis and carbon allocation to roots and shoots. Tinoco-Ojanguren and Pearcy (1995) found that a reduction in the R:FR ratio of the incoming radiation (from 1.25 to 0.26) influences morphogenetic aspects of early-successional species but does not modify steady-state photosynthetic characteristics. The leaf area ratio and the leaf weight ratio of pioneer species (*Heliocarpus appendiculatus* and *Cecropia obtusifolia*) are significantly reduced under low R:FR ratios, but no response was detected in the late-successional species studied (*Rheedia edulis*). On the other hand, carboxylation capacity and dark respiration under low light were not affected by the R:FR ratio. Maximum photosynthetic rate, influenced only weakly in *H. appendiculatus*, was higher under high R:FR ratios. Kitajima (1994) found no influence of the R:FR ratio on gas exchange, allocation patterns, or growth rates in 13 species with different successional status in Panamá. Lee et al. (1996), however, found that species of rain forest trees differing in their successional status respond differently to cultivation under contrasting light intensities and R:FR ratios. Growth rates and architectural characteristics of two early-successional species (*Endospermum malaccense* and *Parkia javanica*) change drastically in response to changes in light intensity. Some properties, particularly the ratio of leaf area to stem length, tend to decrease more in light-demanding than in shade-tolerant species under conditions of lower R:FR ratios. This finding suggests that early-successional species allocate more carbon to stems than to leaves when grown under low R:FR ratios.

We think this question should be addressed for mahogany in a future experiment that uses natural shading or infrared filters. The responses of plant growth and leaf chemical composition we observed were mainly caused by the differences in the amounts of energy in photosynthetically active radiation and not by the spectral composition of the light.

Implications for Growth Under Natural Conditions

The hybrid mahogany seedlings studied showed strong reductions in relative growth rates under the low-light treatment, indicating a significant degree of shade tolerance but also a large plasticity in relative growth rate (hypothesis 1). Mahogany seedlings maintained a positive carbon gain, indicated by continuous biomass accumulation in the low-light treatment over 417 days (hypothesis 2). We obtained evidence that light intensity more markedly affects the growth of those individuals with the largest intrinsic growth capacity, as was particularly noticeable in this experiment because the seed source was a hybrid from parents differing significantly in their growth capacity (Ward and Lugo, Chapter 3, this volume). If seedlings with potentially high growth rates—but suppressed by low light in natural or plantation forests—survive for prolonged periods, they may respond quickly to the opening of medium to large gaps.

Growth under full sunlight resulted in pronounced water stress, particularly in seedlings without a large root system (hypothesis 3). The sudden opening of large gaps may jeopardize survival of seedlings germinated and grown under low light intensity because of the imposition of water stress (Gerhardt 1994). Seedlings that established after the opening of large clearings, however, are much more drought tolerant, both because of the large accumulation of osmotically active substances in their vacuoles, and because they produce organic substances possibly associated with photoprotection of the photosynthetic machinery (hypothesis 4). Mahogany seedlings showed high plasticity in leaf area but not in leaf dry weight (Fig. 8.4), a fact that explains the larger area:weight ratios of seedlings developing under low light. The concentrations per unit of leaf dry weight of P, S, Ca, K, and Mg were very similar among treatments and leaf age, pointing to lower plasticity. The N concentration was significantly larger in low-light seedlings (see Fig. 8.7), but because of the large area:weight ratios of low-light leaves, the N concentration per unit area was higher in the intermediate and high treatments ($\sim70\,mmol\,m^{-2}$) than in the low-light treatment ($55\,mmol\,m^{-2}$). This finding is characteristic of a shade-intolerant species that regulates N allocation to leaves according to the availability of light during growth (Anten et al. 1996).

We did not measure how quickly seedlings reacted to sudden increases in the light regime. Studies on other species, however, indicate that regardless of the shade tolerance of the species, leaves that developed under a low light regime have a low tolerance to sudden increases in light intensity. New leaves developed after a change in light regime are much better adapted to the high light in shade-intolerant species, however (see Kitajima 1996).

Acknowledgments. This work was done in cooperation with the University of Puerto Rico. We thank the following coworkers of the International Institute of Tropical Forestry: M. Alayón, M. Arturi, W. Edwards, J. Francis, E. López, and M.J. Sánchez.

Literature Cited

Anten, N.P.R., Hernández, R., and Medina, E. 1996. The photosynthetic capacity and leaf nitrogen concentration as related to light regime in shade leaves of a montane tropical forest tree *Tetrorchidium rubrivenium. Functional Ecology* **10**:491–500.

Chiariello, N.R., Mooney, H.A., and Williams, K. 1989. Growth, carbon allocation and cost of plant tissues. In *Plant Physiological Ecology*, eds. R.W. Pearcy, J. Ehleringer, H.A. Mooney, and P.W. Rundel, pp. 327–365. Chapman & Hall, London.

Fetcher, N., Strain, B.R., and Oberbauer, S.F. 1983. Effects of the light regime on the growth, leaf morphology, and water relations of seedlings of two species of tropical trees. *Oecologia* (Berlin) **58**:314–319.

Geary, T.F., Barres, H., and Briscoe, C.B. 1972. *Hybrid Mahogany Recommended for Planting in the Virgin Islands.* Research Paper ITF-15. U.S. Department of Agriculture, Forest Service, Institute of Tropical Forestry, Río Piedras, PR.

Gerhardt, K. 1994. *Seedling Development of Four Tree Species in Secondary Tropical Dry Forest in Guanacaste, Costa Rica.* Dissertation, Uppsala University, Sweden.

Gullison, R.E., Panfil, S.N., Strouse, J.J., and Hubbell, S.D. 1996. Ecology and management of mahogany (*Swietenia macrophylla* King) in the Chimanes forest, Beni, Bolivia. *Botanical Journal of the Linnean Society* **122**:9–34.

Harborne, J.B. 1973. Flavonoids. In *Phytochemistry, Vol. 2*, ed. L.P. Miller, pp. 344–380. Van Nostrand Reinhold, New York.

Huang, C.Y., and Schulte, E.E. 1985. Digestion of plant tissue for analysis by ICP emission spectroscopy. *Communications in Soil Science and Plant Analysis* **16**:943–958.

Hunter, A.H. 1982. *International Soil Fertility Evaluation and Improvement: Laboratory Procedures.* Department of Soil Science, North Carolina State University, Raleigh, NC.

Kitajima, K. 1994. Relative importance of photosynthetic traits and allocation patterns as correlates of seedling shade tolerance of 13 tropical trees. *Oecologia* (Berlin) **98**:419–428.

Kitajima, K. 1996. Ecophysiology of tropical tree seedlings. In *Tropical Forest Plant Ecophysiology*, eds. S.S. Mulkery, R.L. Chazdon, and A.P. Smith, pp. 559–596. Chapman & Hall, New York.

Lamb, F.B. 1966. *Mahogany of Tropical America. Its Ecology and Management.* University of Michigan Press, Ann Arbor, MI.

Lee, D.W., Baskaran, K., Manzor, M., Manzor, M., Mohamed, H., and Yap, S.K. 1996. Irradiance and spectral quality affect Asian tropical rain forest tree seedling development. *Ecology* **77**:568–580.

Lehto, T., and Grace, J. 1994. Carbon balance of tropical tree seedlings: a comparison of two species. *New Phytologist* **127**:455–463.

Marrero, J. 1950. Results of forest planting in the insular forests of Puerto Rico. *Caribbean Forester* **11**:107–147.

Marschner, H. 1995. *Mineral Nutrition of Higher Plants*, second edition. Academic Press. London.

Mole, S., Rogler, J.A.M., and Waterman, P.G. 1988. Light-induced variation in phenolic levels in foliage of rain forest plants. I. Chemical changes. *Journal of Chemical Ecology* **14**:1–21.

Nelson, D.W., and Sommers, L.E. 1982. Total carbon, organic carbon, and organic matter. In *Methods of Soil Analysis, Part 2,* pp. 576–571. American Society of Agronomy, Madison, WI.

Newton, A.C., Baker, P., Ramnarine, S., Mesen, J.F., and Leakey, R.R.B. 1993. The mahogany shoot borer: prospects for control. *Forest Ecology and Management* **57**:301–328.

Nobles, R.W., and Briscoe, C.B. 1966. *Height Growth of Mahogany Seedlings, St. Croix, Virgin Islands.* Research Note ITF-10. U.S. Department of Agriculture, Forest Service, Institute of Tropical Forestry, Río Piedras, PR.

Osunkowya, O.O., Ash, J.E., Hopkins, M.S., and Graham, A.W. 1994. Influence of seed size and seedling ecological attributes on shade-tolerance of rain-forest tree species in northern Queensland. *Journal of Ecology* **82**:149–163.

Ramos, J., and Grace, J. 1990. The effects of shade on the gas exchange of seedlings of four tropical trees from México. *Functional Ecology* **4**:667–677.

Snook, L.K. 1996. Catastrophic disturbance, logging and the ecology of mahogany (*Swietenia macrophylla* King): grounds for listing a major tropical timber species in CITES. *Botanical Journal of the Linnean Society* **122**:35–46.

Strauss-Debenedetti, S., and Bazzaz, F.A. 1991. Plasticity and acclimation to light in tropical Moraceae of different successional positions. *Oecologia* (Berlin) **87**:377–387.

Tabatabai, M.A., and Bremner, J.M. 1991. Automated instruments for determination of total carbon, nitrogen and sulfur in soils by combustion techniques. In *Soil Analysis, Modern Instrumental Techniques*, second edition, pp. 261–286. Dekker, New York.

Tinoco-Ojanguren, C., and Pearcy, R.W. 1995. A comparison of light quality and quantity effects on the growth and steady-state and dynamic photosynthetic characteristics of three tropical tree species. *Functional Ecology* **9**:222–230.

Verissimo A., Barreto, P., Tarifa, R., and Uhl, C. 1995. Extraction of a high value natural resource in Amazonia: the case of mahogany. *Forest Ecology and Management* **72**:39–60.

Waterman, P.G., and Mole, S. 1998. Analysis of phenolic plant metabolites. In *Methods in Ecology*. Blackwell, Oxford.

Whitmore, J.L., and Hinojosa, G. 1977. *Mahogany* (Swietenia) *Hybrids*. Research Paper ITF-23. U.S. Department of Agriculture, Forest Service, Institute of Tropical Forestry, Río Piedras, PR.

9. Regeneration, Growth, and Sustainability of Mahogany in México's Yucatán Forests

Laura K. Snook

Abstract. Big-leaf mahogany was studied on nine mixed-species stands that became established naturally between 2 and 75 years ago after catastrophic disturbances (hurricane blowdown, fire, or bulldozer clearing). More than 50% of adult big-leaf mahogany trees had survived a severe hurricane, leaving 2.8 seed trees ha^{-1}. After fire, 29% to 100% of adult mahogany trees survived, leaving an average of 1.4 seed trees ha^{-1}. Thirty or more years later, postdisturbance mahogany trees were found at densities of 18 ha^{-1} after fire, as compared to 6 ha^{-1} after a hurricane. In mixed-species aggregations, mahogany trees grew at densities as great as 47 trees ha^{-1}, accounting for up to 10% of the individuals and 27% of the basal area. A chronosequence of postfire stands 15 to 75 years old revealed annual diameter increments ranging from more than 1 cm yr^{-1} between 15 and 30 years to 0.38 cm yr^{-1} between 45 and 75 years. Assuming constant growth, a big-leaf mahogany requires 122 years, on average, to reach the 55-cm minimum cutting diameter, although the fastest-growing trees may do so in 82 years. The current selective harvesting system, based on a 25-year cutting cycle, cannot be expected to ensure sustainable harvests of big-leaf mahogany because extraction exceeds growth and adequate regeneration conditions are not provided. Harvest rates should be reevaluated and efforts made to increase the harvest of other species and implement silvicultural treatments, or shifting agricultural systems should be integrated into the forest management regime to provide for the regeneration of this valuable shade-intolerant species.

Keywords: Big-leaf mahogany, *Swietenia macrophylla*, Silviculture, Disturbance, Hurricane, Fire, Sustainability, Natural regeneration, Growth, Community forestry

Introduction

For centuries, the extraction of big-leaf mahogany timber has been one of the primary economic activities in the state of Quintana Roo, on México's Yucatán peninsula (Fig. 9.1). Harvests have been maintained over this long period as changes in extraction technology and markets continuously redefined the mahogany resource. Between the seventeenth century and the early 1900s, the successive replacement of manual labor by draft animals and a combination of narrow-gauge railroads and crawler tractors increased the forest resource from a fringe of less than 100 m to a band 60 km wide along the Río Hondo, the perimeter of the Laguna de Bacalar, and other bodies of water in the region, along which logs were floated to ships or processing plants (Chaloner and Fleming 1850; Mell 1917; Record 1924; Lamb 1966; Napier 1973; Konrad 1988). Since then, road building and rubber-tired skidders have permitted logging of big-leaf mahogany from almost every area of the forest (Villaseñor 1958; Medina et al. 1968; Snook 1998).

Changes in markets have also contributed to maintaining big-leaf mahogany harvests. Until the 1940s, only select mahogany trees were harvested for the international log export market (Medina 1948). Huge trees were left standing because they were imperfect. From the 1950s to the 1980s, many of these trees were harvested for a local veneer mill (Medina et al. 1968). Now that big-leaf mahogany timber is being sawn locally into boards, trees of lower quality and smaller diameters can be processed. These changes in markets and transformation technologies have redefined the mahogany resource so that trees left behind in earlier logging operations—

Figure 9.1. The study area on México's Yucatán peninsula.

because they were not considered commercial—are providing the bulk of today's harvests (Snook 1993, 1998).

The framework for sustaining mahogany harvests has changed significantly over the past decades. First, an untapped forest frontier no longer exists. Second, the ancient trees left behind in earlier eras have already been harvested from two-thirds of the 25 annual cutting areas of the 400,000 ha of community forest reserves in the state. These trees have accounted for much of the volume harvested during the past 16 years, providing a one-time windfall. The long-term economic viability of forestry in Quintana Roo depends on assuring sustainable mahogany harvests by providing for regeneration on each cutting area each year and balancing the rate of harvest with the rate of growth. This strategy requires an understanding of the patterns and processes of mahogany regeneration and growth in these forests and the design and implementation of silvicultural management systems based on this knowledge. This study set out to obtain this information and propose silvicultural guidelines for sustaining mahogany harvests into the future.

Big-Leaf Mahogany in the Forests of the Yucatán Peninsula

Big-leaf mahogany grows as a canopy emergent and can reach 70 m tall and 300 cm in diameter (Pennington and Sarukhan 1968). In the course of this study, trees 35 m tall and 150 cm in diameter were measured. Mahogany trees begin to flower and fruit at about 12 years old. Flowers are fragrant and apparently pollinated by as-yet-unidentified insects. Fruits are hard capsules 12 to 18 cm long that mature in 10 to 12 months. Each fruit contains 40 to 50 seeds, 1 to 2 cm long with 6 to 7 cm wings; they are dispersed by the wind during the dry season, when adult mahogany trees are leafless (in Quintana Roo, March and April) (Lamb 1966; Pennington and Sarukhan 1968; Pennington et al. 1981). Mahogany seeds have been observed to land 60 m downwind (northwest and south) from a mother tree 30 m tall (Rodríguez et al. 1994), and they probably fly much farther (Alrasjid and Mangsrud 1973). In Quintana Roo, mahogany seeds germinate between June and October, once the rains begin (Negreros and Snook, unpublished data). Seeds do not maintain their viability from one year to the next (Lamb 1966; Parraguirre 1994; Morris et al. 2000). Observations from both Central and South America show that mahogany seedlings are almost never found in the understory (Finol 1964; Lamb 1966; Snook 1993; Gerhardt 1996) nor do they seem to survive in felling gaps (Stevenson 1927; Quevedo 1986; Veríssimo et al. 1995; Gullison et al. 1996).

Although big-leaf mahogany is more common in seasonal tropical forests like those of Quintana Roo than in any other forest type in México (Pennington and Sarukhan 1968), mahogany trees grow at an average density of only 1 commercial-sized ($\geq$55 cm diameter at breast height, dbh) tree per hectare (Lamb 1966; Medina et al. 1968) and up to 7 trees ha^{-1} of 15 cm or less dbh (Argüelles 1991; Flachsenberg 1993a) in a matrix of 200 to 400 other trees ha^{-1} ($\geq$15 cm dbh) of 60 or more different species (Argüelles 1991; Snook 1993). The most abundant species in these forests, growing at densities of 15 to 60 trees ha^{-1}, is sapodilla or chicozapote (*Manilkara zapota*), the source of the chicle latex used to make chewing gum (Medina et al. 1968; Argüelles 1991; Barrera de Jorgenson 1993, 1994; Flachsenberg 1993a), followed by breadnut or ramón (*Brosimum alicas-trum*), which may also be found at densities greater than 15 trees ha^{-1} ($\geq$15 cm dbh) (Argüelles 1991).

The forests of Quintana Roo are seasonal tropical forests with a dry season 5 to 7 months long with rainfall less than 100 mm month^{-1}. Seasonal forests of this type are the most extensive in Central America (Murphy and Lugo 1986). Annual rainfall in the big-leaf mahogany forests of central

Quintana Roo is 1200 to 1500 mm yr^{-1}, and falls mostly between May and October (Secretaría de Agricultura y Recursos Hidráulicos *in* Snook 1993). During the dry season, which becomes most extreme in March and April, many tree species drop their leaves for a short time (Pennington and Sarukhan 1968; Snook 1993). Soils in the region are derived from limestone, and the topography is flat to slightly rolling.

For millennia, the forests of Quintana Roo have been affected by a spectrum of drastic natural and human-caused disturbances. Almost every year, in August or September tropical cyclones or hurricanes (Wilson 1980; Escobar 1981) bring heavy rains and winds as high as 300 km h^{-1} from the south, southeast, or east (Jauregui et al. 1980; Whigham et al. 1991). Hurricanes usually measure about 600 km in diameter (Jauregui et al. 1980; Wilson 1980). Periodically, they defoliate or knock down thousands of hectares of forest, as in 1942, 1955 ('Janet'), 1974 ('Carmen'), 1988 ('Gilbert'), and 1995 ('Opal' and 'Roxanne') (Medina 1948; Miranda 1958; Lindo et al. 1967 *in* Johnson and Chaffey 1973; Escobar 1981; López-Portillo et al. 1990; García et al. 1992; author's observations). The effects of a sixteenth-century hurricane on the Yucatán forests were described as follows: "There came a storm that grew into a hurricane. The storm blew down all the high trees. The land was left so treeless that those of today look as if planted together and thus all grown of one size. To look at the country from heights, it looks as if all trimmed with a pair of shears" (de Landa 1566).

Forest fires have also been frequent in Quintana Roo. During particularly dry years, forest fires have been caused by lightning (Wolffsohn 1967), but more typically they spread from shifting agricultural fields. Extensive fires are typical in the years after hurricanes, when fallen foliage, branches, and trees provide abundant fuel, and may burn hundreds of thousands of hectares of forest. Forest fires were extensive in the posthurricane years 1945, 1975, and 1990 (Lamb 1966; Pérez 1980; López-Portillo et al. 1990; Whigham et al. 1991; García et al. 1992).

Shifting agriculture has been practiced in the forests of Quintana Roo since at least 2000 B.C., when the early Maya became established there (Hammond 1982). In this system, patches of forest of 0.5 to 3 ha or more (Murphy 1990) are cleared and burned, planted, and cultivated for 1 or more years, then abandoned and recolonized by forest species. In today's forest, the density of crumbling pyramids and other Mayan structures, currently overgrown by trees, reveals that much of today's forest grew up on abandoned agricultural lands and urban centers after the collapse of the Mayan empire. The process of depopulation began about 900 years ago and continued through the period of Spanish conquest and the establishment of Mexican control (de Landa 1566; Gates 1937; Hammond 1982; Edwards 1986). More recent additions to the disturbance regime in Quintana Roo are felling gaps, skid trails, and log yards opened by bulldozers, produced by commercial timber harvesting.

Methods

The frequency and range of disturbances that affect the forests of Quintana Roo make it an ideal setting for studying the processes of forest regeneration and growth. Historical events have established a variety of "natural experiments" (Diamond 1986), where different kinds of catastrophic disturbance, or treatments, have taken place in the past, giving rise to new forest stands. Because of the frequency of disturbance, a chronosequence could be identified of stands that became established at different times in the past after the same kind of disturbance. By comparing the sizes of trees at different stages along the chronosequence, their development and growth over time could be analyzed.

To determine the ages of trees not known to produce annual growth rings, it was necessary to know the ages of sample stands. Fortunately, the forests of Quintana Roo have been and continue to be inhabited and used by chicle tappers, mahogany loggers, and hunters who remember what has happened in the past on different parts of their forests. In response to questions, they described and led me to stands where they knew what kind of disturbance had occurred, and when. I confirmed stand histories by evaluating site evidence (Lorimer 1985). On stands affected in the past by fire, I found fire scars on some standing trees and charred trunks on the ground; on posthurricane stands, evidence included broken branches and stems, sometimes resprouted, and the remains of uprooted trees; old log yards had scattered sawn-off hollow logs that had been cut off and left behind when log trucks were loaded.

To avoid confusing the effects of time with the effects of soil, sample stands were selected for study only if they were growing on the red soils known in the Mayan classification as "kankab," described as chromic cambisols in the Food and Agriculture Organization system (Flachsenberg 1993a). These soils cover 52% of the 20,000-ha forest reserve of the community of Noh Bec (Argüelles 1991), where the study was carried out (see Fig. 9.1). Nine stands were sampled, ranging in age from 2 to 75 years since the most recent catastrophic disturbance and regeneration event (Table 9.1).

Data Collection

Two sampling systems were used. In each stand of more than 1 ha, one or more transects of 1 km by 10 m or 20 m wide (depending on stand age and tree size and visibility) were established using a compass and tape. Within these transects, all big-leaf mahogany trees were measured and evidence of damage noted. Where mahogany trees had already been harvested, their stumps were measured and a formula relating stump diameter to diameter at breast height (dbh) on standing trees was used to determine their diameters. In all stands, plots were established to include mahogany trees and

Table 9.1. Age, Disturbance History and Estimated Area of Sample Stands, and Sampling Method Used in Each Stand

	Sample Stands		Sampling Method		
Age	Certainty[a]	Disturbance	Area (ha)	Plot Size (m)	Transects (ha)
2	C	Bulldozer[b]	1	9	—
8	C	Bulldozer[c]	<1/2	25	—
15	C	Bulldozer[c]	<1/2	25	—
15	E	Bulldozer[d]	<1/2	25	—
15	C	Fire	200	25	1
30	C	Fire	200	314	1
34	C	Hurricane	>200	314	4
45	C	Fire	>200	1000	3
75	NC	Fire	200	1000	4

[a] C, age and history confirmed independently by more than one informant; NC, age obtained from only one informant; E, age estimated.
[b] Clearing established as a helicopter landing pad.
[c] Log yard.
[d] Road edge.

their associates. Plot size varied with stand age and tree size, so that each plot contained about 40 trees (Table 9.1). Within each plot, all trees larger than a minimum size (2 m tall in stands up to 15 years old; 15 cm dbh in older stands) were identified, and their diameters and heights measured. In addition, any damage to the stem or the crown and its probable cause were recorded.

During the first field season, increment cores were collected from big-leaf mahogany trees on plots to see whether ring numbers correlated with historical data. Ten cross-sectional slabs were also cut from mahogany stumps of known age in a thinned plantation. In a confirmation of previous observations (Rodríguez 1944; Medina et al. 1968), initial counts showed that the number of rings did not correspond to the age of the stand or the plantation, nor did ring numbers correspond among neighboring trees. As a result, no further samples were collected.

Data Analysis

Neither hurricanes nor fires destroyed all trees in the affected area. To analyze the patterns of regeneration and growth required differentiating two cohorts of trees in each stand: those that had survived the stand-initiating disturbance, and those that became established afterward. The ages of the latter group could be assumed to correspond closely to the number of years since the disturbance; those of the former group could not be determined. A tree was considered to have survived the stand-initiating disturbance if it was damaged in a way that corresponded to that type of disturbance and if it was larger than other trees in the stand (Lorimer 1985).

In stands affected by fire, damage to surviving trees consisted of fire scars at the base. Trees that had survived a hurricane were likely to have bent stems or broken branches, which might have resprouted. Unlike fires, which typically destroy smaller individuals, hurricanes are more likely to damage canopy trees. Because small trees and saplings in the understory typically escape wind damage, these trees were compared to conspecifics of known age to determine whether they were survivors. In cases of doubt, they were included in the postdisturbance cohort. Data were analyzed by using Systat and Sygraph (Wilkinson 1988).

Results and Discussion

The Survival of Big-Leaf Mahogany After Fire and Hurricane

Big-leaf mahogany trees and stumps found in transects through the different stands were classified, based on their sizes and evidence of damage, into three categories: those that died as a consequence of the disturbance (mortality); those that survived the disturbance (survivors); and those that became established after the disturbance (regeneration). Some trees may have been entirely destroyed by fire, leaving no evidence, although mature mahogany trees are considered extremely fire resistant when alive, as is their timber (Chaloner and Fleming 1850: 53).

After the hurricane of 1955, adult mahogany trees survived at an average density of 2.8 trees ha^{-1}, a rate of survival greater than 50% (Table 9.2). These results are similar to Lamb's observation (1966: 111) that 6 years after the hurricane of 1942 in Belize, 3 mahogany trees ha^{-1} survived. Mahogany is wind resistant because of its strong, flexible wood (Kukachka 1959), its buttresses, and its few, heavy branches (Snook 1993). In transects through burned areas, surviving adult mahogany trees were found at a density of 0.5 to 2.0 ha^{-1}, representing 29% to 100% survivorship (Table 9.2). Mature individuals have thick bark and survive fire well, but large trees with fire scars were observed. The lower density and percentage of surviving mahogany trees in older stands may indicate that some survivors had died in the subsequent years, perhaps after having suffered damage. Because mahogany seeds are not viable beyond one rainy season and mahogany seedlings are typically rare or absent from the understory seedling bank, the survival of mature mahogany trees through fires and hurricanes is necessary to ensure colonization of open areas after such disturbances.

Natural Regeneration of Big-Leaf Mahogany After Disturbance

The density of big-leaf mahogany trees that became established after fire and hurricane (Table 9.2) revealed two phenomena. First, the density of mahogany trees on the 15-year-old postfire stand was more than eight times the density in stands 30 years and older. The lower density of mahogany

Table 9.2. Density (ha^{-1}) of Adult Mahogany Trees That Died or Survived After a Hurricane or a Fire, and Percentage Surviving; and Density (ha^{-1}) of Mahogany Regeneration Established After the Disturbance

Disturbance	Time Since Disturbance	Mortality (number)	Trees Surviving (number)	Surviving (percentage)	Postdisturbance Big-Leaf Mahogany Regeneration[a] (number)
Fire	15	0	2.0	100	133
Fire	30	3.0	2.0	40	22
Fire	45	1.2	1.1	48	15
Fire	75	1.2	0.5	29	16
Hurricane	34	2.5	2.8	53	6.3

[a] Densities were determined from transects through sample stands.

trees on older stands probably reflected the process of self-thinning or stem exclusion, natural mortality in proportion to the increase in size of each individual (Westoby 1984; Oliver and Larson 1990). Second, the average density of postdisturbance mahoganies that became established after fire was three times higher (18 ha^{-1}) than the number that became established after a hurricane (6 ha^{-1}), although the density of surviving adult seed trees was twice as high after hurricane (2.8 ha^{-1}) as after fire (1.4 ha^{-1}). This finding indicated that the conditions for mahogany establishment were more favorable or more extensive, or both, after a fire had destroyed all seedlings and most adult trees, than after a hurricane, which produces a series of treefall gaps (Whigham et al. 1991). Opportunities were limited for mahogany seedlings to become established among the understory trees and juvenile stages of other species that typically survive disturbances such as hurricanes and logging that create canopy gaps. On stands more than 30 years old, other tree species were found at densities averaging 7,000 sapling-sized individuals ha^{-1} (>2 m tall and <15 cm dbh) and more than 110,000 seedlings ha^{-1} (<2 m tall) (Snook 1993).

Among the younger stands that became established after clearing or fire, big-leaf mahogany trees were found at densities greater than 1000 ha^{-1} (Table 9.3). The highest density of mahogany trees, equivalent to 5600 ha^{-1}, was found along the edge of a logging road, where the soil was mounded up. Mahogany trees averaged 5 m tall, and some reached 7 m. They were the most abundant species, accounting for 38% of all individuals. These patterns paralleled observations made by Wolffsohn (1961) in Belize, where he noted that mahogany regenerated most abundantly on areas cleared by fire, agriculture, or machinery.

In stands 30 years old or older, big-leaf mahogany trees represented up to 10% of the trees and 27% of the basal area in the mixed-species aggregations where they were growing (Table 9.3). Within these aggregations of 200 to 400 individuals ha^{-1} of about 40 tree species, postdisturbance

Table 9.3. Density on Plots of Mahogany Trees in Mixed-Species Aggregations Established After Disturbance

Stand	Age (years)	Big-Leaf Mahogany (trees ha^{-1})	As a Percentage of All Trees	Mahogany (m^2 ha^{-1})	Mahogany (percentage of total BA[a])
				Mahogany	
				Basal Area (BA)	
Landing pad	2	1400	11	—	—
Log yard	8	4200	23	—	—
Log yard	15	2000	21	—	—
Road edge	15	5600	38	—	—
Fire	15	1700	16	—	—
Fire	30	32	6	2.2	7
Hurricane	34	37	8	2.5	7
Fire	45	47	10	4.7	18
Fire	75	43	10	7.4	27

[a] For trees ≥15-cm diameter.

mahogany trees were found at a density equivalent to 32 to 47 individuals ha^{-1}. In the 75-year-old stand, the crowns of the 42 mahogany trees ha^{-1} were close to touching each other, creating a supercanopy over the main canopy. As stand age increased, mahogany represented a progressively higher proportion of stand basal area because it grows more rapidly than almost all associated tree species (Snook 1993).

Mahogany Growth in Mixed-Species Stands

Big-leaf mahogany growth rates were found to change with tree age (Table 9.4). The most rapid growth, exceeding 1 cm dbh yr^{-1}, was observed in trees between 15 and 30 years old. The diameter increment of mahogany determined by comparing the size of trees on the 30-year-old stand with those on the 45-year-old stand (0.44 cm yr^{-1}) was almost identical to the increment of 0.43 cm yr^{-1} calculated by Juárez (1988) on the basis of three remeasurements at 5-year intervals of mahogany trees between 25 and 35 cm dbh, and corresponds to the average growth rate determined from two subsequent remeasurements of mahogany trees in Noh Bec (Whigham et al. 1998).

Table 9.4. Average Diameters and Periodic Annual Increments (PAI) of Mahogany Trees Established After Fire in Stands of Known Age[a]

Age (years)	15	30	45	75
dbh + SE (cm)	3.1 ± 0.2	19.4 ± 2.8	25.9 ± 2.3	37.2 ± 1.4
PAI (cm yr^{-1})[a]	0.20	1.09	0.44	0.38
n (trees)[b]	131	18	30	74

[a] PAI = (dbh t_2 – dbh t_1)/(t_2 – t_1).
[b] n, number of trees measured in each age-class.

The average diameter of big-leaf mahogany trees in the oldest sample stand was 37 cm, considerably less than the legal cutting limit of 55 cm. To estimate the time required for a mahogany to reach a diameter of 55 cm, a formula was derived from the periodic annual increments (PAI) of mahogany trees of known age. If the oldest stand was 75 years old, and if mahogany continued to grow in the future at the same annual rate of diameter growth as it did between 45 and 75 years, the time required for an average mahogany tree to reach a larger diameter can be calculated by using the following formula, derived from the data in Table 9.4: year = [(diameter − 25.9)/0.38] + 45. If these suppositions are true, a mahogany tree growing at an average rate in these forests would reach the diameter limit of 55 cm in 122 years. This growth rate means that the trees harvested during the course of this study, which averaged more than 80 cm dbh, became established about 200 years earlier, and that the largest trees measured were about 350 years old.

Some trees grow more rapidly than the average and others grow more slowly. When big-leaf mahogany age-cohorts were divided into thirds by diameter, cumulative annual increments ranged from $0.32 \, cm \, yr^{-1}$ to age 75, for the slowest-growing third, to $0.67 \, cm \, yr^{-1}$ to age 75, for the fastest-growing third. If these trees were to continue growing at the same rate into the future, the fast-growing 33% could be expected to reach 55 cm at about 82 years (Fig. 9.2). The range of diameters between fast- and slow-growing individuals in a single age-cohort explains how Olmsted and Álvarez (Plan Piloto Forestal 1987) found that 2 of the 57 mahogany trees ha^{-1} growing in a log yard (<4% of the total), had reached 55 cm diameter at age 34. The average diameter of the mahogany trees they measured on that stand was

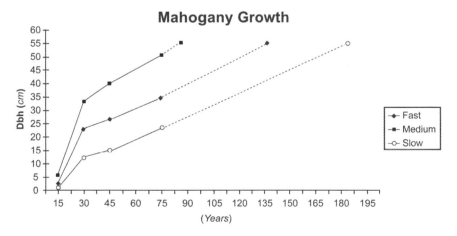

Figure 9.2. Diameters of mahogany trees at different ages, subdivided into thirds by size/growth rate, derived from trees on stands of known age and extrapolated to 55-cm diameter.

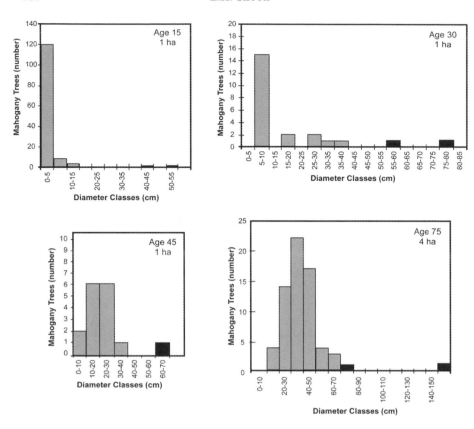

Figure 9.3. Mahogany trees by diameter class in four essentially single-cohort stands established after fire. *Black bars* represent trees that survived the stand-initiating fire.

25.4 cm, a size that falls on the average for that age as determined in this study (Table 9.4).

The diameter–class frequency distribution of big-leaf mahogany in the stands sampled reflects this variation in growth rates, changing from an inverse J-shaped curve, as faster-growing individuals move out of the sapling class, to a bell-shaped curve (Fig. 9.3). This growth pattern is typical of even-aged cohorts (Oliver and Larson 1990) and has been described for mahogany stands elsewhere (Finol 1964; Gullison and Hubbell 1992).

Conclusions and Implications for Sustainable Silviculture

In the tropical forests of Quintana Roo, the regeneration strategy of big-leaf mahogany is characterized by the capacity of adult trees to survive hurricanes and fires that periodically destroy most other trees; the production, by those survivors, of winged seeds capable of dispersing by wind to those recently opened areas; the capacity of mahogany seedlings to become established in open clearings; and the capacity of mahogany to develop and grow in essentially single-cohort mixed-species stands. In postdisturbance aggregations, mahogany trees more than 30 cm dbh can be found at densities approaching 50 trees ha^{-1}. Data from this study suggest that, although the fastest-growing mahogany trees may reach 55 cm dbh in about 82 years, two-thirds of the trees will take well over a century, and some closer to two (see Fig. 9.2).

How do these ecological parameters of big-leaf mahogany regeneration and growth relate to current forest management, and what do they imply for sustaining future mahogany harvests? In Quintana Roo, 46 forest ejidos[1] control a total of nearly 400,000 ha of commercial forest in blocks that range from 1,000 to 30,000 ha per ejido (Argüelles 1993). The forest management plan for each ejido is designed to assure continuous yields from its particular forest by controlling the rate and spatial distribution of harvesting, based on a polycyclic system with a 25-year cutting cycle and a minimum diameter limit of 55 cm. Whatever the size of the ejido forest or the standing volume of commercial-sized mahogany trees calculated from forest inventories, these figures are divided by 25 to determine how much area and how much volume can be harvested each year. All mahogany trees larger than the minimum diameter are harvested from 1 of the 25 cutting areas each year. Harvests on a particular area are scheduled to recur at 25-year intervals.

Sustaining Yields from Existing Trees

For each of the first 25 years of the cutting cycle, 1/25 of the existing stock of big-leaf mahogany trees larger than the diameter limit is harvested. According to current harvesting guidelines, all these trees will be logged during the first cutting cycle, although many of the largest have been felled and left in the woods because of poor wood quality (Argüelles 1991; author's observations). If inventories of trees in commercial size-classes accurately described the trees on the whole forest reserve of each ejido, this selective, diameter-limit harvesting system could be expected to assure continuous and relatively constant yields of timber over the 25 years of the first cutting cycle (of which 9 years remained in 2001). Where inventories do not cover the full forest reserve, however, annual harvests do not reflect timber availability on the whole forest, or over the whole cutting cycle. This discrepancy has led to unexpected fluctuations in timber availability from one cutting area, and one year, to the next.

Beginning with the second cutting cycle, in year 26 (2010), the forest reserve will be cut over again, one parcel each year. No gigantic big-leaf mahogany trees, centuries old, will remain; thus, the trees harvested on this second cut will be those currently in the 35- to 54-cm size-classes, the so-

---

[1] An ejido is a communal landgrant and its residents.

called reserve. During the 25 years of the first cutting cycle, they are expected to have grown to commercial diameters. The third cutting cycle, beginning in year 51, will cut over the 25 annual cutting areas, one by one, for the third time. Harvesting will focus on trees currently in the 15- to 34-cm size-classes, the so-called recruits, or repoblado, which will have had 50 years to grow into commercial size-classes (Argüelles 1991).

The abundance and proportion of different age-classes of big-leaf mahogany on any area is a function of the timing and characteristics of catastrophic disturbances in the past. In some areas, the number of trees currently in precommercial size-classes may be greater than the number in commercial size-classes. Trees in these categories, however, may be fast-growing young trees or slow-growing older trees. Based on extrapolations in Figure 9.2, most trees 45 cm in diameter today can be expected to attain 55 cm in 25 years, but even the fastest-growing trees that measure 35 cm dbh today could not be expected to attain 55 cm in 25 years. Additional data on growth are needed, but both minimum diameters and total volumes of harvests in the second and third cutting cycles are likely to be lower than expected. The 25-year cutting cycle used in Quintana Roo was not derived from growth data but from the 25-year duration of the forest concession granted to a veneer company in the 1950s (Rodríguez 1944; Medina et al. 1968; Snook 1998). Across the border in Belize, where forests are very similar but may benefit in some areas from deeper soils and more rainfall, mahogany forests are managed on a 40-year cutting cycle.

Sustaining Yields from New Trees

The long-term sustainability of big-leaf mahogany harvests, beyond the 75 years of the first three cutting cycles that focus on existing trees, depends on establishing mahogany regeneration on each parcel after each harvest. Regeneration requires seeds or seedlings and a favorable environment for their survival and growth. Opportunities for regeneration depend on the ecological requirements of the species, the calendar and characteristics of harvesting operations, and the design and implementation of silvicultural techniques.

The practices of big-leaf mahogany logging in Quintana Roo are defined by both ecological and economic parameters. A key ecological factor is the fact that commercial-sized mahogany trees grow at relatively low densities in stands comprising hundreds of individuals of dozens of other species (Argüelles 1991; Snook 1993). The economic context for logging is that mahogany is by far the most valuable species, with unlimited market demand. Few associated species have commercial value, and those that do have limited demand and low value (Flachsenberg 1993b). As a consequence of these combined factors, mahogany is selectively logged while associated species are left standing. Although mahogany typically regenerates in clearings produced by the destruction of trees of other species, selective harvesting of mahogany inverts those conditions. This practice short-circuits the regeneration ecology of mahogany in two ways: by reducing the availability of mahogany seed, and by perpetuating conditions unfavorable to the establishment and growth of mahogany seedlings.

Because big-leaf mahogany seeds do not retain their viability beyond one season (Lamb 1966; Rodríguezy Pacheco and Barrio 1979; Parraguirre 1994), natural regeneration requires that seed sources be retained within dispersal distance of an appropriate clearing. The current harvesting schedule, in which mahoganies are felled during January and February, just before their seeds are dispersed (March–April), does not permit the seeds from harvested trees to contribute to natural regeneration. Furthermore, if the less-valuable species are harvested, they are normally felled and extracted after mahogany, and after the period of mahogany seed dispersal. This schedule means that mahogany seeds are not able to colonize felling gaps during that year. Because mahogany regenerates in essentially even-aged aggregations, in old stands all or nearly all mahogany trees have attained commercial size. In such stands, diameter-limit harvesting can deplete seed sources over a large area. If some precommercial trees in smaller-diameter classes remain, they could serve as seed sources, although their rate of seed production is a fraction of that of trees greater than 75 cm (Gullison et al. 1996; Camara and Snook 1998).

The lack of seeds can be overcome, at some cost, by sowing or planting. A more difficult challenge is the fact that postharvest conditions are inhospitable for the survival of mahogany seedlings because of their intolerance of shade and competition. Selectively harvesting big-leaf mahogany creates relatively minor effects on the forest and was calculated to reduce canopy cover by only 2% (Whitman et al. 1997); this represents a problem for a species that typically regenerates after catastrophic disturbance and requires direct sunlight to grow. The selective harvest of mahogany trees in Quintana Roo produces an average of one treefall gap ha^{-1} that is 10 m or less to 20 m in diameter, plus skid trails less than 5 m wide and a few 0.2- to 1-ha log yards per cutting area (Argüelles 1991). In one evaluation, only 3.6% of a cutting area was calculated to have been disturbed by mahogany harvesting, and none of that area was considered favorable for regeneration. When more species were extracted, 5.4% of the area was affected, leaving 1.8% of the cutting area in a condition considered favorable for regeneration (Flachsenberg 1993a). Although log yards provide a favorable environment for mahogany regeneration, they cover only 0.5% to 5% of each cutting area (Snook 1993; Argüelles 1991), so even locating log yards near mahogany seed trees will not be adequate to sustain the forest's potential for mahogany production.

Given the practices and calendar of harvesting and the characteristic stand structures and regeneration dynamics of big-leaf mahogany, natural regeneration will not be sufficient to sustain mahogany in these forests. Silvicultural management is necessary. To date, the only silvicultural activ-

ities have consisted of planting to compensate for the lack of mahogany
seed sources. Seeds are collected from felled or standing mahoganies, sown
in nurseries, and the seedlings transplanted into felling gaps, skid trails, and
log yards. An evaluation of plantings 1 to 3 years later, however, showed
that only 22% of mahogany seedlings had survived, in part because of poor
seedling quality and poor planting technique (Negreros 1995), but probably also because of competition from other species in both the overstory
and understory. An evaluation of plantings in log yards showed a survivorship of 25% to 75% after 5 to 6 years, and the stocking of survivors was
considered adequate (Synnott 1995).

Sustaining Big-Leaf Mahogany Harvests Through Silvicultural Management

If the objective of forest management in Quintana Roo is to sustain yields
of big-leaf mahogany over time, current practices must be modified to take
into account what is known about the growth and regeneration of this
species. In the short term, the 25-year cutting cycle and its implicit 75-year
rotation must be recognized as shorter than the time required for mahogany
trees to grow to the 55-cm commercial diameter. Diameter limits and
annual harvests should be reevaluated so that harvests are balanced with
the rate of growth. Because most logs are now being sawn into boards
rather than peeled for veneer, trees could be processed at 40 cm rather than
55 cm dbh, and they could be harvested at about 80 years. Even if trees were
harvested younger and smaller, however, harvesting smaller trees would
yield lower annual harvest volumes. This reduction would have important
economic implications for the ejidos that depend on mahogany harvests for
a significant portion of their income.

Furthermore, sawing yields the lowest return per unit of timber
processed. The relatively small scale of ejido forestry operations represents
a significant constraint to making adjustments to increase the value of each
meter harvested. Volumes of big-leaf mahogany per ejido ($0-1588 \text{m}^3 \text{yr}^{-1}$)
(Flachsenberg 1993b) are typically too low to provide a constant supply of
logs to any kind of transformation industry, and at that scale it is not eco-
nomical to invest in the technology necessary to transform mahogany into
the products with the highest value per unit of volume.

Over the longer term, efforts to sustain or enhance big-leaf mahogany
production will have to focus on improving the opportunities and conditions for establishing mahogany regeneration by increasing the number of
openings created by harvesting and reducing competition for mahogany
seedlings. At a minimum, however, two relatively minor modifications in
the calendar of harvesting could enhance the potential for natural regeneration of mahogany: harvesting mahogany after its seeds are dispersed and
harvesting other species first, before mahogany seeds fall. These changes
would require that mahogany be harvested between April and the begin-

ning of the rains in June or July, and it would delay the influx into ejido economies of capital from early mahogany sales. It would also require that markets be further developed for species other than mahogany. In addition to providing a greater range of silvicultural options, however, this strategy would greatly increase the yield from each hectare. Incorporating into the management plan the harvesting of other products including railroad ties and poles and thatch for construction (Murphy 1990, 1994) would also enhance regeneration opportunities. These currently haphazard activities could be concentrated in time (before mahogany seeds fall) and space (each year's cutting area) to maximize the openings available for mahogany regeneration and minimize competition for seedlings from midcanopy and understory species.

Additional treatments would probably increase the success of regeneration. During the 1920s, in Belize, silvicultural practices that included the poisoning and girdling of undesirable species to open up the canopy and understory cleaning were implemented to encourage the establishment of big-leaf mahogany regeneration before logging and foster its subsequent development. These methods led to the establishment of about 100 mahogany seedlings ha^{-1} (Stevenson 1927). Although these treatments were abandoned during the depression (Johnson and Chaffey 1973), experiments with canopy opening, combined with understory cleaning, have been undertaken in Quintana Roo. Three years after treatment, mahogany seedlings of natural origin were found at a density of 1000 ha^{-1} (Negreros and Mize 1994). These seedlings may require periodic cleanings to successfully compete with the sprouts of understory seedlings and saplings of other species.

Line plantings of big-leaf mahogany seedlings were widely established in the forests of Quintana Roo in the 1950s (Miranda 1958; Medina et al. 1968; Weaver 1987). They were apparently unsuccessful (Miranda 1958), probably because of canopy shade, root competition, and understory sprouts. Line planting of mahogany has been considered successful in Puerto Rico, however, where seedlings have been carefully tended (Bauer 1991).

More intensive silviculture would be required to duplicate the conditions that have favored the establishment of high-density stands of big-leaf mahogany in the past. Probably the most common disturbance pattern to have given rise to high-density mahogany stands in Quintana Roo has been hurricanes followed by fire. Very similar ecological conditions are produced by slash-and-burn agriculture, still the mainstay of the subsistence farmers who make up the rural and forest-dwelling population of Quintana Roo (Murphy 1990, 1994). The regeneration of mahogany and associated tree species was probably favored in the past by shifting agriculture, yet, ironically, the first step in organizing forestry in Quintana Roo was to create permanent forest reserves where agriculture is prohibited (Snook 1991). There were logical reasons for doing so: where population densities are high, fallow periods become so short that forests never regrow beyond a scrubby stage, and significant logistical difficulties and risks can be incurred when agricultural activities and their associated fires take place in a commercial forest.

Nonetheless, farmers have noticed that where big-leaf mahogany seed trees occur nearby, seedlings become successfully established on their abandoned fields (huamils). Some farmers are reluctant to clear a fallow field where mahogany trees have regenerated. Currently, researchers are working with members of forest ejidos to establish experimental slash-and-burn fields where mahogany seeds are being sown along with corn, beans, and squash to evaluate the costs and success of this agrosilvicultural system. Experimental patch clear-cuts and mechanical clearings imitating log yards are also being established to permit a comparison of the relative costs and results of creating mahogany regeneration conditions by using these three intensive methods. These systems are expected to result in the establishment of even-aged, mixed-species stands with 50 to 100 mahogany trees ha^{-1} at the end of the rotation.

Beyond Big-Leaf Mahogany: Sustaining the Forest

Maintaining future yields of big-leaf mahogany is only one rather simple criterion for determining the sustainability of forestry practices in the forests of Quintana Roo. The larger objectives of forest management in this area are to sustain both the forest-based economies of local communities and the forest itself, part of the Mayan forest ("Selva Maya"), the largest continuous block of tropical forest north of the Amazon basin. Although big-leaf mahogany is the most valuable timber species, for both economic and ecological reasons many other species should be taken into account in designing silvicultural management plans for these forests. For example, chicle latex provides half or more of some people's annual incomes, so a healthy population of productive chicle trees must be maintained. Many species that do not reach commercial diameters are used for building houses and for fencing (Snook 1993; Murphy 1990, 1994; Snook and Barrera de Jorgenson 1994). Several animal species are also important to the subsistence of local people. In the process of refining the forest management system, consideration should be given to the fact that sapodilla and breadnut fruits are major food sources for pacas and deer, two of the most important game species in the forest (Jorgenson 1993, 1994, 1998).

Even beyond their utilitarian values, in a tropical forest where little is known about pollination and dispersal, maintaining species diversity should be considered a safeguard for the future of many species. For example, it is not yet known what species pollinate big-leaf mahogany flowers and what other species these creatures may require to survive. Both food supplies and habitat needs for fauna should be considered. Intensive silviculture for the production of mahogany timber need not reduce the diversity of tree species because patchy catastrophic disturbance events have sustained the existing mixture of species for centuries. The tree species associated with mahogany either regenerate successfully under the same open, cleared conditions as mahogany or in the kinds of small canopy gaps produced by logging and wind (Snook 1993), so if thought is given to maintaining the full spectrum of species, this can be achieved as part of a silvicultural regime designed to sustain mahogany harvests. Nonetheless, if more species are harvested, more knowledge is required to ensure that their populations are not depleted.

As part of the effort to sustain biodiversity, attention should also be paid to sustaining a mixture of ages and sizes of trees, including some ancient ones. If the current diameter-limit harvesting plan is followed, by the end of the first cutting cycle the age and size structure of the forest will have been altered significantly from what it has been over the past 800 years or more. Big-leaf mahogany trees attain larger sizes than other trees in this forest. The harvest of huge, old mahoganies is reducing the availability of nesting cavities for parrots and toucans, important seed dispersers and inhabitants of the forest.

Big-leaf mahogany has been an important product of the forests of Quintana Roo since the Maya used these large trees for making canoes (Hammond 1982). New knowledge of its regeneration ecology and growth rates can be integrated into silvicultural management plans to sustain its production into the future. Unusual opportunities for silvicultural management are provided by the diverse economies of the local forest communities. Shifting agricultural techniques used to produce the local staples of corn, beans, and squash mimic natural disturbance patterns. The fact that local people use a wide range of resources, from building poles and thatch to chicle latex and game, also provides a broader framework than a simple industrial timber economy for developing forest management plans. Silviculture in this region could integrate ecological understanding with local patterns of resource use, focusing on intensive management of small areas for multiple products produced in a mosaic of time and space. Because land is abundant, mahogany could conceivably be sustained, along with a mixture of other species, as part of the shifting agricultural system, in a mosaic of age structures that would simultaneously provide wildlife habitat, chicle supplies, and building materials. A diversified peasant economy may provide the best framework for a kind of silviculture that works with the complexities of these species-diverse tropical forests.

Acknowledgments. This study was carried out as a doctoral dissertation for Yale's School of Forestry and Environmental Studies, New Haven, CT, USA. Financial support was provided by a Fulbright Doctoral Dissertation grant from the U.S. Department of Education; the Charles A. Lindbergh Fund; the Tropical Resources Institute (TRI) of Yale's School of Forestry and Environmental Studies; and by the German-Mexican Forestry Agreement of the German GTZ. Additional support was provided by México's National Institute for Research on Forestry, Agriculture and Animal Husbandry (INIFAP), through their San Felipe Bacalar Field Station, and México's former National Institute for Research on Biotic Resources (INIREB). Research was carried out in Noh Bec, Quintana Roo, México. I thank all those institutions, as well as the following people: Abel Rodríguez Tun, of Noh Bec, whose knowledge of the forest was crucial to every day of data collection during 13 months of fieldwork; Bernaldo Blanco and Francisco Tadeo of Noh Bec, who provided much of the historical information that allowed me to find my sample stands; and Javier Chavelas Polito of INIFAP, for his botanical knowledge and other support, and his family, for their hospitality. Thanks also to three anonymous reviewers for their helpful comments.

Literature Cited

Alrasjid, H., and Mangsud. 1973. [Natural regeneration trials with mahogany (*Swietenia* spp.) in the Ngraho and Tonbo forest circles, East Java.] Laporan, Lembaga Penelitian Hutan 165, 25 p. *Forestry Abstracts* **36**:190.

Argüelles, S.L.A. 1991. *Plan de Manejo Forestal para el Bosque Tropical de la Empresa Ejidal Noh Bec*. Tesis, Ing. Agrónomo Esp. en Bosques, División de Ciencias Forestales, Universidad Autónoma de Chapingo, México.

Argüelles, S.L.A. 1993. Conservación y manejo de selvas en el estado de Quintana Roo, México. In *Conservación y Manejo de Selva en el Estado de Quintana Roo, México*. Ponencia presentada en el I Congreso Forestal Centroamericano, III Congreso Forestal de Guatemala, Petén, Guatemala, 29 ag.–4 sept.

Barrera de Jorgenson, A. 1993. *Chicle Extraction and Forest Conservation in Quintana Roo, México*. M.S. thesis, University of Florida, Gainesville, FL.

Barrera de Jorgenson, A. 1994. La extracción de chicle y la conservación del chicozapote (*Manilkara zapota*) en las selvas de Quintana Roo. In *Madera, Chicle, Caza y Milpa: Contribuciones al Manejo Integral de las Selvas de Quintana Roo, México*, eds. L. Snook and A. Barrera de Jorgenson, pp. 47–66. PROAFT/ INIFAP/USAID/WWF-US. (Available from L. Snook.)

Bauer, G.P. 1991. Line planting with mahogany (*Swietenia* spp): experiences in the Luquillo Experimental Forest, Puerto Rico, and opportunities in tropical America. Proceedings, Humid Tropical Lowlands Conference: Development Strategies and Natural Resource Management **5**:45–64. DESFIL Project TRD/USAID and the U.S. Department of agriculture, Forest Service Tropical Forestry Program, Panama City, Panama, June 17–21, 1991.

Camara, L., and Snook, L. 1998. Fruit and seed production by mahogany (*Swietenia macrophylla*) trees in the natural tropical forests of Quintana Roo, México. Journal of the Tropical Resources Institute, Yale School of Forestry and Environmental Studies, New Haven, CT, pp. 18–21.

Chaloner, E., and Fleming. 1850. *The Mahogany Tree*. Rockliff and Son, Liverpool.

de Landa, Fr. Diego. 1566/1937/1978. *Yucatán Before and After the Conquest*. Dover, New York.

Diamond, J. 1986. Overview: Laboratory, field experiments and natural experiments. In *Community Ecology*, eds. J. Diamond and T.J. Case, p. 322. Harper & Row, New York.

Edwards, C.R. 1986. The human impact on the forest in Quintana Roo, México. *Journal of Forest History* **30**:120–127.

Escobar, N.A. 1981. *Geografía General del Estado de Quintana Roo*. Fondo de Fomento Editorial del Gobierno del Estado de Quintana Roo, Chetumal.

Finol, H. 1964. Silvicultural study of some commercial species in the university forest of Caimital, Barinas. *Revista Forestal Venezolana* **7**(10/11):17–63.

Flachsenberg, H. 1993a. Aspectos socioculturales, técnicos, económicos y financieros en el manejo del bosque tropical. In *Conservación y Manejo de Selva en el Estado de Quintana Roo, México*, pp. 1–27. Ponencia presentada en el I Congreso Forestal Centroamericano, III Congreso Forestal de Guatemala, Petén, Guatemala, 29 ag.–4 sept.

Flachsenberg, H. 1993b. Descripción general. Unpublished internal report, Plan Piloto Forestal/Acuerdo México-Alemania, Chetumal, Quintana Roo, México.

García C., X., Rodríguez S., B., and Chavelas P., J. 1992. Regeneración natural en sitios afectados por el huracán Gilberto e incendios forestales en Quintana Roo. *Revista Ciencia Forestal en México* **17**(72):75–99.

Gates, H. 1937, 1978. Introduction. In *Yucatán Before and After the Conquest*, Fr. Diego de Landa, pp. 1–15. Dover, New York.

Gerhardt, K. 1996. Germination and development of sown mahogany (*Swietenia macrophylla* King) in secondary tropical dry forest habitats in Costa Rica. *Journal of Tropical Ecology* **12**:275–289.

Gullison, R.E., and Hubbell, S. 1992. Regeneración natural de la mara (*Swietenia macrophylla*) en el bosque Chimanes, Bolivia. *Ecología en Bolivia* **19**:43–56.

Gullison, R.E., Panfil, S.N., Strouse, J.J., and Hubbell, S.P. 1996. Ecology and management of mahogany (*Swietenia macrophylla* King) in the Chimanes Forest, Beni, Bolivia. *Botanical Journal of the Linnean Society* **122**:9–34.

Hammond, N. 1982. *Ancient Maya Civilization*. Rutgers University Press, New Brunswick, New Jersey.

Jauregui, E., Vidal, J., and Cruz, F. 1980. Los ciclones y tormentas tropicales en Quintana Roo durante el período 1871–1978. In *Quintana Roo: Problemática y Perspectiva*, pp. 47–64. Memorias del Simposio, Instituto de Geografía, UNAM & Centro de Investigaciones de Quintana Roo, Cancún, Quintana Roo, octubre de 1980.

Johnson, M.S., and Chaffey, D.R. 1973. *An Inventory of Chiquibul Forest Reserve, Belize. Land Resource Study 14*. Foreign and Commonwealth Office, Overseas Development Administration, Land Resources Division, Surbiton, Surrey, England.

Jorgenson, J.P. 1993. *Gardens, Wildlife Densities and Subsistence Hunting by Maya Indians in Quintana Roo, México*. Ph.D. dissertation, University of Florida, Gainesville, FL.

Jorgenson, J.P. 1994. La cacería de subsistencia practicada por la gente Maya en Quintana Roo. In *Madera, Chicle, Caza y Milpa: Contribuciones al Manejo Integral de las Selvas de Quintana Roo, México*, eds. L. Snook and A. Barrera de Jorgenson, pp. 19–46. PROAFT/INIFAP/USAID/WWF-US. (Available from L. Snook).

Jorgenson, J.P. 1998. The impact of hunting on wildlife in the Maya forest of México. In *Timber, Tourists and Temples: Conservation and Development in the Mayan Forest of México, Belize and Guatemala*, eds. R. Primack, D. Bray, H. Galletti, and I. Ponciano, pp. 179–194. Island Press, Washington, DC.

Juárez B., C.J. 1988. *Análisis del Incremento Periódico de Caoba (*Swietenia macrophylla *King) y Cedro (*Cedrela odorata*) en un Relicto de Selva en el Estado de Campeche*. Tesis, Ing. Agrónomo Esp. en Bosques, División de Ciencias Forestales, Universidad Autónoma de Chapingo, Texcoco, México.

Konrad, H.W. 1988. De la subsistencia forestal tropical a la producción para exportación: La industria chiclera y la transformación de la economía maya de Quintana Roo de 1890 a 1935. In *Etnohistoria e Historia de las Américas*.

Memorias, pp. 161–182. 45 Congreso Internacional de Americanistas. Ediciones Uniandes, Bogotá, Colombia.

Kukachka, B.F. 1959. Mahogany (*Swietenia macrophylla* King) Meliaceae. Foreign Wood Series 2167. U.S. Department of Agriculture, Forest Service, Forest Products Laboratory, Madison, WI.

Lamb, F.B. 1966. *Mahogany of Tropical America: Its Ecology and Management.* University of Michigan Press, Ann Arbor, MI.

López-Portillo, J., Keyes, M.R., González, A., Cabrera, E.C., and Sánchez, O. 1990. Los incendios de Quintana Roo. Catástrofe ecológica o evento periódico? *Ciencia y Desarrollo* **16**(91):13–57.

Lorimer, C.G. 1985. Methodological considerations in the analysis of forest disturbance history. *Canadian Journal of Forest Research* **15**:200–213.

Medina, R.B. 1948. *La Explotación forestal en el territorio de Quintana Roo.* Tesis. Ing. Agrónomo Esp. en Bosques. Escuela Nacional de Agricultura, Chapingo, México.

Medina, R.B., Cuevas, A.L., and de los Santos, M.V. 1968. Ajuste al proyecto de ordenación forestal. UIEF MIQRO, Chetumal, México.

Mell, C.D. 1917. True mahogany. Bulletin 474. U.S. Department of Agriculture, Forest Service, Washington, DC.

Miranda, F. 1958. Estudios acerca de la vegetación. In *Los Recursos Naturales del Sureste y su Aprovechamiento*, ed. E. Beltran, pp. 213–272. Instituto Mexicano de Recursos Naturales Renovables, México City, México.

Morris, M.H., Negreros-Castillo, P., and Mize, C. 2000. Sowing date, shade, and irrigation affect big-leaf mahogany (*Swietenia macrophylla* King). *Forest Ecology and Management* **132**:173–181.

Murphy, J. 1990. Indigenous forest use and development in the "Maya Zone" of Quintana Roo, México. Master's paper, Graduate Program in Environmental Studies, York University, Ontario, Canada.

Murphy, J. 1994. Aprovechamiento forestal y la agricultura de milpa en el ejido de X-Maben, Quintana Roo, México. In *Madera, Chicle, Caza y Milpa: Contribuciones al Manejo Integral de las Selvas de Quintana Roo, México*, eds. L. Snook and A. Barrera de Jorgenson, pp. 3–18. PROAFT/INIFAP/USAID/WWF-US. (Available from L. Snook).

Murphy, P.G., and Lugo, A.E. 1986. Ecology of a tropical dry forest. *Annual Review of Ecology and Systematics* **17**:67–88.

Napier, I.A. 1973. A brief history of the development of the hardwood industry in Belize. *Coedwigwr* **26**:36–43.

Negreros Castillo, P. 1995. Enrichment planting as a silvicultural technique for sustaining Honduras mahogany (*Swietenia macrophylla*) and Spanish cedar (*Cedrela odorata*) production: an evaluation of experiences in Quintana Roo, México. Paper presented at the conference Conservation and Community Development in the Selva Maya of Belize, Guatemala and México, Chetumal, Quintana Roo, November 8–11.

Negreros, C.P., and Mize, C. 1994. El efecto de la abertura del dosel y eliminación del sotobosque sobre la regeneración natural de una selva de Quintana Roo. In *Madera, Chicle, Caza y Milpa: Contribuciones al Manejo Integral de las Selvas de Quintana Roo*, eds. L. Snook and A. Barrera de Jorgenson, pp. 107–126. INIFAP/PROAFT/AID/WWF-US, Merida, México. (Available from L. Snook).

Oliver, C.D., and Larson, B. 1990. *Forest Stand Dynamics.* Biological Resource Management Series. McGraw-Hill, New York.

Parraguirre L., C. 1994. Germinación de las semillas de trece especies forestales comerciales de Quintana Roo. In *Madera, Chicle, Caza y Milpa: Contribuciones al Manejo Integral de las Selvas de Quintana Roo, México*, eds. L. Snook and

A. Barrera de Jorgenson, pp. 67–80. PROAFT/INIFAP/USAID/WWF-US. (Available from L. Snook).

Pennington, T.D., and Sarukhan, J. 1968. *Árboles Tropicales de México.* INIF/FAO, México.

Pennington, T.D., Styles, B.T., and Tayler, D.A.H. 1981. Meliaceae. *Flora Neotropica Monograph* **28**:1–472.

Pérez, V.G. 1980. El clima y los incendios forestales en Quintana Roo. In *Quintana Roo: Problemática y Perspectiva,* pp. 65–80. Memorias del Simposio, Instituto de Geografía, octubre de 1980. UNAM & Centro de Investigaciones de Quintana Roo, Cancun, Quintana Roo, Mexico.

Plan Piloto Forestal. 1987. El Remate, Ejido Noh Bec. Unpublished internal document. Sociedad de Productores Forestales Ejidales de Quintana Roo. Chetumal, Quintana Roo, Mexico.

Quevedo H., L. 1986. *Evaluacíon del Efecto de la Tala Selectiva Sobre la Renovación de un Bosque Humedo Subtropical en Santa Cruz, Bolivia.* Masters thesis, Universidad de Costa Rica, CATIE, Turrialba, Costa Rica.

Record, S.J. 1924. *Timbers of Tropical America.* Yale University Press, New Haven, CT, pp. 348–356.

Rodríguez C., R. 1944. *La Explotación de los Montes de Caoba en el Territorio de Quintana Roo.* Tesis. Ing. Agr. en Bosques. Escuela Nacional de Agricultura, Chapingo, México.

Rodríguez S., B., Chavelas P., J., and García C., X. 1994. Dispersión de semillas y establecimiento de caoba (*Swietenia macrophylla*) después de un tratamiento mecánico del sitio. In *Madera, Chicle, Caza y Milpa: Contribuciones al Manejo Integral de las Selvas de Quintana Roo, México,* eds. L. Snook and A. Barrera de Jorgenson, pp. 81–90. PROAFT/INIFAP/USAID/WWF-US. (Available from L. Snook.)

Rodríguez y Pacheco, A.A., and Barrio Chavira, J.M. 1979. Desarrollo de caoba (*Swietenia macrophylla* King) en diferentes tipos de suelos. *Ciencia Forestal* **4**(22):45–64.

Snook, L.C. 1991. Opportunities and constraints for sustainable tropical forestry: lessons from the Plan Piloto Forestal, Quintana Roo, México. Proceedings, Humid Tropical Lowlands Conference: Development Strategies and Natural Resource Management **5**:65–83. DESFIL Project, TRD/USAID and the U.S. Department of Agriculture, Forest Service Tropical Forestry Program, Panama City, Panama, June 17–21, 1991.

Snook, L.K. 1993. *Stand Dynamics of Mahogany* (Swietenia macrophylla) *and Associated Species After Fire and Hurricane in the Tropical Forests of the Yucatán Península, México.* Doctoral dissertation, Yale School of Forestry and Environmental Studies, New Haven, CT. (University Microfilms International 9317535, Ann Arbor, MI.)

Snook, L.K. 1998. Sustaining harvests of mahogany from México's Yucatán forests: past, present and future. In *Timber, Tourists and Temples: Conservation and Development in the Mayan Forest of México, Belize and Guatemala,* eds. R. Primack, D. Bray, H. Galletti, and I. Ponciano, pp. 61–80. Island Press, Washington, DC.

Snook, L.K., and Barrera de Jorgenson, A., eds. 1994. *Madera, Chicle, Caza y Milpa: Contribuciones al Manejo Integral de las Selvas de Quintana Roo, México.* PROAFT/INIFAP/USAID/WWF-US, Mérida, México. (Copies available from L. Snook.)

Stevenson, N.S. 1927. Silvicultural treatment of mahogany forests in British Honduras. *Empire Forestry Journal* **6**:219–227.

Synnott, T.J. 1995. Practices for sustainable silviculture at the Plan Piloto Forestal in Quintana Roo, México. Unpublished final report to the Biodiversity

Support Program. Biodiversity Support Program, WWF/TNC/WRI, Washington, DC.

Verissimo, A., Barreto, P., Tarifa, R., and Uhl, C. 1995. Extraction of a high-value natural resource in Amazonia: the case of mahogany. *Forest Ecology and Management* **72**(1):39–60.

Villaseñor A., R. 1958. Los bosques y su explotación. In *Los Recursos Naturales del Sureste y su Aprovechamiento*, ed. E. Beltran, pp. 273–326. Instituto Mexicano de Recursos Naturales Renovables, México City, México.

Weaver, P.L. 1987. Enrichment planting in tropical America. In *Management of the Forests of Tropical America: Prospects and Technologies*, eds. J. Figueroa, F. Wadsworth, and S. Branham, pp. 259–277. U.S. Department of Agriculture, Forest Service, International Institute of Tropical Forestry, Río Piedras, PR.

Westoby, M. 1984. The self-thinning rule. *Advances in Ecological Research* **14**:167–225.

Whigham, D.F., Olmsted, I., Cano, E.C., and Harmon, M.E. 1991. The impact of hurricane Gilbert on trees, litterfall and woody debris in a dry tropical forest in the northeastern Yucatán península. *Biotropica* **23**(4a):434–441.

Whigham, D.F., Lynch, J.F., and Dickinson, M.B. 1998. Dynamics and ecology of natural and managed forests in Quintana Roo, México. In *Timber, Tourists and Temples: Conservation and Development in the Maya Forest of Belize, Guatemala and México*, eds. R.B. Primack, D.B. Bray, H.A. Galletti, and I. Ponciano, pp. 267–281. Island Press, Washington, DC.

Whitman, A., Brokaw, N.V.L., and Hagan, J.M. 1997. Forest damage caused by selection logging of mahogany (*Swietenia macrophylla*) in northern Belize. *Forest Ecology and Management* **92**:87–96.

Wilkinson, L. 1988. *Systat. The System for Statistics for the PC*. SPPSS, Inc., Chicago, IL.

Wilson, E.M. 1980. Physical geography of the Yucatán península. In *Yucatán: A World Apart*, eds. E.H. Mosley and E.D. Terry. University of Alabama Press, Tuscaloosa, AL.

Wolffsohn, A.L.A. 1961. An experiment concerning mahogany germination. *Empire Forestry Review* **40**(1):71–72.

Wolffsohn, A.L.A. 1967. Post-hurricane forest fires in British Honduras. *Commonwealth Forestry Review* **46**:233–238.

10. Regeneration of Big-Leaf Mahogany in Closed and Logged Forests of Southeastern Pará, Brazil

James Grogan, Jurandir Galvão, Luciana Simões, and
Adalberto Veríssimo

Abstract. The regeneration status of big-leaf mahogany, after selective extraction remains controversial. This study, conducted at three sites in the forest-savanna woodland (cerrado) transition zones of Brazil's southeastern Amazon River basin, describes regeneration patterns in undisturbed and logged forests. Experimental and observational studies of seed and seedling ecology under closed forest conditions indicated that annual seed-production patterns may be overlain by supra-annual masting cycles. Germination rates in closed forest were high, and seedlings were established at highest densities within 40 m of parent trees. Survivorship of natural regeneration was 18.2% after 10 months. We found seedlings in 65% of 40 treefall gaps created by mahogany extraction 2 or 3 years before sampling. Of these seedlings, 63% were less than 70 cm tall; that is, they were suppressed under cover of more vigorous, competing vegetation. We suggest that management systems for natural forests must account for seed-production cycles and that management should occur at spatial and temporal scales mimicking disturbance regimes that regulate natural distribution patterns of adult trees.

Keywords: Meliaceae, Population ecology, Seedling ecology, Amazon River basin, Sustainable management

Introduction

Many demographic studies of higher plants have stressed the importance of seed and seedling ecology—the most perilous phases in life history cycles (Grubb 1977; Harper 1977)—toward understanding density and distribution patterns of adult populations (Bormann 1953; Clark and Clark 1985; Howe and Schupp 1985; Peters 1989; de Steven 1991a,b). Beyond theoretical issues associated with population and community ecology, thorough understanding of regeneration dynamics is prerequisite to effective silviculture in natural forest ecosystems. We present data from ongoing regeneration studies of big-leaf mahogany in the southeastern Amazon River basin. By describing patterns of seed production, dispersal, germination, predation, and seedling survivorship in three forested sites spread across 250 km of Pará State, Brazil, we provide a natural history framework for interpreting regeneration patterns found during random surveys in undisturbed closed forest and after merchantable stems were selectively extracted. We expect that these studies—placed in the context of a multidisciplinary research program examining genetic structure, reproduction, growth, and population ecology at stand, landscape, and regional scales—will orient management systems designed to ensure the continued presence of mahogany in natural forests of this region.

Despite big-leaf mahogany's extraordinary commercial value, which has driven more than four centuries of exploitation across Central and South America (Record and Hess 1943; Lamb 1966), relatively little is known about its seed and seedling dynamics in natural forests. Big-leaf mahogany probably produces seed annually, although production may be subject to masting cycles of unknown periodicity (Lamb 1966). Dispersal of relatively bulky, winged seeds has been shown to be largely restricted to within 80 m of adult trees, with median dispersal distances between 32 and 36 m from adult trees (Gullison et al. 1996) or within downwind funnel-shaped "shadows" covering 2 to 4 ha when hurricane winds extend dispersal ranges (Lamb 1966; Rodríguez et al. 1992 *in* Snook 1993).

Germination rates of fresh seeds in nursery settings are typically greater than 85% (Chinte 1952; Campbell de Araujo 1971; Rodríguez and Barrio 1979 *in* Snook 1993). Germination under closed canopies in semievergreen forests has been reported at rates between 35% and 60%, with moisture apparently triggering response (Gerhardt 1996). Parrots, agoutis, pacas, and other forest rodents are known animal predators (Lamb 1966; Snook 1993), but their effects on population dynamics remain undocumented. Wolffsohn

(1961) cited insect predation of dispersed seeds as a mechanism probably important in regulating Central American populations. Gullison et al. (1996) attributed postdispersal seed mortality to insects and fungus. Gullison and Hubbell (1992) reported up to 300 seedlings ha^{-1} established in closed forest in the immediate vicinity of adult trees. Seedlings have been described as shade tolerant (Stevenson 1927) and as shade intolerant or light demanding (Smith 1942; Lamb 1966; Gullison and Hubbell 1992). Seedling densities in gaps and along skid trails and extraction roads are often high compared to closed forest (Stevenson 1927; Wolffsohn 1961; Lamb 1966; Snook 1993). Gullison et al. (1996) indicated that seedlings may survive for years in a suppressed state under closed forest canopies.

Regeneration after logging is generally reported as poor to nonexistent. Stevenson (1927) described large areas of "primary association" forest on moist, well-drained sites in British Honduras (Belize) as having little or no regeneration after logging. Regeneration was nearly sufficient for replacement in drier forests where merchantable stems were "well distributed." Stevenson noted that, where found, initial high seedling densities around exposed stumps declined rapidly in competition with advance regeneration of other species, often fading to zero within 1 year. Quevedo (1986), working in Bolivia, encountered regeneration in 3-year-old logging gaps created by mahogany extraction, but not in 9-year-old gaps. Gullison and Hubbell (1992) found an average of 48 seedlings in 18 treefall gaps created by recent big-leaf mahogany extraction in Bolivia. In the same study area, however, Gullison et al. (1996) found regeneration attributable to release or recruitment after logging in only 2 of 28 gaps 18 to 20 years old. In south Pará, Brazil, Veríssimo et al. (1995) found regeneration in 31% of 70 plots centered on stumps in logging gaps 3 to 9 years old. High correlation with remnant seed trees indicated that many of these seedlings established after mahogany extraction. Lamb (1966) stated that on favorable sites, that is, where soil moisture and drainage are good, mahogany fares poorly in competition with advance regeneration, except after large-scale disturbances. Snook (1992, 1993) demonstrated that mahogany might establish successfully after hurricanes and fires at intervals of decades to centuries on México's Yucatán Península. She concluded that single-treefall gaps created by selective extraction of widely spaced adult trees present insufficient (being too small) or inappropriate (lacking soil disturbance) growing conditions for mahogany regeneration. Gullison et al. (1996), attributing even-aged population structures to periodic recruitment after landscape-scale hydrologic disturbances, reached similar conclusions.

Considering the regeneration cycle as that period spanning seed production to seedling recruitment into sapling classes, we begin here to quantify regeneration in south Pará at various spatial and temporal scales, and to discover at what stage or stages in this cycle mahogany experiences significant mortality. Our ultimate objective is to address a fundamental question about big-leaf mahogany behavior in south Pará that remains unanswered for any region within its enormous range: Does it depend, on average, on advance regeneration in closed forest or on postdisturbance establishment to complete its life cycle? Ideally, the intervention stages of silvicultural management systems would be designed with the answer to this question in mind.

The Study Region and Sites

The study region is located between 6.5° and 8° south of the equator, equidistant between Belém and Brasília, on the southeastern edge of the Amazon Basin, where forest meets savanna woodland (cerrado) on a broad, uneven front (Figs. 10.1, 10.2). The study region to the Araguaia River marks the easternmost extension of big-leaf mahogany's natural range, which continues north in a broad arc to Altamira and then west and south into Amazonas, Rondônia, Acre, and Bolivia. Rainfall in the study region is highly seasonal, with 1700 to 2000 mm falling almost entirely between November and May. Annual totals increase and seasonality decreases moving west and north into the interior basin (Salati et al. 1978). Although early wet-season storms are often accompanied by strong winds, hurricane-scale storms like those striking Central America—shown by Snook (1993) to play a key role in big-leaf mahogany population dynamics in Quintana Roo, México—are not a feature of this landscape.

The Brazilian Shield, a Precambrian crystalline bedrock yielding shallow, well-structured, generally acidic soils, underlies the study region (Irion 1984; Klammer 1984). Aside from low mountains that rise intermittently across the landscape, topographic relief is gentle, typically about 5 to 20 vertical meters across slopes hundreds of meters or kilometers wide. Even so, soil stratification according to slope position is predictable, with red, brown, or orange Oxisols and Ultisols on upper slopes, and gray or white sandy Entisols at the bases of slopes where water flows during the rainy season. Large-scale flooding with silt deposition such as that reported from Bolivia, the disturbance mechanism attributed by Gullison and Hubbell (1992) as primarily responsible for mahogany recruitment there, is not a feature of this landscape.

Big-leaf mahogany distribution patterns in the study region demonstrate strong correlation with low ground, as first reported by Veríssimo et al. (1995). Mature trees typically occur—or occurred, before extraction—typically growing in irregular lines and clumps along the banks of seasonal streams or in the relatively flat expanses to either side of them. Mahogany rarely grows on high ground. From the air, if mahogany crowns could be distinguished from the background forest matrix, the pattern of the regional watershed could be imagined in broad outline.

Reductions in forest cover associated with cattle ranching, smallholder agriculture, mahogany extraction, and gold mining have transformed south-

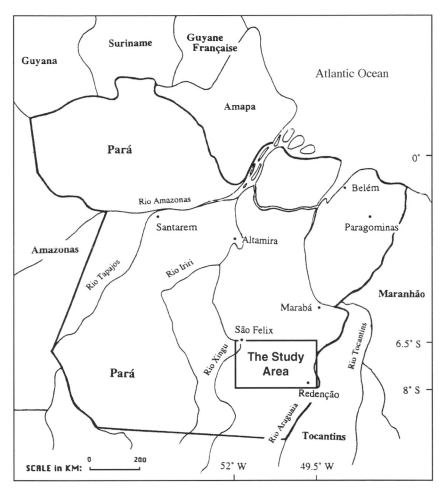

Figure 10.1. Pará State, Brazil (in *dark outline*), showing the study area.

eastern Pará since paved roads penetrated the region in the early 1970s (Schmink and Wood 1992; Veríssimo et al. 1995). Outside the Kayapó Indigenous Area, merchantable stands of mahogany are now rare east of the Iriri River. Even so, big-leaf mahogany remains a common feature of this landscape as smaller size-classes in logged forest. During the forest

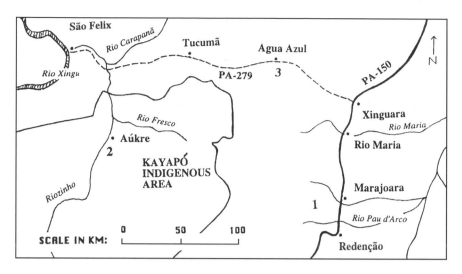

Figure 10.2. The study area, showing field sites: *1*, Marajoara Management Project; *2*, Pinkaití; *3*, Agua Azul.

conversion that frequently follows extraction of merchantable mahogany stems, trees less than 40 cm dbh are often left standing during clear-felling for pasture or agriculture. These trees usually die in fires set to clear debris, however.

Our principal research site is 75 km northwest of Redenção in a 4400-ha managed forest called the Marajoara Management Project, established in 1992 by the Serraria Marajoara (SEMASA) logging company, according to requirements set by the Brazilian Ministry for the Environment (see Fig. 10.2). Population, growth, and fruit production studies are also underway at the Agua Azul Management Project run by the Peracchi logging firm, and at the Pinkaití Research Station run by Conservation International in cooperation with Kayapó Amerindians from the village of Aúkre. Subsections of the Marajoara site have been logged by SEMASA since 1993. The Agua Azul site is a remnant stand of small trees left by loggers in the late 1980s. The forest at Pinkaití is unlogged but is surrounded by logged areas.

The Regeneration Puzzle

Our primary objective, on arriving in June–July 1995, was to quantify regeneration in undisturbed and logged forests as a precursor to framing questions that would shape the design of a broader research program. Our findings appeared consistent with reports from Central America

(Stevenson 1927; Lamb 1966; Snook 1993): Seedling densities in the immediate vicinity of what we assumed were reproductively mature trees appeared low to nonexistent at the start of the long dry season. We asked: Is this a problem of seed production, dispersal, germination, seedling establishment, or high seedling mortality? Given production and dispersal, what are the primary agents of seed and seedling mortality? How long can seedlings be expected to persist under closed canopies? In reviewing the observations that follow, bear in mind the narrow timeframe on which they are based: only one year for most data sets.

Methods

We counted fruit capsules on all trees larger than 20 cm dbh at the Agua Azul ($n = 92$, in about 225 ha) and Pinkaití ($n = 53$, in about 200 ha) sites in June–July 1996, during early to mid dry season when leaf shedding left capsules in stark outline against the sky. A subsample of 23 trees was selected for the count at Marajoara, nonrandomly insofar as these were the trees within easy walking distance (a 1.5-km radius) of camp. We estimated seed production by multiplying capsules by 50, a conservative figure in view of counts by Gullison et al. (1996) and our own observations that each of five compartments per capsules typically contains 12 to 15 seeds, 2 to 4 of which are often malformed.

With seedling distribution patterns as a proxy for seed-dispersal patterns, our preliminary surveys indicated that dispersal declines sharply beyond a 40-m radius drawn from parent stems. We therefore sampled seedling distribution patterns around 16 trees at Pinkaití (these being larger stems around which we deemed regeneration most likely to be present) and 12 trees at Marajoara (all known to have regeneration beneath their crowns) in three 12°-wide wedge transects extended to 40 m in randomly stratified directions. This design represented a 10% sampling intensity in 0.5 ha per tree.

We followed germination of 2681 seeds collected from five sites spread across south Pará; they were planted in polyvinyl bags filled with gray sandy soils from Marajoara, under partial shade cast by loose palm-leaf thatching. Three provenances—Jaú, Jabá, and Toco Sapo—were near the Xingu River, 100 to 200 km south or north of São Felix do Xingu, with seeds taken at each site from 24 to 30 trees in less than 1000 ha. Marabá seeds were collected from a clump of 20 trees of unknown origin planted in the city of Marabá. Seeds from Castelo do Sonho came from hundreds of trees in the Serra do Cachimbo region, 1200 km due west of Marajoara. Because of logistical constraints on the number of seeds that could be planted, seeds from Jaú, Toco Sapo, Marabá, and Castelo do Sonho were sorted and selected for large size and apparent health. By contrast, seeds of all sizes from Jabá were planted. We also followed germination of 1475 preselected (that is, large) seeds from Castelo do Sonho planted in two experiments in closed forest at Marajoara.

We checked seeds planted in closed forest at 2-week intervals to record germination rates and fates of individual seeds. With experience, the telltale signs of different mortality agents became legible, for example, the size and type of entry hole identifying different insect predators, or the color and spreading pattern distinguishing different types of fungi. We estimated seedling densities in closed forest at Marajoara and Pinkaití from data collected in seedling distribution studies. Densities were calculated for the 0.5-ha area within 40-m radii of adult trees.

Relative effects of different mortality agents during the 10 months after germination under closed forest were derived from monthly censuses of 1961 naturally occurring seedlings around eight parent trees at Marajoara and of 821 seedlings that established in three planting experiments, also at Marajoara.

To assess density and vigor of postlogging regeneration, we sampled 40 single-treefall gaps created by mahogany extraction randomly stratified across an area of 1400 ha at Marajoara. These gaps were divided equally between 2 and 3 years old. At each treefall gap, we searched high-light zones, where observational work suggested a strong likelihood of encountering surviving seedlings: around stumps to a 5-m radius, across the interiors and around the perimeters of the crown zone, and to 50 m along the exiting skid trail. Heights and crown depths of all seedlings were measured to estimate growth and vigor.

Results

Of 168 adult trees included in the census, 72 (43%) bore fruit, producing 1 to more than 100 capsules and dispersing from 50 to more than 5000 seeds (Table 10.1). Higher average production values at Pinkaití, where sampled trees were large compared to the other two sites, indicated rising fecundity with stem size, as reported from Bolivia by Gullison et al. (1996). Only one tree in the survey produced capsules at rates comparable to those recorded in Bolivia for stems larger than 80 cm dbh, however. Of trees producing capsules, 75% had fewer than 10. Most seedlings sampled in this survey were found between 5 and 10 m from the parent tree. Densities fell more or less linearly to 40 m (Fig. 10.3).

Germination rates in the nursery ranged from 86% to 95% for the five provenances. In experiments in closed forest, 67.5% of planted seeds germinated; termites, beetles, and ants accounted for 75% of pregermination mortality. Pathogens and disappearances accounted for the other 25%.

Most big-leaf mahogany trees at all three sites had no regeneration in their immediate vicinity. Eight to 10 months after germination, average seedling densities in the 0.5 ha immediately surrounding trees with regen-

Table 10.1. Fruit Capsule and Seed Production at Three Sites in South Pará, Brazil

Site	Trees[a]	Percentage[a]	Capsules[a]	Seeds[a]
Marajoara	23	60.9	7.0 (6.6)	350
Agua Azul	92	40.2	6.2 (5.3)	300
Pinkaití	53	39.6	16.1 (21.2)	800
Total	168	42.9	9.2 (13.0)	450

[a] Trees, number of trees censused; percentage, percentage of censused trees fruiting; capsules, number of capsules per fruiting tree (SD); seeds, estimated number of seeds per fruiting tree.

eration at two sites, 9 of 16 sampled trees at Pinkaití and 12 trees at Marajoara, were 52.5 at Pinkaití (range, 10–130) and 76.7 at Marajoara (range, 10–240) (Table 10.2). At Pinkaití, a higher proportion of surveyed seedlings than at Marajoara were older than 1 year. This difference implied that production, dispersal, germination, or survivorship was low during the previous production cycle at Pinkaití (that is, for the seeds dispersed during the 1995 dry season).

Survivorship of naturally occurring seedlings was 18.2% at the end of 10 months (Table 10.3). Survivorship of seedlings established in planting

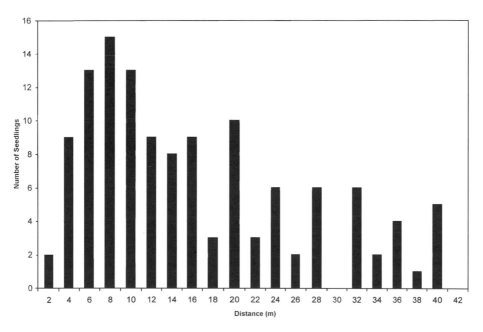

Figure 10.3. Seedling distribution patterns at Marajoara and Pinkaití; distance (m) is from parent trees.

Table 10.2. Seedling Density to 40 m of Parent Trees at Marajoara and Pinkaití

Site	Trees[a]	Seedlings[a]	Range[a]	Percentage[a]
Pinkaití	9	52.5 (44.9)	10–130	39
Marajoara	12	76.6 (70.7)	10–240	6

[a] Trees, number of trees with regeneration; seedlings, estimated number of seedlings in 0.5 ha per tree with regeneration (SD); range, range of estimated number of seedlings; percentage, percentage of seedlings older than 1 year (that is, one growing season).

experiments was higher, 28.3%. The principal causes of seedling death included fungal pathogens, seedling defoliation by caterpillars of an unidentified nocturnal moth, disappearances (mostly from the tramplings and scufflings of peccaries and armadillos), and moisture stress associated with the onset of the dry season (Table 10.4).

The average sampled area per mahogany treefall gap was 536 m² (Table 10.5). The sampling zone most likely to contain seedlings was the area immediately adjacent to the stump. Twenty-six of 40 treefall gaps, or 65% of the total, showed regeneration. Four treefalls, at four widely separate locations, accounted for 55% of all regeneration encountered, or 17 seedlings per treefall. The remaining 22 treefalls with regeneration averaged 2.6 seedlings. Treefalls with and without regeneration were randomly distributed throughout the sampling area.

Gaps created by mahogany extraction 3 years before the sampling period accounted for 34% of all seedlings encountered, and these gaps had slightly

Table 10.3. Survivorship of Wind-Dispersed (Natural) and Planted Seedlings in Closed Forest Around Eight Parent Trees at Marajoara, November 1995 to September 1996

Tree	Natural			Planted	
	Seedlings[a]	%[b]	Exp[c]	Seedlings[a]	%[b]
1	35	37.1	1	191	34.6
2	34	38.2	2	507	23.9
3	31	45.2	3	123	36.6
4	118	26.3			
5	127	13.4			
6	94	34.0			
7	737	19.5			
8	785	11.7			
Total	1961	18.2		821	28.3

[a] Seedlings, number of seedlings per tree or experiment.
[b] Percentage, percentage survivorship.
[c] Exp, experimental plantings.

Table 10.4. Mortality Agents for Seedlings Established from Wind-Dispersed (Natural) and Planted Seeds in Closed Forest at Marajoara, November 1995 to September 1996, as a Percentage of Total Deaths

Origin	P	L	C	O	D	M
Natural	19.2	36.2	1.2	8	19.2	16.2
Planted	37.8	14.6	9.5	13.1	6.3	18.7

P, fungal pathogens; L, larvae of an unidentified moth; C, stem cut, usually by grasshoppers; O, other causes (such as insects, falling debris); D, seedling disappeared; M, dry season moisture stress.

higher average height and range of heights than the 2-year-old gaps (Table 10.5). Of the seedlings encountered in the survey, 63% were less than 70 cm tall; that is, they performed poorly at given ages of 2 years or older, compared to potential height growth in high-light environments (up to 6 m in 2 years; Fig. 10.4). Only 25% of all seedlings were more than 100 cm tall; of these, only two demonstrated truly robust height growth, 260 and 340 cm tall, and form. Competing vegetation, including both advance regeneration of canopy trees and a diverse community of "exploitative gap-invaders" (Bormann and Likens 1979), consistently outgrew and overtopped mahogany seedlings in the single-treefall gaps created by extraction.

Discussion

These data require cautionary notes before discussion:

- Low fruit production rates must be viewed in light of the generally small size-classes of trees comprising the study, especially at Agua Azul and Marajoara. Sampled trees at Agua Azul were all between 20 and 60 cm dbh. At Marajoara, a logged site, only 6 of 23 sampled trees were larger

Table 10.5. Postlogging Seedling Regeneration in Single-Treefall Gaps

Gaps				Seedlings		
Year[a]	Gaps[b]	Present[c]	Area[d]	Seedlings[e]	Avg.[f]	Ht.[g] (range)
2	20	12	531	83	4	61.1 (23–260)
3	20	14	539	43	2	65.3 (33–340)

[a] Year, gap age.
[b] Gaps, number sampled.
[c] Present, number of gaps with seedlings present.
[d] Area, average area sampled per gap (in m²).
[e] Seedlings, total number of seedlings encountered in all gaps.
[f] Avg., average number of seedlings for all gaps.
[g] Ht., average seedling height (in cm) (with range).

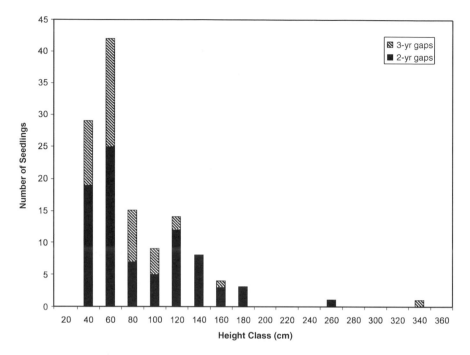

Figure 10.4. Postlogging seedling regeneration height classes (cm) for seedlings in or near 2-year and 3-year single-treefall gaps at Marajoara.

than 60 cm dbh. Low fruit production could also be a function of masting cycles operating at undetermined intervals.

- Seedling distributions only approximate seed-dispersal patterns, especially if density- or distance-dependent mortality agents regulate population structures (Augspurger 1984; Becker and Wong 1985; Howe and Schupp 1985; Sork 1987; Schupp 1988). The unidentified moth whose larvae defoliate and kill large numbers of seedlings may prove to be one such agent.
- Germination rates for seeds planted in closed forest may overestimate natural rates because these seeds were selected for large size, and the experiments were established in late October, 1 month before the full onset of the rainy season and 2 to 3 months after seeds were naturally dispersed, a period during which predators likely increase mortality rates.
- Seedling density estimates should be viewed with the highly dynamic nature of seedling populations in mind. Presented here is a point estimate for a given moment: middle (at Pinkaití) to late (at Marajoara) dry season, whereas population densities are known to oscillate through the year, peaking at the onset of each rainy season.
- Higher survivorship of seedlings established from planted seeds may derive from greater vigor and hardiness associated with larger average

seed size (Spurr 1944; Chinte 1952). Observational and experimental studies in 1997 and 1998 have addressed these considerations.

With these caveats in mind, the following points can be made:

- High germination rates and relatively high survivorship of dispersed seeds indicate that if a tree produced and dispersed seeds, a seedling bank was a likely consequence.
- Seedlings may occur at high densities in 0.5-ha circles drawn around adult trees, tapering to zero or near zero beyond 100 m. For example, we estimated 0.75 seedlings m^{-2} in the 0.5 ha surrounding a pair of large adults at Marajoara shortly after the 1995 germination period, for a total of more than 10,000 seedlings in a 100-m radius (in 3 ha).
- Seedlings appeared remarkably persistent, repeatedly resprouting after defoliation and breakage and tolerating low understory light intensities in closed forest.
- About 10% to 20% of seedlings could be expected to survive through 1 year.
- Survivorship of 52% of 25 naturally occurring seedlings older than 1 year suggested persistence in closed forest beyond 2 years, once established.

The fact that in 1996 most adult trees had little or no regeneration within dispersal distance indicated that seed production during 1995 was minimal or failed for most trees. Observations from earlier surveys in 1995 indicated that seed production was likewise low in 1994. With seed-production data from 1996, these observations indicate that if trees produce seeds in masting cycles—and local informants insist that trees are highly productive in some years compared to others—then these cycles may be 3 or more years long. Seed production may therefore represent an important bottleneck in natural regeneration cycles by lowering the odds that seeds or seedlings will be ready for dispersal or in place when gaps form. The implication is that management plans for natural forests may need to account for masting cycles at whatever temporal and spatial scales they occur. Alternatively, the generally small size of trees in this study may guarantee low production because of size-dependent fecundity rates, a potential management problem in logged forests cited by Gullison and associates from their Bolivian studies.

Scarce or failed regeneration after logging is less surprising if pulsed seed production, the supercycle within which annual or perennial seedling banks bloom and fade, stacks the odds against recruitment after selective extraction. For example, if masting occurs at 4-year intervals, then surveys will document low to nearly zero postlogging regeneration rates in 3 of 4 years. A second temporal issue is the seasonal timing of felling: the longer the interval between the germination period and felling, the fewer seedlings are likely to survive. At Marajoara, trees were extracted in both 1994 and 1993 (our 2- and 3-year-old gaps) during the mid- to late dry season, when seedling banks (where they existed) would have been sharply reduced compared to initial densities. Occasional gaps with high seedling densities, 4 of 40 treefalls sampled, suggested that fruiting cycles of individual trees may be asynchronous with proposed forest-wide masting cycles. An important point is that seedlings were present, albeit usually at low rates, in 65% of sample treefalls. This finding quantitatively confirms our qualitative impression that mahogany seedlings and saplings can be common in the post-logged forest. On the other hand, competing vegetation consistently outgrew these seedlings, and most seedlings appeared suppressed, as indicated by height growth. Without further disturbance, natural mortality under closing canopies could reduce seedlings and saplings to densities approximating those reported by Quevedo (1986) after 9 years and Gullison et al. (1996) after 18 to 20 years.

These rates of natural mortality raise an important question, currently being addressed experimentally: Do seedlings and saplings retain sufficient ability to respond, after one to several years of suppression, to sudden gap formation, such as happens at the time of extraction? If not, then natural recruitment would primarily be through postdisturbance colonization by dispersed seeds, and the notion that preestablished seedlings could be moved into recruitment classes through canopy manipulations would need to be reconsidered.

Conclusions

From the results of this study, we concluded:

- Natural regeneration is commonly present and may even be locally abundant 2 to 3 years after logging in mahogany's range in the southeastern Amazon basin. Where seedlings are absent, low seed production during the 2 to 3 years before extraction may be the cause.
- If management strategies are to be keyed to natural regeneration, then better understanding of seed-production patterns over many years at individual tree, watershed, and regional scales may be necessary to ensure that silvicultural manipulations optimize use of available seeds.

Corollaries include the following:

- Understanding how extraction, at whatever intensity, affects fruit production by remaining trees is important; for example, variation in the timing of flowering on the scale observed at Marajoara in 1996 could lead to temporal isolation of remnant flowering trees (Loveless and Grogan, unpublished data). In addition, given that mahogany seedlings are not likely to grow beyond 100 m of adult trees even in low-ground areas, retaining seed trees in the larger size-classes serves the dual function of generating high seedling densities in their immediate vicinity and ensuring that seeds are available for artificial dissemination to other sites.

- Given that adult trees generally appear restricted to low ground in south Pará, and perhaps to specific microtopographic or edaphic positions within these drainage areas, then the most effective management strategies will likely be those designed to mimic disturbance regimes at the scale of the lowest-order watershed. We are currently testing hypotheses about disturbance regimes posited to shape distribution patterns of mahogany on this landscape.

Acknowledgments. Our institutional affiliation in Brazil is through the Belém-based, nongovernmental organization The Institute of People and the Environment of the Amazon (IMAZON). Collaborators include researchers from IMAZON, Pennsylvania State University, the College of Wooster, and the National Institute for Amazonian Studies (INPA) in Manaus. Funding support has come from the USDA Forest Service's International Institute of Tropical Forestry, the IKEA Foundation, and a private donor. Generous infrastructural support has been provided by the SEMASA and Peracchi timber export companies, Conservation International-Brasil, and the Kayapó from the village of Aúkre. We thank Chris Uhl and Marilyn Loveless for reviewing a draft of this paper.

Literature Cited

Campbell de Aravjo, V. 1971. Sôbre a germinação do mogno (aguano) *Swietenia macrophylla* King. *Acta Amazonica* **1**:59–69.

Augspurger, C.K. 1984. Seedling survival of tropical tree species: Interaction of dispersal distance, light-gaps, and pathogens. *Ecology* **65**(6):1705–1712.

Becker, P., and Wong, M. 1985. Seed dispersal, seed predation, and juvenile mortality of *Aglaia* sp. (Meliaceae) in lowland dipterocarp rainforest. *Biotropica* **17**:230–237.

Bormann, F.H. 1953. Factors determining the role of loblolly pine and sweetgum in early old-field succession in the Piedmont of North Carolina. *Ecological Monographs* **23**:339–358.

Bormann, F.H., and Likens, G.E. 1979. *Pattern and Process in a Forested Ecosystem.* Springer-Verlag, New York.

Chinte, F.O. 1952. Trial plantings of large leaf mahogany in the Philippine Islands. *Caribbean Forester* **13**(2):75–84.

Clark, D.B., and Clark, D.A. 1985. Seedling dynamics of a tropical tree: impacts of herbivory and meristem damage. *Ecology* **66**:1884–1892.

de Steven, D. 1991a. Experiments on mechanisms of tree establishment in old-field succession: seedling emergence. *Ecology* **72**:1066–1075.

de Steven, D. 1991b. Experiments on mechanisms of tree establishment in old-field succession: seedling survival and growth. *Ecology* **72**:1076–1088.

Gerhardt, K. 1996. Germination and development of sown mahogany (*Swietenia macrophylla* King) in secondary tropical dry forest habitats in Costa Rica. *Journal of Tropical Ecology* **12**:275–289.

Grubb, P.J. 1977. The maintenance of species-richness in plant communities: the importance of the regeneration niche. *Biology Review* **52**:107–145.

Gullison, R.E., and Hubbell, S.P. 1992. Regeneración natural de la mara (*Swiete-nia macrophylla*) en el bosque Chimanes, Bolivia. *Ecología en Bolivia* **19**:43–56.

Gullison, R.E., Panfil, S.N., Strouse, J.J., and Hubbell, S.P. 1996. Ecology and management of mahogany (*Swietenia macrophylla* King) in the Chimanes Forest, Beni, Bolivia. *Botanical Journal of the Linnean Society* **122**:9–34.

Harper, J.L. 1977. *Population Biology of Plants*. Academic Press, London.

Howe, H.F., and Schupp, E.W. 1985. Early consequences of seed dispersal for a Neotropical tree (*Virola surinamensis*). *Ecology* **66**:781–791.

Irion, G. 1984. Clay minerals of Amazonian soils. In *The Amazon: Limnology and Landscape Ecology of a Mighty Tropical River*, ed. H. Sioli, pp. 537–579. Junk, Boston.

Klammer, G. 1984. The relief of the extra-Andean Amazon Basin. In *The Amazon: Limnology and Landscape Ecology of a Mighty Tropical River*, ed. H. Sioli, pp. 47–83. Junk, Boston.

Lamb, F.B. 1966. *Mahogany of Tropical America. Its Ecology and Management*. University of Michigan Press, Ann Arbor, MI.

Peters, C.M. 1989. *Reproduction, Growth and the Population Dynamics of* Brosimum alicastrum *Sw. in a Moist Tropical Forest of Central Veracruz, México*. Dissertation, Yale University, New Haven, CT.

Quevedo, L.H. 1986. *Evaluación del Efecto de la Tala Selectiva Sobre la Renovación de un Bosque Húmedo Subtropical en Santa Cruz, Bolivia*. Thesis, Universidad de Costa Rica, San José, Costa Rica.

Record, S.J., and Hess, R.W. 1943. *Timbers of the New World*. Yale University Press, New Haven, CT.

Salati, E., Marques, J., and Molion, L.C. 1978. Orígem e distribuição das chuvas na Amazônia. *Interciencia* **3**:200–206.

Schmink, M., and Wood, C. 1992. *Contested Frontiers in Amazonia*. Columbia University Press, New York.

Schupp, E.W. 1988. Seed and early seedling predation in the forest understory and in treefall gaps. *Oikos* **51**:71–78.

Smith, J.H.N. 1942. The formation and management of mahogany plantations at Silk Grass Forest Reserve. *Caribbean Forester* **3**:75–77.

Snook, L.K. 1992. Mahogany and logging in the forests of Quintana Roo, México: why silvicultural management is necessary to sustain *Swietenia macrophylla*. Paper presented at the Mahogany Workshop: Review and Implications of CITES, Tropical Forest Foundation, Washington, DC, Feb. 3–4.

Snook, L.K. 1993. *Stand Dynamics of Mahogany (*Swietenia macrophylla *King) and Associated Species After Fire and Hurricane in the Tropical Forests of the Yucatán Península, México*. Dissertation, Yale School of Forestry and Environmental Studies, New Haven, CT.

Sork, V.L. 1987. Effects of predation and light on seedling establishment in *Gustavia superba*. *Ecology* **68**:1341–1350.

Spurr, S.H. 1944. Effect of seed weight and seed origin on the early development of eastern white pine. *Journal of Ecology* **25**:467–480.

Stevenson, N.S. 1927. Silvicultural treatment of mahogany forests in British Honduras. *Empire Forestry Journal* **6**:219–227.

Veríssimo, A., Barreto, P., Tarifa, R., and Uhl, C. 1995. Extraction of a high-value natural resource in Amazonia: the case of mahogany. *Forest Ecology and Management* **72**(1):39–60.

Wolffsohn, A.L.A. 1961. An experiment concerning mahogany germination. *Empire Forestry Review* **46**:71–72.

11. Effects of Large-Scale Flooding on Regeneration of Big-Leaf Mahogany in the Bolivian Amazon

Raymond E. Gullison, Corine Vriesendorp, and Agustín Lobo

Abstract. The role that episodic flooding and deposition of alluvial sediments might play in maintaining big-leaf mahogany and associated timber species was studied in Bolivian Amazonian forests. Big-leaf mahogany, Spanish cedar, *Hura crepitans*, and *Calophyllum brasiliense* had significantly higher densities near active rivers than at sites distant from rivers. Alluvial deposition at one site resulted in a median mortality of 12% of the basal area of the forest (all tree species ≥5 cm dbh), but mortality ranged up to 100%. This reduction in basal area caused large and significant increases in canopy openness. Big-leaf mahogany and *H. crepitans* showed strong increases in height growth with increasing canopy openness. Growth of the mahogany saplings projected by using the growth curves and light environments quantified in this study suggest that this type of fluvial disturbance was large enough to explain their regeneration. Stand structures and growth curves for all three of the associated species suggested that conditions created by episodic flooding and deposition also might improve their regeneration.

Keywords: Mahogany, *Swietenia macrophylla*, Ecology, Regeneration, Flooding, Deposition, Seedling demography, Bolivia, Disturbance, River

Introduction

The stand structures, life histories, and population dynamics of tree species are determined in part by the temporal and spatial patterns of the regeneration sites they require (Clark 1991a,b). Certain Neotropical tree species, such as big-leaf mahogany, require large disturbances to regenerate (Lamb 1966; Snook 1996). The disturbances—fires and hurricanes—that maintain big-leaf mahogany in forests in its northern range are absent from the southwest Amazonian forests where big-leaf mahogany grows at its southern limits. However, the big-leaf mahogany in these forests suggests that some large-scale disturbances do happen. Recent research suggests that these disturbances may be hydrologic. Foster et al. (1986) have shown how the migration of river bends (point bar migration) can create large areas of primary succession that can maintain some emergent tree species (hereafter referred to as "disturbance species") that require large-scale disturbances to regenerate. Gullison et al. (1996), based on the stand structures of big-leaf mahogany populations, argued that the hydrologic disturbances of erosion, episodic flooding, and deposition caused by logjams are agents of large-scale forest disturbance in the Bolivian Amazon, and that they might be responsible for maintaining mahogany in these forests.

We investigated the role that episodic flooding and deposition might play in maintaining species that require large-scale disturbances to regenerate in Bolivian Amazonian forests. Our hypotheses were that flooding and deposition of alluvial sediments can cause large disturbances to Bolivian Amazonian forest, that these disturbances are sufficiently large to maintain mahogany in these forests, and the mechanism to explain mahogany regeneration in disturbed areas is elevated light in the understory caused by the death of large trees.

From these three hypotheses, we tested the following predictions:

From the first, that alluvial deposition should cause significant tree mortality, the rate of mortality should be similar to that caused by other agents of large-scale forest disturbance, and past episodic flooding should be evident throughout the forest.

From the second, that mahogany should survive better than other tree species through hydrological disturbances, allowing its seeds to disperse to disturbed areas, as Snook (1993) found for mahogany in México, and mahogany should be most dense in areas of past or recent disturbance.

From the third, that light is higher in the flooded forest because of tree mortality caused by deposition of sediments, and based on the growth of mahogany in different light regimes, the light environments of disturbed areas are sufficient to explain the regeneration of mahogany in these areas, but the available light in undisturbed forest is not.

We tested these predictions with data from a long-term study in the Chimanes Forest, Beni, Bolivia. Although the primary focus of this chapter is big-leaf mahogany, we used available data on three other commercial timber species with similar ranges to look for evidence that the regeneration of these species also depends on, or at least is facilitated by, fluvial disturbances.

Methods

Site Description

The Chimanes Permanent Timber Production Forest (66°00′ to 67°00′W and 14°30′ to 16°00′S) is in the Chimanes Region in the state of Beni, Bolivia (Burniske 1994). Botanical plots have documented a species-richness diversity of 66 to 92 tree species of 10 cm or more in diameter at breast height per hectare (dbh ha^{-1}) (Gullison et al. 1996). Annual mean temperature is 26°C. Median precipitation ranges from 2021 to 2316 mm yr^{-1}. A single, well-defined dry season lasts from May to September.

The Chimanes Forest has two main forest types of interest to this study. Nonflooded alluvial plain forest (high forest) grows on high terraces and areas of low hills (Government of Bolivia 1993). It has a standing volume of 150 to 180 m^3 ha^{-1}. The topography of these forests shows that erosion was much more prevalent here in the past and that the existing mahogany stands have regenerated in these areas of erosion (Gullison et al. 1996). We found little recent mahogany regeneration in high forest.

Temporarily flooded, alluvial plain forest (low forest) grows on low terraces that may or may not be flooded during the wet season. It has a standing volume of about 100 m^3 ha^{-1}. Inventories of mahogany in low forest found a lack of regeneration similar to that in high forest, with the exception of the Cuberene River site, which has abundant regeneration (Gullison et al. 1996). For this reason, the majority of plots in this study are near the Cuberene River (Fig. 11.1). The Cuberene is relatively small, up to 60 m wide and 6 m deep.

According to the local people, a smaller river joined the Cuberene upstream from our sites about 12 years ago, and the enlarged Cuberene became more active. Nine years ago, a logjam began to form downstream,

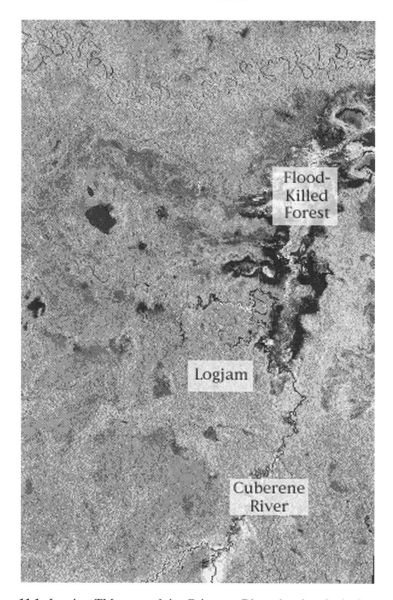

Figure 11.1. Landsat TM scene of the Cuberene River showing the logjam, newly formed savanna, and four study plots.

continued to grow, and is now 2 to 3 km long. The blockage of flow at the logjam and the diversion of the river have caused an area of about 44,000 ha of forest to become savanna, which can be seen in the satellite photo (Fig. 11.1). The logjam is also increasing flooding and deposition of sediment on the forest upstream, and these areas of recently killed forest are where mahogany is regenerating well (Gullison et al. 1996). Only areas of alluvial sediment deposition have substantial tree mortality. Mortality is not elevated in temporarily flooded forest, in the absence of deposition mortality (Gullison, personal observation).

Study Species

The four species considered in our study are big-leaf mahogany (Meliaceae), Spanish cedar (Meliaceae), *Hura crepitans* (Euphorbiaceae), and *Calophyllum brasiliense* (Guttiferae). This study is part of a larger project with the goal of designing a management plan for mahogany. The three additional species were included because their association with mahogany (Gullison, personal observation) suggests a similar life history and because their harvest is likely in the near future.

Big-leaf mahogany ranges from the Gulf Coast of México to the southern Amazon (Lamb 1966). It is confined to somewhat seasonal rain forest climates and is absent from the central Amazon. The trees are emergents, attaining heights up to 50 m and diameters to 3 m in the Chimanes Forest (Gullison et al. 1996). Wind disperses the winged seeds in the dry season (Lamb 1966). This species is considered one of the premier cabinet woods in the world, and it has almost exclusively been the focus of selective logging in the Chimanes Forest (Goitía 1990).

Spanish cedar ranges from the Pacific Coast of México (26°N) to Argentina (28°S) in moist and seasonally dry subtropical and tropical life zones (Cintrón 1990). It is often associated with *Calophyllum* sp. and mahogany species, although at a lower density than the other species. It releases its wind-dispersed seeds in the dry season. Trees can reach heights of 40 m. Like mahogany, this species is reported to regenerate after fire in its northern range. Its primary uses are in household furniture. Small amounts have been cut in the Chimanes Forest (Goitía 1990).

Hura crepitans ranges from Costa Rica (24°N) to the Bolivian Amazon (19°S) (Francis 1990). It requires moist soils and often grows in alluvial floodplains. Its fruits are 15 celled (one large seed per cell) and explosively dehiscent; individual seeds can be thrown 8 m or more. The trees can grow to heights of 60 m, with diameters of 3 m. The saplings can grow up through a light overstory of early secondary species, but they normally require openings for successful regeneration. This species has some resistance to light fires and hurricanes. The wood, used in general carpentry and joinery, is not in much demand locally in the Beni, but large amounts are cut closer to La Paz (Claure 1991).

Calophyllum brasiliense ranges from México to Bolivia. In northern South America, it grows on, but is not limited to, riverbanks and stream valleys (Weaver 1990). In subtropical moist forest, it is frequently associated with mahogany and Spanish cedar. It can grow to 45 m and reach diameters greater than 2 m. Its fruits are one-seeded drupes dispersed by rodents, bats, and birds. The species is intermediate in tolerance to shade. It is a widely used wood suitable for general construction, flooring, bridges, and cabinet making. One of the concessions in the Chimanes Forest has started to harvest *C. brasiliense* now that their supplies of mahogany are exhausted (Gullison, personal observation).

Data Collection Overview

Our study used four distinct data sets. The first set is data from natural saplings of the four study species growing in selected focal quadrats on four plots along the Cuberene River. In September 1993, to augment the sample size of naturally occurring mahogany, we planted 10 seeds and 10 bare-rooted saplings (50–100 cm tall) in the focal quadrats for the second data set. These saplings were remeasured in June and November of 1994. In 1992, 120 naturally occurring mahogany saplings 20 to 250 cm in height outside the focal quadrats were identified as the third set for study. Growth and mortality were measured in the dry season in 1993 and 1994. The final data set is the only one to include data from outside the Cuberene study area; 100-ha inventory plots were installed at three other sites (an additional plot in low forest, and two in high forest) to compare the density of focal species among different sites and types of forest.

The Plot Design at Cuberene

Four plots were installed in 1992 at the Cuberene site (Figs. 11.1 and 11.2). Three of these plots (B134, B5, and Pichi) are adjacent to the Cuberene river and include river bends (Fig. 11.2). Plots Pichi and B5 were on relatively high terrain and had little or no flooding. Plot B134 included a wide range of flooding conditions. High-water marks on trees indicated that water reached more than 1 m in the wet season. The Esperanza plot was across a small elevational gradient 300 to 400 m away from the river. The northern end of the plot received substantial flooding (water marks on the trees were present to heights of almost 90 cm); the southern end was not flooded in the wet season. The plots were installed by cutting a baseline roughly parallel to the river, and then lightly cutting perpendicular trails every 20 m to the river bank. Access was thus provided to at least one edge of all 10- by 10-m quadrats in each plot. The quadrats in the three river plots were stratified by their distance from the river (Fig. 11.2). The distance categories are labeled for the B5 plot in Figure 11.2 to illustrate the stratification scheme. The distance categories were point bar successional vegetation, 0 to 10 m from river, 10 to 30 m, 30 to 50 m, 50 to 70 m, and more than 70 m from the river. We then chose a random sample of 5% to 10% of the quadrats from each distance category. The Esperanza plot, far from the river, was simply divided into four before quadrats were randomly chosen, so that the selection was reasonably well distributed across the plot.

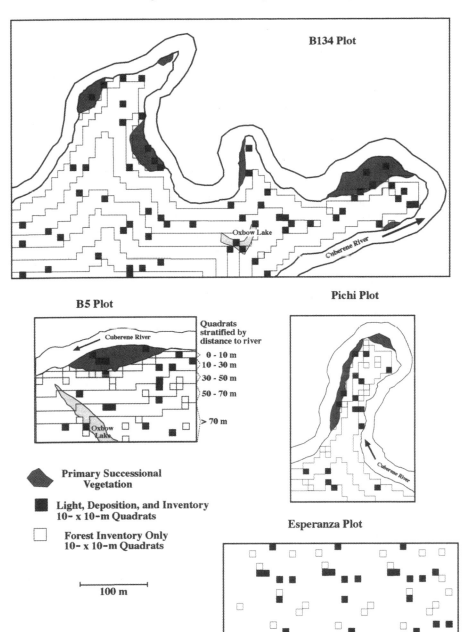

Figure 11.2. The 10- by 10-m focal quadrats in four study plots at the Cuberene site. See text for explanation of sampling design.

Testing the Predictions

Deposition and Tree Mortality

Forest structure in the focal quadrats at Cuberene was determined by identifying and measuring height and diameter of all trees (dead or alive)
5 cm dbh or less. The identity of most standing dead trees could be determined by bark and wood characteristics. Tree mortality was calculated for
each quadrat as the percentage of total basal area that was standing dead.

Alluvial deposition during 1992–1993 was measured in 1993 by digging
through the sediment to the buried leaf layer from the previous dry season
(the forests are semideciduous) at three random points in each quadrat. The
accuracy of this technique was verified the next year by comparing estimates of deposition measured from the depth of leaf layers with deposition
measured by notched stakes. The average deposition ($n = 3$) for each focal
quadrat was used in the following analyses.

Quadrats were divided into four categories by the amount of deposition
they received (none, 0 to 5 cm, 5 to 10 cm, and >10 cm). We used a
Kruskall–Wallis test to determine if the deposition categories differed in
the proportion of dead basal area and amount of live basal area they contained. To test whether mortality on the flooded plots was independent of
tree size, we compared, using a Mann–Whitney U test, the size of trees that
died with those that survived.

Magnitude of Fluvial Disturbance

We compared the amount of tree mortality we found in this study with
literature values of mortality caused by other types of large-scale forest
disturbance.

Evidence for Flooding at Other Sites

We dug pits at Cuberene to determine whether the recent deposition of
sediments in the soil profiles could be identified, and at other sites, under
mature mahogany stands, to look for evidence of past episodic deposition
events. We dug five pits (1 × 1 m and about 1.5 m deep) at Cuberene,
two in currently flooding sites and three in unflooded sites, and five
more at three other sites. In each pit, soil was sampled at 10-cm intervals
and the samples sent to Rutgers University soil laboratory for particle-size
analysis.

Survival of Disturbance Species

The inventory of all trees greater than 5 cm dbh in the focal quadrats
allowed us to compare the mortality of the commonest species with the
mortality of the four study species. We combined the abundance of all individuals in the two flooded plots and those in the two unflooded plots. In

each group, we compared mortality of the four study species with mortality of the five most common species with a chi-square test to determine whether mortality is independent of species. The study species were rare, and their numbers had to be combined to meet the conditions of the chi-square test. If the chi-square test was significant, a binomial test was conducted to compare mortality.

We also present data on the effects of deposition on the mortality of mahogany seedlings and saplings. We tested for an effect of deposition on the mortality rates of the planted saplings (from seeds and bare-root stock) in the focal quadrats, by dividing the quadrats into the same deposition categories as above, and then conducting a Kruskall–Wallis test to determine if mortality is independent of the amount of alluvial deposition received by a quadrat. We used a Mann-Whitney U test to determine if deposition affected the mortality rates of the 120 naturally occurring mahogany saplings outside the focal quadrats. We compared the rates of deposition, and deposition scaled by sapling height, between the saplings that died in 1992–1993 and the saplings that survived.

Density of Focal Species

To compare the density of focal species with other sites, we installed four 100-ha inventory plots. Two plots were in low forest (one at Cuberene, the second more than 1 km from the nearest river), and two plots in high forest. The plots, each 1 by 1 km, contained 10 randomly located 500- by 20-m subplots placed off a central baseline, yielding a sampling intensity of 10%. All mahogany more than 2.5 cm dbh (>10 cm dbh for *H. crepitans*, Spanish cedar, and *C. brasiliense*) were tagged, mapped, and measured. Each randomly located 1-ha transect represented an independent estimate of density at a site. We used an analysis of variance (ANOVA) to compare total density (all individuals >10 cm dbh), and density of trees in the smallest size-classes (2.5–10 cm for mahogany, 10–20 cm for *H. crepitans*, Spanish cedar, and *C. brasiliense*) among the sites. If density was not independent of site, we used a Student–Newman–Keuls test to identify those sites that differed in density.

Light Intensities in Disturbed and Undisturbed Forest

We took hemispherical photographs to quantify light intensities in the focal quadrants at Cuberene. The camera had a Nikkor 8-mm fish-eye lens and used Kodak Tri-X 400 ASA film. We took the pictures only when the sun was hidden but light was still sufficient to expose the photos. Normally, our photography was limited to about 1 h each at dusk and dawn, but some pictures were taken under cloudy skies. Photos were exposed for the sky, without a filter, yielding pictures with enough contrast for further analysis when taken under this restricted light. The photographs were taken above three randomly located stakes in each quadrat—at 30 cm, 1 m, 2 m, and 4 m high—in the dry season (late August to early September 1993) in all four plots, and again in the wet season (late November to early December) only in the quadrats plot B134.

The negatives were digitized by using a system developed by R. Chazdon (University of Connecticut, personal communication). One of us (A. Lobo) developed a computer program to process the digitized images (see Appendix). The program objectively chooses a threshold pixel value to distinguish between sky and vegetation on the image, thus eliminating intra- and interobserver differences. The gray-scale image is then turned into a binary image by using the threshold value, from which the percentage of canopy openness can be calculated simply as the percentage of the total image that is white. We used the three percentage openness values to calculate the average openness at four heights for each focal quadrat.

Here we used data only from plot B134 because it is the most complete. We divided the quadrats based on the amount of deposition they received, and used a three-factor ANOVA to test for effects on canopy openness of season (wet or dry), height (30 cm, 1 m, 2 m, and 4 m), and deposition.

Growth of Focal Species Saplings in Different Light Environments

Three data sets had sample sizes large enough to test for an effect of canopy openness on seedling and sapling growth. These sets were the small, naturally occurring *H. crepitans* saplings in the focal quadrats, the planted mahogany saplings in the focal quadrats, and the naturally occurring mahogany saplings outside the focal quadrats. The height growth of *H. crepitans* saplings less than 50 cm high in the quadrats was first converted to average percentage height increase per quadrat and then regressed against the average percentage openness at 30 cm for the quadrat.

The growth of the planted mahogany saplings in the wet and dry seasons was correlated with their respective light intensities for the quadrats in plot B134. Average and maximum height growth (in centimeters and percentage) was correlated with average light intensities to determine which measures were most strongly influenced by light. The amounts of light at 30 cm in height were used in the regression for the saplings originating from seeds and amounts at 100 cm were used for the saplings planted as bare-rooted stock.

The naturally occurring mahogany saplings outside of the quadrats were divided into three size categories: category I, less than 50 cm; category II, 50 to 100 cm; and category III, 100 to 300 cm. The growth of individual saplings in the preceding year in these categories was then converted to the percentage of maximum height growth observed in that category. The maximum growth observed in each category was 51.5 cm in category I, 116 cm in category II, and 197 cm in category III, during the 1-year period. Standardizing growth in this manner was necessary because of the large range in initial sapling heights. Standardized growth was then regressed against percentage canopy openness measured from a photo taken directly above the sapling in the 1993 dry season.

Tukey Box Graphs

In this chapter, we present many data sets in the form of Tukey box graphs, or box plots, because they show graphically that distributions are not normal. Here, we provide a brief explanation of these box graphs because this form of display may not be familiar to the reader. The median (50th percentile) of a distribution is shown as a dark bar. A box is drawn around the median in which the lower and upper edges represent the 25th and 75th percentiles of the data. Vertical lines extend above and below the box, to the 10th and 90th percentiles. The first and last 10% of the values are plotted individually. Tukey (1977) can be consulted for more information.

Results

Description of Plots at Cuberene

The four plots showed markedly different patterns of tree mortality (Table 11.1). The unflooded plots (B5 and Pichi) had the highest basal area and lowest mortality; B134 had the highest mortality. The focal species (with the exception of Spanish cedar) are locally common enough to be among the five most abundant species on the plots (by basal area).

About two-thirds of the focal quadrats received some alluvial deposition (Table 11.2). The B5 plot received no deposition. Pichi and Esperanza plots received low to intermediate amounts of deposition in some areas, but many of the quadrats escaped deposition entirely. In the following analyses, Pichi is treated as an unflooded plot because deposition has only recently started, and it has not caused any mortality. All the quadrats in B134 received some deposition, sometimes as much as 35 cm in 1 year.

Deposition and Tree Mortality

Deposition significantly increased the mortality of trees of 5 cm or greater dbh (Kruskall–Wallis test: $H = 23.2$, $p < 0.0001$). Nonparametric multiple contrasts indicated that the mortality rates were different in all deposition categories ($p < 0.05$). Quadrats receiving more than 5 cm of deposition had a median mortality of about 12%, ranging up to 100% (Fig. 11.3).

In quadrats without deposition, the median mortality was zero. As a result, live basal area was significantly reduced by deposition (Fig. 11.4) (Kruskall–Wallis test: $H = 11.5$, $p = 0.009$). Nonparametric multiple contrasts indicated that the mortality rates were different in all deposition categories ($p < 0.05$), meaning that mortality increased with increasing

Table 11.1. Description of the Four Plots at the Cuberene River Study Site

Plot	Size (ha)	Quadrats Sampled (number)	Average Basal Area (m²) of Trees ≥5 cm dbh (per 10-m × 10-m quadrat)	Total Tree Species (number)	Total Trees Sampled	Mortality (% basal area)	Mortality (% individuals)	Five Most Common Species by Basal Area (average basal area in m² per 10-m × 10-m quadrat)
Esperanza (Flooded)	4.6	48	0.318	52	493	15.0	15.6	*Unonopsis floribunda*, Annonaceae (0.078) *Hura crepitans*, Euphorbiaceae (0.0301) *Ficus* sp., Moraceae (0.0168) *Sloanea* cf. *guianensis*, Eleaoceceae (0.0156) *Guarea macrophylla*, Meliaceae (0.0140)
B134 (Flooded)	10.8	72	0.215	74	626	30.0	27.5	*Hura crepitans*, Euphorbiaceae (0.0316) *Astrocaryum macrocalyx*, Palmae (0.0145) *Scheelea princeps*, Palmae (0.0102) *Pseudolmedia laevis*, Moraceae (0.0097) *Calophyllum brasiliensi*, Clusaceae (0.0095)

							Dominant species (importance value)
B5 (Unflooded)	3.3	37	0.249	51	336	1.0	1.2	*Unonopsis floribunda*, Annonaceae (0.0780) *Hura crepitans*, Euphorbiaceae (0.0544) *Sapium marmierii*, Euphorbiaceae (0.0227) *Ocotea* sp., Lauraceae (0.0176) *Scheelea princeps*, Palmae (0.0142)
Pichi (Unflooded)	2.7	30	0.259	57	269	1.0	1.0	*Ficus* sp., Moraceae (0.0537) *Swietenia macrophylla*, Meliaceae (0.0194) *Unonopsis floribunda*, Annonaceae (0.0188) *Spondias mombin*, Anacardiaceae (0.0144) *Guarea macrophylla*, Meliaceae (0.0119)

Table 11.2. Deposition of Alluvial Sediments on the Four Plots During the 1992–1993 Flood

Amount of Alluvial Deposit Received in Quadrat (cm³)	Percentage of 10-m × 10-m Quadrats				
	B134	Esperanza	B5	Pichi	Total
0	0	10	16	8	34
0–5	9	15	0	4	28
5–10	16	1	0	0	17
>10	21	0	0	1	22
Total	46	26	16	13	101

deposition. The mortality of individual trees in areas of deposition was independent of tree size (normal approximation to the Mann–Whitney test: B134 ... $n_{dead} = 175$, $n_{live} = 283$, $z = -0.123$, $p = 0.902$; Esperanza. ... $n_{dead} = 75$, $n_{live} = 248$, $z = -0.881$, $p = 0.378$).

Evidence for Flooding at Other Sites

Current alluvial deposition at Cuberene is marked by a high percentage of sand in the upper soil profiles (Fig. 11.5: see profiles of Esperanza flooding and B134). The old forest floor (about 9 years old) is still visible in both pits, at about 60 cm deep in the Esperanza pit and 90 cm deep in the B134 pit. Soil pits in unflooded forest at Cuberene (Fig. 11.5: B5, Esperanza

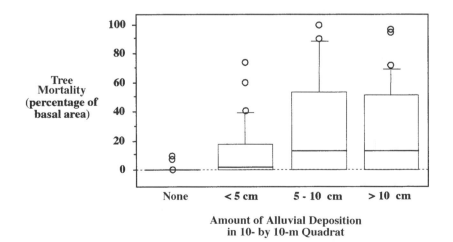

Figure 11.3. Tukey box graph of the percentage mortality (per m² basal area) caused by the deposition of alluvial sediments of all trees ≥5 cm dbh.

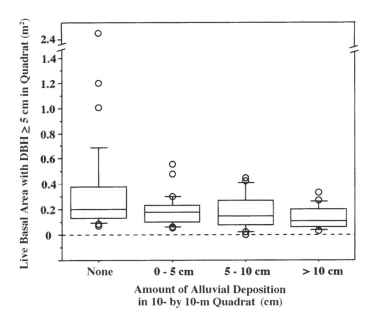

Figure 11.4. Tukey box graph of live basal area of trees ≥5 cm dbh in quadrats with different amounts of alluvial sediments.

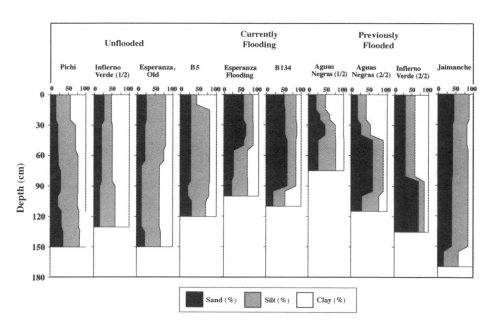

Figure 11.5. Particle-size soil profiles from pits dug on the four plots at the Cuberene site that were the subject of intensive study and from two additional sites with commercial mahogany stands.

Old, and Pichi) did not have a high sand content at the top of the soil profiles.

The soil pits dug at three other sites with mature mahogany stands—Infierno Verde (two pits), Aguas Negras (two pits), and Jaimanche (one pit)—each had at least one pit with a soil profile showing a section with high sand content, suggesting previous episodic flooding and deposition.

Survival of Focal Species

Tree mortality was not independent of tree species for the trees on the flooded plots (Table 11.3: $\chi^2 = 74.2, p < 0.001$). A Scheffé *post hoc* multiple comparison of mortality (Zar 1984) among species on the flooded plots showed that the group containing big-leaf mahogany, Spanish cedar, and *C. brasiliense* had significantly lower mortality than all the common species and that *H. crepitans* had lower mortality than at least one of the common species. Mortality and sample sizes were too low on the unflooded plots to compare with a chi-square test (Table 11.3).

Median mortality of saplings planted as bare-rooted stock was about 50%, independent of the amount of deposition received (Fig. 11.6. Kruskall–Wallis test: $H = 3.08, p = 0.38$). For seedlings planted as seeds,

Table 11.3. Comparison of Mortality Rates of the Focal Species with the Five Most Common Tree Species in the Two Flooded and Two Unflooded Plots[a]

Species	Trees 5 cm dbh (number)	Dead (number)	Mortality (proportion)
Flooded plots			
Psuedolmedia laevis	72	44	0.61
Symphonia globulifera	46	9	0.20
Unonopsis floribunda	168	29	0.17
Inga sp.	46	8	0.17
Guarea macrophylla	57	6	0.11
Hura crepitans	23	2	0.09
Calophyllum brasiliense			
Spanish cedar	13	0	0
Big-leaf mahogany			
Unflooded plots			
Psuedolmedia laevis	27	1	0.04
Unonopsis floribunda	110	3	0.03
Guarea macrophylla	54	0	0
Ocotea sp.	31	0	0
Rheedia cf. *madruno*	30	0	0
Hura crepitans			
Calophyllum brasiliense	4	0	0
Big-leaf mahogany			

[a] Vertical indicate species that do not significantly differ in their mortality rates at $p = 0.05$ level.

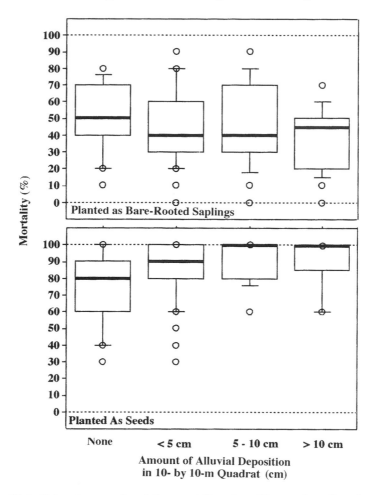

Figure 11.6. Tukey box graphs of the mortality of seedlings and saplings (originating from seeds on bare-rooted saplings) in quadrats with different amounts of deposition.

however, the amount of deposition did influence mortality (Kruskall–Wallis test: $H = 12.9$, $p = 0.005$). Nonparametric multiple contrasts indicated that the mortality rates were different in all deposition categories ($p < 0.05$). In unflooded quadrats, median mortality was 80%, but on the most heavily flooded quadrats, median mortality was 100%.

Between 1992 and 1993, 76 of the 120 naturally occurring big-leaf mahogany saplings died. No significant difference was found in the amount of deposition received by the seedlings that died and those which lived (Table 11.4). When deposition was scaled by the heights of the saplings, however, a highly significant difference in deposition appeared

Table 11.4. Comparison of the Amount of Deposition, and Deposition/Seedling Height, Between Naturally Occurring Big-Leaf Mahògany Seedlings That Died Between 1992 and 1993 and Those That Survived

	n	Median	Minimum	Maximum	
Alluvial deposition (cm)					
Seedlings that died	76	10.2	2	43.3	
Seedlings that lived	44	9.3	0	48.4	
Mann–Whitney					$p = 0.564$
U test					
Deposition/Seedling height					
Seedlings that died	76	0.42	0.05	3.33	
Seedlings that lived	44	0.18	0	1.04	
Mann–Whitney					$p = 0.0001$
U test					

between the two groups. Taken together, the results suggested that smaller mahogany saplings were more susceptible to alluvial deposition than larger ones.

Density of Focal Species

Significant differences were found in the densities of all species in almost all size categories among the four sites (Table 11.5). *Post hoc* comparisons by Student–Newman–Keuls tests showed that rather than having significant differences between low- and high-forest sites, the Cuberene site (Low Forest 1 in Table 11.5) had significantly higher densities of commercial tree species in most size-classes than did both the other low-forest site and the two high-forest sites.

Light Intensities in Disturbed and Undisturbed Forest

Canopy openness values for quadrats receiving less than 5 cm, 5 to 10 cm, or more than 10 cm deposition were distributed along a height gradient from 30 to 400 cm (Fig. 11.7: $n = 13$ for quadrats receiving less than 5 cm deposition; $n = 16$ for quadrats receiving 5 to 10 cm deposition; and $n = 23$ for quadrats receiving more than 10 cm deposition). A three-factor ANOVA found that openness increased significantly with both increasing alluvial deposition ($df = 2$, $SS = 4899$, $MS = 2449$, $F = 51$, $p = 0.0001$) and height ($df = 3$, $SS = 4507$, $MS = 1502$, $F = 31$, $p = 0.0001$), but that no significant difference was found between wet and dry season light intensities ($df = 1$, $SS = 6$, $MS = 6$, $F = 0.1$, $p = 0.72$). None of the interactions was statistically significant. Median canopy openness at 30 cm was about 2%, and

Table 11.5. Results from ANOVA Comparisons of Species Densities at Four Sites in the Chimanes Forest

Species	Diameter Class (cm)	Cuberene (low forest) 1[a]		Low forest 2		High forest 1		High forest 2		p
		Density (/ha)	Standard Error	Density (/ha)	Standard Error	Density (/ha)	Standard Error	Density (/ha)	Standard Error	
Big-leaf mahogany	2.5–10	0.60	0.22	0.00	0.00	0.10	0.10	0.10	0.10	0.0113
	10–20	1.00	0.26	0.20	0.13	0.00	0.00	0.00	0.00	0.0001
	>10	3.80	0.51	0.70	0.21	0.30	0.15	0.20	0.13	0.0001
Spanish cedar	10–20	0.60	0.31	0.00	0.00	0.10	0.10	0.30	0.21	0.1503
	>10	0.90	0.28	0.00	0.00	0.60	0.16	0.90	0.23	0.0083
Calophyllum brasiliensi	10–20	0.60	0.22	0.00	0.00	0.00	0.00	0.00	0.00	0.0006
	>10	3.90	0.82	0.10	0.10	0.20	0.13	0.00	0.00	0.0001
Hura crepitans	10–20	1.60	0.56	1.20	0.57	0.30	0.21	0.20	0.20	0.0693
	>10	9.10	1.35	13.40	1.87	1.20	0.77	1.90	0.48	0.0001

ANOVA, analysis of variance.

[a] Sites that do not differ significantly in density are grouped by underlining.

[b] Low forest 1 is the Cuberene River site.

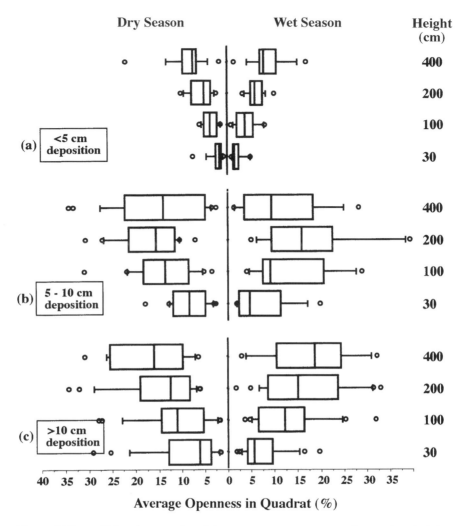

Figure 11.7a–c. Tukey box graphs of the percentage openness values at four heights in the wet and dry seasons for quadrats receiving (**a**) <5 cm deposition, (**b**) 5–10 cm deposition, and (**c**) >10 deposition, in the flooded B134 plot.

at 400 cm about 7.5% in the sites receiving the least deposition. In contrast, median canopy openness at 30 cm was about 7% and at 400 cm about 14%, in the quadrats receiving the most deposition.

Growth of Study Species in Different Light Environments

Only *H. crepitans* was abundant enough to test for a relation between growth of naturally occurring saplings in the quadrats with percentage openness (Fig. 11.9). A significant relation was found between average

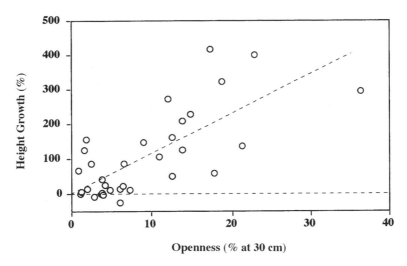

Figure 11.8. One-year growth of *Hura crepitans* seedlings and saplings as a function of percentage canopy openness ($y = 11.29 * x$, $r^2 = 0.73$, $n = 34$, $p < 0.0001$).

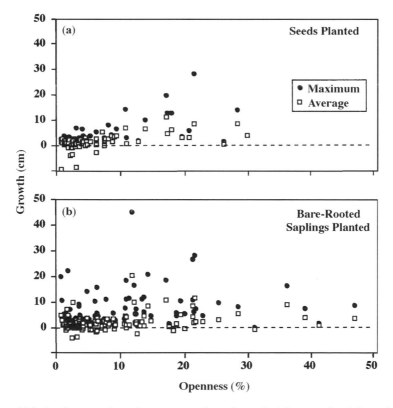

Figure 11.9a,b. Scatter plots of average and maximum height growth of planted bigleaf mahogany saplings originating from (**a**) seeds or (**b**) bare-rooted saplings versus percentage of canopy openness.

percentage height growth and canopy openness measured at 30 cm in the quadrats (regression equation: $y = 11.29 * x$, $r^2 = 0.73$, $n = 34$, $p < 0.0001$).

Growth of the planted mahogany seedlings and saplings was influenced by canopy openness. Both average and maximum growth of seedlings was significantly correlated with canopy openness (average growth, $\rho = 0.48$, $p = 0.001$; maximum growth, $\rho = 0.50$, $p = 0.0001$). Similarly, average and maximum growth of saplings planted as bare-rooted stock was also significantly correlated with canopy openness, although not as strongly (average growth, $\rho = 0.23$, $p = 0.013$: maximum growth, $\rho = 0.33$, $p = 0.0004$).

Finally, the standardized growth of the naturally occurring mahogany saplings selected for detailed study also had a significant positive regression against canopy openness ($y = 0.017 * x$, $r^2 = 0.72$, $n = 36$, $p < 0.0001$) that explained much of the variation in growth (Fig. 11.10).

Discussion

The data presented in this chapter supported the hypothesis that episodic flooding and the corresponding deposition of alluvial sediment caused by logjams can cause large-scale forest disturbance. The median tree mortality caused by alluvial deposition in the Cuberene site was 12% of the basal area but ranged up to 100% in some quadrats. Soil profiles suggested episodic flooding in many places in the Chimanes Forest, although we do not know when.

The amount of mortality caused by flooding is similar to the amount caused by hurricanes in Central American and Caribbean forests. Hurricanes can cause mortality of 3% to 13%, although the rates of nonlethal damage, such as defoliation and branch breaking, can be much higher (Brokaw and Grear 1991; Walker 1991; Basnet et al. 1992; Foster and Boose 1992). Estimates of the mortality of trees in tropical forest fires are scarce, but anecdotal reports suggest it can approach 100%.

One possible difference among these disturbance agents is how they affect different size-classes of trees. We found that for trees more than 5 cm dbh the mortality caused by deposition was independent of tree size. Mortality was greatest for small mahogany saplings. Hurricane mortality is apparently biased toward larger trees (Basnet et al. 1992), although not always (Zimmerman et al. 1994). The effects of fire are more like effects from flooding than hurricanes: fire kills all saplings and seedlings, and most of the larger trees (Snook 1993).

For species to be maintained in tree communities by large-scale disturbances, they must regenerate better than the other species there do. For

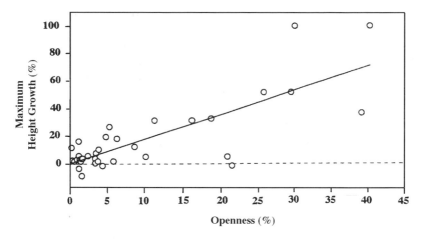

Figure 11.10. Height growth of naturally occurring big-leaf mahogany seedlings and saplings (standardized by the maximum growth observed in each size category) in relation to canopy openness ($y = 0.017 * x$, $r^2 = 0.72$, $n = 36$, $p < 0.0001$).

species with little or no seed dormancy, when disturbances destroy much of the seedling and sapling bank successful regeneration depends on the seeds dispersing into the sites after the disturbance. Seed input will be increased for species that survive the disturbances better than other species do, as appears to be true for mahogany in its northern range, where it and Spanish cedar have been observed to survive fires and hurricanes better than all other tree species (Lamb 1966; Snook 1993).

The sample sizes of big-leaf mahogany and the other focal species were very small on the study plots, but they had significantly lower mortality when grouped together than did the common species on the flooded plots. Gullison et al. (1996), in a complete inventory of mahogany on the most heavily flooded plot (B134), found that mortality was only 4.9% (1 of 19), but mortality of the most common species ranged from 24% to 80%.

In addition to the better ability of adults to survive disturbances, disturbance species need to successfully recruit in these sites by outcompeting other species there. To show whether mahogany and the other focal species can compete with all the other species present would mean comparing their growth by following succession for many years. Obviously, we have not done that; instead, we projected the growth of mahogany based on the percentage openness we quantified in areas of deposition (see Fig. 11.7), and the mahogany growth curve in response to different percentages of openness (see Fig. 11.10). Assuming that mortality is a function of sapling height and deposition as our data (see Table 11.4) showed (deposition/height = 0 to 0.4, mortality = 42%; 0.4 to 1.6, mortality = 90%; >1.6, mortality = 100%), we started cohorts of 1000 seedlings growing from an initial height of 30 cm, and continued annual cycles of growth and mortality until either the surviving saplings reached 4 m in height (the extent of our light data), until all seedlings had died, or for 12 years. When growth was projected for the range of canopy openness values found at the Cuberene site, saplings were unable to attain heights of 4 m at median light intensities in any deposition category. They were able to reach 4 m in 7 to 8 years at maximum intensities in the two highest deposition categories, however (Fig. 11.11). This finding suggested that light intensities were sufficient to explain regeneration of mahogany on these sites, and that light intensities in undisturbed or lightly disturbed forest were insufficient to allow regeneration.

This study presents data from a short period and from a single episodic deposition event, which may not even be complete. The outcomes of this particular flooding event, and the general frequency and scale of this type of disturbance, are unknown. Hydrologic disturbances did appear to be the major disturbance force structuring the tree communities in the Chimanes Forest, however. Further study of the role of these disturbances in tree regeneration should help provide an understanding of maintaining disturbance species in these forests and a rational basis for managing species like big-leaf mahogany.

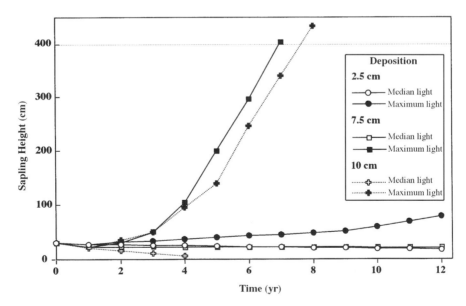

Figure 11.11. The projected growth of saplings in three different flooding regimes and their corresponding light intensity. Saplings can grow rapidly under the maximum light regimes in the areas receiving either 7.5 or 10 cm of alluvial sediments annually.

Acknowledgments. Johnny Galloso, Lucio Isita, Valentin and Rudolfo García, and The Grand Chimanes Council deserve special thanks for help in the field. Miguel del Aguila conducted the forest inventories. Steve Panfil helped either directly or indirectly on most aspects of the project. Robin Chazdon's laboratory lent support in setting up the digitizing equipment. Dan Schragg and Arielle Levine assisted in soil sampling and analysis. Kyle Harms, Steve Panfil, and three anonymous reviewers commented on earlier versions of this manuscript. Steve Hubbell provided indispensable supervision throughout. This material is based on work supported by the National Science Foundation and by the Office of Forestry, Environment and Natural Resources, Bureau of Science and Technology, of the U.S. Agency for International Development (USAID) under National Science Foundation (NSF) grant BSR-9100058. In addition, the project was supported by BOLFOR, Bolivia's Sustainable Forest Management Project funded by USAID and the Government of Bolivia through the Ministry of Sustainable Development and the Environment. We also thank the following companies and organizations for financial support: the Thompson Mahogany Company, Industria Maderera "San Francisco" S.R.L.; Hermann Miller, Inc.; the USDA Forest Service, the U.S. Agency for International Development/Bolivia, and the International Tropical Timber Organization. R. Gullison was supported by a 1967 Centennial National Science and Engineering Research Council Doctoral Fellowship.

Literature Cited

Basnet, K., Likens, G.E., Scatena, F.N., and Lugo, A.E. 1992. Hurricane Hugo: damage to a tropical rain forest in Puerto Rico. *Journal of Tropical Ecology* **8**:47–55.

Brokaw, N.V.L., and Grear, J.S. 1991. Forest structure before and after Hurricane Hugo at three elevations in the Luquillo Mountains, Puerto Rico. *Biotropica* **23**:386–392.

Burniske, G.R. 1994. *Final Evaluation—Phase 1. Conservation, Management, Utilization, Integral and Sustained Use of the Forests of the Chimanes Region of the Department of El Beni, Bolivia.* International Tropical Timber Organization, Yokohama, Japan.

Cintrón, B.B. 1990. *Cedrela odorata* L. In *Silvics of North America*, pp. 250–257. U.S. Department of Agriculture, Forest Service, Washington, DC.

Clark, J.S. 1991a. Disturbance and life history on the shifting mosaic landscape. *Ecology* **72**:1102–1118.

Clark, J.S. 1991b. Disturbance and population structure on the shifting mosaic landscape. *Ecology* **72**:1119–1137.

Claure, H. 1991. *Política Económica de Exportación de Maderas en Bolivia.* Unpublished report produced for Conservation International, Washington, DC.

Foster, D.R., and Boose, E.R. 1992. Patterns of forest damage resulting from catastrophic wind in central New England, USA. *Journal of Ecology* **23**:386–392.

Foster, R.B., Arce, B.J., and Wachter, T.S. 1986. Dispersal and the sequential plant communities in Amazonian Peru flood plain. In *Frugivores and Seed Dispersal*, eds. A. Estrada and T.H. Fleming, pp. 357–370 Junk, Dordrecht.

Francis, J.K. 1990. *Hura crepitans* L. In *Silvics of Tropical Trees*. U.S. Department of Agriculture, Forest Service, Washington, DC. International Institute of Tropical Forestry, Río Piedras, PR.

Goitía, L. 1990. *Informe Técnico-Económico: Proyecto de Acciones Forestales Iniciales (Preliminar)*. Unpublished project report available from the Bolivian Forest Service.

Government of Bolivia. 1993. *Conservation, Management, Harvesting, and Integrated and Sustained Use of Forests in the Chimanes Region, Beni, Bolivia*. International Tropical Timber Organization Project Proposal PD 33/93, Yokohama, Japan.

Gullison, R.E., Panfil, S.P., Strouse, J.J., and Hubbell, S.P. 1996. Ecology and management of mahogany (*Swietenia macrophylla* King) in the Chimanes Forest, Beni, Bolivia. *Botanical Journal of the Linnean Society* **122**:9–34.

Lamb, F.B. 1966. *Mahogany of Tropical America. Its Ecology and Management*. University of Michigan Press, Ann Arbor, MI.

Snook, L.K. 1993. *Stand Dynamics of Mahogany* (Swietenia macrophylla *King*) *and Associated Species after Fire and Hurricane in the Tropical Forests of the Yucatán Península, México*. Dissertation, Yale School of Forestry and Environmental Studies. University Microfilms International 9317535, New Haven, CT.

Snook, L.K. 1996. Conservation biology of mahogany (*Swietenia macrophylla* King): An ecological strategy coevolved with periodic catastrophic disturbance is maladaptive for selective logging. *Botanical Journal of the Linnean Society* **112**:35–46.

Tukey, J.W. 1977. *Exploratory Data Analysis*. Addison-Wesley, Reading, PA.

Walker, L.R. 1991. Tree damage and recovery from Hurricane Hugo in Luquillo Experimental Forest, Puerto Rico. *Biotropica* **23**:379–385.

Weaver, P.L. 1990. *Calophyllum calaba* L. In *Silvics of North América*, Vol 2 Hardwoods, pp. 172–178. U.S. Department of Agriculture, Forest Service, Washington, DC.

Zar, J.H. 1984. *Biostatistical Analysis*. Prentice-Hall, Englewood Cliffs, NJ.

Zimmerman, J.K., Everham, E.M.R. III, Waide, B., Lodge, D.J., Taylor, C.M., and Brokaw, N.V.L. 1994. Responses of tree species to hurricane winds in subtropical wet forest in Puerto Rico: implications for tropical tree life histories. *Journal of Ecology* **82**:911–912.

Appendix

The thresholding algorithms use the histogram of the gray-scale pixel values from a digitized hemispherical photograph, which represents the number of pixels per gray level (0–255). Images can be processed as either negatives or positives, but we discuss the images as negatives. The algorithms assume that the images consist of two types of objects, gaps and vegetation, and that the contrast between them is good. Contrast is optimized by taking the hemispheric pictures with a red filter that blocks the green light reflected by vegetation and proportionally reduces a much larger amount of light from vegetation than from the sky. Under these conditions, histograms are bimodal, with one mode corresponding to gaps and the other to vegetation.

We experimented with two different computer algorithms for thresholding the images. We ultimately used only the second method discussed here,

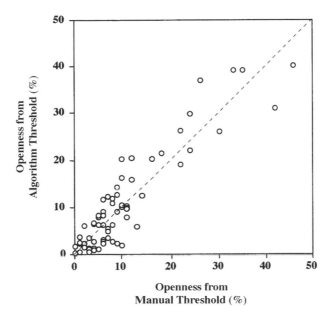

Figure 11.12. Scatterplot of percentage openness values calculated from manual and algorithm thresholds ($r = 0.92$, $n = 70$).

but we present both for the benefit of other researchers interested in using this technique. The first algorithm selects the threshold value as the minimum of a smoothing spline fitted to the log transform of the histogram between the two modes. This strategy produced excellent results with reasonably good digital images that had histograms that spanned most of the 0 to 255 range of digital values. Unfortunately, images of this quality were uncommon. Most of the images spanned about two-thirds of the range, and the video input software did not allow us to stretch the histogram to occupy the full range. Additional problems were that, in some photos, noise was relatively high, the smaller mode (usually the one corresponding to gaps) was very short, and the valley between the modes was often very flat. Under these conditions, the algorithm often located the minimum at the base of the Gaussian corresponding to the highest mode (usually the one corresponding to vegetation), producing a threshold that severely overestimated the amount of gap.

A second, simpler algorithm performed better with the lower-quality images. It assumes that, although the part of the histogram between the modes contains data from the gap and sky, the part between the gap's mode and the lower extreme of the histogram contains only data from the gap. The algorithm also assumes that the distribution of pixel values of the gap is symmetrical. The algorithm simply computes the threshold as twice the distance between the gray level that corresponds to the mode and the minimum. For example, if the mode is at gray level 45 and the minimum is 28, the threshold will be 62 [45 + (45 − 28)]. Most of the images had histograms that were too noisy for a threshold estimation based on a better parameterization of the gap's distribution of gray-scale values. The algorithm included a function that identified images with suspiciously high thresholds (beyond the first third of the gray-scale range), so that these images could be checked manually and thresholded, if necessary.

Performance of the second algorithm was tested against 70 manually thresholded photos (Fig. 11.12). Correlation of the manual with the automatically thresholded values was high ($r = 0.92$). The algorithm inaccurately thresholded (but flagged) about 10% of the images, and these were thresholded manually. The use of this algorithm, or an improved version, should help reduce the subjectiveness of hemispherical photographs to quantify openness and light intensities. Although individual observers may become highly precise in classifying pictures manually, the thresholding algorithm avoids the problem of interobserver differences and should facilitate the comparison of studies done by different researchers.

12. Regeneration After Hurricane Disturbance of Big-Leaf and Hybrid Mahogany Plantations in Puerto Rico

Hsiang-Hua Wang and Frederick N. Scatena

Abstract. A 1-year study of natural mahogany regeneration was conducted in three plantations 8 years after they received different intensities of hurricane-induced disturbance. These plantations were established in 1931, 1938, and 1963 with combinations of big-leaf mahogany and a hybrid of big-leaf and small-leaf mahogany. All the plantations had natural regeneration of mahogany before Hurricane Hugo in 1989. Eight years after the hurricane, lack of seed rain and competition with pioneer species limited regeneration in the most severely damaged plantation. In the plantations with moderate and low damage, seed rain and regeneration were abundant, with an inverse J-shape size distribution of mahoganies. Even in these plantations, however, regeneration was inversely related to the density of pioneer species and the amount of coarse woody debris on the forest floor. In general, both seed rain and natural regeneration were low in areas with more than 500 pioneers per hectare and with less than $20\,m^2\,ha^{-1}$ basal area of mahogany. These findings suggested that mahogany is not competitive in areas colonized by pioneer species, but it regenerated well in areas with moderate canopy disturbance. The success of these plantations indicated that establishing small plantations can be effective in maintaining mahogany populations in disturbed areas.

Keywords: Mahogany, Plantations, Natural regeneration, Disturbance ecology, Puerto Rico

Introduction

For centuries, big-leaf mahogany has been one of the world's most valued timbers (Lamb 1966; Keay 1996). In natural forests, the trees are long lived and typically grow in even-aged stands where regeneration has been related to several types of catastrophic disturbances including hurricanes, fires, and floods (Lamb 1966; Verissimo et al. 1995; Snook 1996; Gullison et al. 1996). The species close association with disturbances suggests that they only regenerate after catastrophic disturbances (that is, with >5% stand mortality *sensu* Lugo and Scatena 1996). Mahoganies, however, do not have many traits typically associated with early-successional, pioneer species (*sensu* Swaine and Whitmore 1988). Moreover, they are long lived; have relatively short-lived seeds, low rates of seed rain, and relatively small seedling banks; do not regenerate well in large gaps; and can survive for years under partial light (Lamb 1966; Johnson and Chaffey 1973; Snook 1996; Gullison et al. 1996). Furthermore, many older mahogany plantations have abundant nondisturbance-related natural regeneration and inverse J-shape diameter distributions (see Chapter 17 by Wadsworth et al. and Chapter 13 by Mayhew et al., this volume).

Our chapter presents results of a 1-year study of natural regeneration of big-leaf mahogany and a hybrid of big-leaf and small-leaf mahogany in three plantations damaged by a hurricane 8 years earlier. The purpose of the study was to quantify the influences of natural disturbance on the regeneration of mahogany in settings where disturbance intensity varied but seed and seedling availability were abundant before the disturbance.

Study Site and Plantation History

Between 1931 and 1985, the USDA Forest Service planted about 1381 ha of mahoganies across Puerto Rico (Marrero 1947; Bauer 1987). We studied three mature mahogany plantations resulting from those plantings (Table 12.1). In similar physical environments and previously studied (Ewel 1963; Lugo 1992; Rodríguez Pedraza 1993; Fu et al. 1996), the plantations were

Table 12.1. Description of Study Plantations in the Luquillo Experimental Forest, Puerto Rico[a]

Date of establishment
  El Verde   1931
  Sabana   1963
  Bisley   1938

Planted species
  El Verde   Mahogany hybrid
  Sabana   Mahogany hybrid
  Bisley   Big-leaf mahogany

Elevation, annual rainfall,[b] aspect
  El Verde   200–300 m, 3150 mm yr^{-1}, west
  Sabana   180–260 m, 3060 mm yr^{-1}, west
  Bisley   220–300 m, 3180 mm yr^{-1}, north

Soil
  El Verde   Catalina stony clay, clayey, oxidic, isohyperthermic, Typic
              Topohumults
  Sabana   Humatas clay, clayey, kaolinitic, isohyperthermic, Typic
              Tropohumults
  Bisley   Humatas clay, clayey, kaolinitic, isohyperthermic, Typic
              Tropohumults

Silvicultural thinnings
  El Verde   1945, 1954, 1958, 1965
  Sabana   Never
  Bisley   1949, 1954, 1964

Average diameter growth since establishment (average diameter/plantation age)
  El Verde   0.8 cm yr^{-1}
  Sabana   1.1 cm yr^{-1}
  Bisley   1.0 cm yr^{-1}

Hurricane influence since establishment
  El Verde   2 within 30-km radius, 7 within 60-km radius
  Sabana   1 within 30-km radius, 4 within 60-km radius
  Bisley   1 within 30-km radius, 5 within 60-km radius

[a] All areas are in the subtropical wet forest zone (Ewel and Whitmore 1973).
[b] Annual rainfall was estimated using a forest-based relation between rainfall and elevation (García Martinó et al. 1996).

established at different times with different combinations of big-leaf mahogany and the hybrid. The plantations also received slightly different treatments during establishment and were affected to different degrees by a hurricane in 1989.

In September of 1989, the western part of Hurricane Hugo passed over eastern Puerto Rico (Scatena and Larsen 1991). Storm-facing slopes in the northeast corner of the Luquillo Experimental Forest (the Forest) and within 15 km of the hurricane's center received the most damage (Scatena and Larsen 1991; Boose et al. 1994). According to site visits immediately after the storm (Scatena, personal observervation) and a forest-wide damage map produced from aerial photos and field surveys (Boose et al. 1994), the Bisley plantation received the most damage. The Sabana site received intermediate damage and the El Verde plantation was least affected.

At the Bisley site, the mature natural forest adjacent to the study plantations lost 50% of its aboveground biomass during the storm (Scatena et al. 1996). Within 5 years, regeneration and growth of survivors had increased aboveground biomass to 86% of the prehurricane amount. A comparative study of the effects of Hurricane Hugo on the Sabana and El Verde plantations (Rodríguez Pedraza 1993) showed that in the Sabana plantation all stems had some form of damage and 52% were severely damaged (stem snap, standing dead, or tipover). In the El Verde plantation, 3.5% of the stems were not damaged and 37% were severely damaged. The Sabana plantation lost 54% of its aboveground biomass and suffered 51% mortality during the hurricane, and the El Verde plantation had 33% mortality and a 41% reduction in biomass. Both plantations suffered greater damage than paired natural forest of similar age. A 12-year history of the El Verde plantation and its paired natural forest showed that, because the hurricane opened the plantation's canopy to a greater degree, more early-successional, pioneer species established in the plantation (Fu et al. 1996). Posthurricane fluctuations in species composition were also greater in the plantation than in the adjacent natural forest.

Before Hurricane Hugo, each of the plantations had experienced the passage of other hurricanes (see Table 12.1). The most severe was in 1932 (Scatena and Larsen 1991), 1 year after the El Verde plantation was planted and before the other plantations were established. Although the 1932 storm severely affected surrounding forests, the small saplings of this plantation showed little damage (F.H. Wadsworth, International Institute of Tropical Forestry, personal communication). In 1955, another hurricane passed near the Forest and killed 29% of the mahoganies in the Bisley area (Wadsworth, unpublished Institute report). The Sabana plantation had not yet been established, and the El Verde plantation suffered only minor damage during this storm (F.H. Wadsworth, personal communication).

During their first 6 years, mahogany plantations in Puerto Rico had average diameter growths between 0.6 and 1.4 cm yr^{-1}, average height growth between 0.7 and 1.0 m yr^{-1}, and volume increments of 0.3 to 1.6 m^3 yr^{-1} (Marrero 1947). The average diameter growth of the mahoganies in the study plantations, as estimated from the average diameter of adults and the plantation age, is similar to that for other plantations in Puerto Rico (see Table 12.1). These growth rates were faster than those reported for large, mature trees in natural forests (Gullison et al. 1996; Snook 1996), but they are within the range reported for other mahogany plantations (Lamprecht 1989). Nevertheless, mahoganies in Puerto Rico grow much slower than native pioneer species in hurricane-disturbed areas, where height growth of

4 to 5 m yr^{-1} and diameter growth of 4 cm yr^{-1} are common (Scatena et al. 1996).

All three plantations had annual seed crops and abundant seedlings and saplings before Hurricane Hugo. In the El Verde plantation, regenerated seedlings of hybrid mahogany were observed 24 years after establishment (Marrero and Wadsworth 1955, unpublished Institute report). Moreover, the authors noted that, "A fair amount of natural regeneration was seen in the sample plot . . . for the first time during March of this year. This regeneration apparently came as a result of heavy fruiting last winter after recent thinning . . . This is the first crop which has appeared in the forest plantation."

Methods

In January and February of 1997, three 4- × 50-m-transect plots were randomly located on ridge, slope, and valley topographic positions in each plantation, following methods previously used in the Bisley area (Basnet et al. 1992). In each of the nine transects, forest structure, seedling dynamics, seed rain, and the phenology of adult mahogany were monitored for 1 year (Fig. 12.1). In each transect, all live trees (2.5 cm in dbh or with heights ≤50 cm) were tagged and identified to species following the nomenclature of Liogier

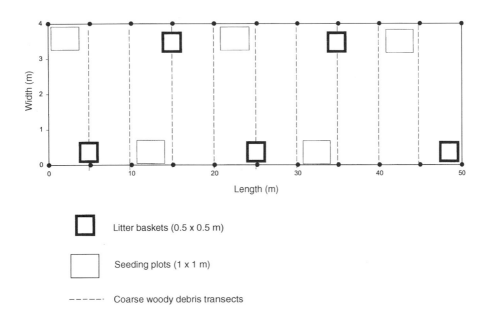

Litter baskets (0.5 x 0.5 m)

Seeding plots (1 x 1 m)

‐ ‐ ‐ ‐ ‐ Coarse woody debris transects

Figure 12.1. Schematic design of transect plots located in three mahogany plantations in the Luquillo Experimental Forest, Puerto Rico (not drawn to scale).

and Martorell (1982). We divided these mahoganies into five categories: seedling (height <50 cm); young juvenile (height 50 cm, <2.5-cm dbh); old juvenile (>2.5-cm dbh but <10-cm dbh); small tree (>10-cm dbh but <20-cm dbh); and adult tree (20-cm dbh). Differences in population structure among plantations were compared using analysis of variance (ANOVA) and Duncan's multiple range test.

Two indices were used to quantify the amount of hurricane disturbance in each transect: an importance value (half the sum of the percentage density and percentage basal area) of the most common pioneer species in the area, *Cecropia peltata* (also known as *C. schreberiana*), and the percentage of forest floor covered by coarse (≥20-cm-diameter) woody debris. We calculated this percentage by measuring the intersecting length of all coarse woody debris in 10 line transects in each plot (Fig. 12.1; percent = 100*length of line transect occupied by debris/total length of line transects).

Five plots, 1 × 1 m, were established in each transect to monitor mahogany seedling dynamics over the course of 1 year. All mahogany seedlings inside the plots were tagged and revisited monthly between February 1997 and July 1997 and once every 2 months between July 1997 and February 1998. The seedlings were grouped into two categories, old and new: "old" seedlings were those present when the plots were established in February 1997, and "new" seedlings germinated during the study period. During each sampling, we tallied the number of new seedlings, old seedlings, and old seedlings without leaves in February 1997 but with leaves produced during the period of investigating. The number of leaves on each seedling was also recorded.

Five 50- × 50-cm seed rain baskets were placed in each transect (Fig. 12.1) and collected monthly. The phenology of five mature mahogany trees in each transect was also observed monthly. For each tree, the presence or absence of the following traits was noted: abscissing leaves, new leaves, seed pods, new flowers, abscissing flowers, and aborted fruits.

We tested seed germination from both the Sabana and El Verde plantations in two ways. First, 200 mature, recently fallen seeds were collected from the understory of the plantations in April 1997. The seeds were sown in germinating plates with sterilized soil (50 seeds per plate, four plates per plantation) at the Institute's nursery. The plates were watered daily and the number of germinated seeds recorded. Second, 180 seeds were collected from the Sabana plantation and immediately sown in the plantation in 1- × 1-m plots in one of three treatments: seeds sown on bare soil after the litter layer was removed, seeds sown on the surface of the litter layer, and seeds sown under 4 cm of litter. Twenty seeds were sown in each plot, with three plots per treatment.

Photosynthesis photon flux density (PPFD) at 1 m above the ground was measured in each transect by using a Sunfleck Ceptometer (Model SF-80, Decagon Devices, Pullman, Washington, USA). Measurements were made before (March 13, 1997) and after (April 25, 1997) the mahoganies had dropped their leaves. Fifty PPFD measurements were made in each transect and averaged to compare light intensity before and after the mahoganies shed their leaves. All measurements were made on clear days when the sun was directly overhead (1100–1300).

Results

Forest Structure and Composition

Before Hurricane Hugo, adult mahoganies dominated the canopy at all three sites with heights ranging between 25 and 35 m. Eight years after Hurricane Hugo, basal area, stem density, and species diversity were similar in the three plantations (Tables 12.2, 12.3). Large trees of the pioneer species *C. peltata* dominated the canopy at Bisley, and the adult mahoganies still dominated the canopies at El Verde and Sabana. The two indices of disturbance (the percentage of forest floor covered by coarse woody debris and the importance value of *C. peltata*) confirmed earlier observations that the Bisley plantation had suffered the greatest damage, the Sabana site had intermediate damage, and El Verde plantation had the least damage (Table 12.3).

Although the basal area, stem density, and species diversity were similar in the three plantations, their composition and population structure were different (Tables 12.2, 12.3, 12.4). In the El Verde plantation, hybrid mahogany and understory *Prestoea montana* comprised 67% of the individuals (see Table 12.2). At the Sabana site, hybrid mahogany and three common secondary species (*Casearia arborea*, *Cecropia peltata*, and *Syzygium jambos*) comprised about 65% of the individuals. At Bisley, mahogany comprised only about 1% of the individuals and the common pioneer species *Cecropia peltata* and *Psychotria berteriana* had their highest densities. In all the plantations, the densities of young juveniles were less than 1% of seedling density (see Table 12.4). The densities of young and old juveniles were significantly higher in El Verde, and the number of adult mahoganies was significantly less in Bisley. The number of small trees varied greatly within each site, but it did not vary among the three sites. Because of the age of the plantation and the spacing between individual trees, we believe that many, if not most, of the individuals in the small-tree category in the Sabana plantation were not natural regeneration. Instead, they were slow-growing individuals planted when the plantation was established. Considerable within-stand variations in height and volume increments have also been reported for nearby plantations of hybrid mahogany (Weaver and Bauer 1986).

Within individual plots, the densities of mahogany seedling, young juveniles, and old juveniles were negatively related to the density of the pioneer species *C. peltata* (Fig. 12.2). The densities of mahoganies in the smaller

Table 12.2. Density, Basal Area, and Important Value Index [IVI = (%BA + % density)/2] in Three Plantation Forests in the Luquillo Experimental Forest, Puerto Rico, in 1998 for Plants with Diameter ≥2.5 cm at 1.3 m Aboveground

	El Verde			Sabana			Bisley		
	Density (ha⁻¹)	BAᵃ (m²ha⁻¹)	IVIᵇ (%)	Density (ha⁻¹)	BAᵃ (m²ha⁻¹)	IVIᵇ (%)	Density (ha⁻¹)	BAᵃ (m²ha⁻¹)	IVIᵇ (%)
Alchornea latifolia	0	0	0	33.30	0.73	1.1	16.70	0.01	0.2
Alchorneopsis portoricensis	0	0	0	0	0	0	33.30	0.21	0.6
Aniba bracteata	0	0	0	100.00	0.16	1.5	0	0	0
Andira inermis	0	0	0	0	0	0	33.30	0.15	0.5
Buchenavia capitata	0	0	0	33.30	5.09	4.8	0	0	0
Byrsonima spicata	0	0	0	16.70	0.45	0.6	0	0	0
Casearia arborea	33.30	0.21	0.9	533.30	1.75	8.9	33.30	0.11	0.5
Casearia guianensis	66.70	0.09	1.4	100.00	0.22	1.6	0	0	0
Casearia sylvestris	0	0	0	0	0	0	33.30	0.18	0.6
Cecropia peltata	66.70	1.56	3.1	683.30	6.69	15.2	1800.0	13.88	33.1
Cordia borinquensis	0	0	0	0	0	0	33.30	0.06	0.5
Cyathea arborea	0	0	0	16.67	0.07	0.3	0	0	0
Cyathea arborea, understory	0	—	—	0	—	—	50.00	—	—
Dacryodes excelsa	16.70	0.01	0.4	0	0	0	416.70	1.69	6.3
Dendropanax arboreus	33.30	0.04	0.7	0	0	0	16.70	0.22	0.4
Faramea occidentalis	33.30	0.07	0.7	0	0	0	0	0	0
Guarea guidonia	33.30	0.20	0.9	0	0	0	33.30	13.60	12.5
Henrietella fascicularis	0	0	0	16.67	0.04	0.3	0	0	0
Heterotrichum cymosum	0	0	0	0	0	0	16.70	0.08	0.3
Homalium racemosum	16.70	0.15	0.5	16.67	0.26	0.5	0	0	0
Inga fagifolia	100.00	0.52	2.5	83.33	0.51	1.6	100.00	0.41	1.5
Inga vera	0	0	0	33.33	0.16	0.6	0	0	0
Manilkara bidentata	16.70	0.01	0.4	50.00	4.75	4.7	33.30	0.22	0.6
Miconia prasina	0	0	0	83.33	0.10	1.3	0	0	0
Miconia racemosa	33.30	0.02	0.7	166.70	0.54	2.8	0	0	0

Species									
Miconia tetrandra	0	0	0	16.70	0.04	0.3	0	0	0
Myrcia deflexa	33.30	0.17	0.9	0	0	0	0	0	0
Myrcia splendens	150.00	0.28	3.2	116.70	0.38	2.0	0	0	0
Ocotea leucoxylon	0	0	0	216.70	0.75	3.7	150.00	1.19	2.8
Ormosia krugii	0	0	0	0	0	0	16.67	0.04	0.3
Palicourea riparia	116.70	0.08	2.3	16.67	0.01	0.3	0	0	0
Prestoea montana	0	0	0	0	0	0	350.00	8.11	11.3
Prestoea montana, understory	1633.33	—	—	0	—	—	400.00	—	—
Psychotria berteriana	150.00	0.34	3.3	16.67	0.03	0.39	16.70	1.20	11.7
Sapium laurocerasus	0	0	0	0	0	0	33.30	0.05	0.05
Schefflera morototoni	50.00	0.31	1.3	133.30	2.10	3.6	133.30	1.05	2.5
Sloanea berteriana	0	0	0	0	0	0	16.70	0.04	0.3
Swietenia macrophylla	0	0	0	0	0	0	50.00	13.16	12.3
Swietenia macrophylla × S. mahagoni	1217.00	29.08	56.8	550.00	32.50	35.0	0	0	0
Syzygium jambos	83.30	0.09	1.7	550.00	0.93	8.4	0	0	0
Tabebuia heterophylla	366.70	10.16	18.7	16.67	1.11	1.2	0	0	0
Tetragastris balsamifera	0	0	0	0	0	0	50.00	0.49	1.0
Trichilia pallida	0	0	0	0	0	0	16.70	0.04	0.3
Total (without understory Prestoea montana, Cyathea arborea)	2617	43.38	100	3550	59.38	100	4333	56.2	100
Total with understory (Prestoea montana, Cyathea arborea)	4250	43.38	—	3600	59.38	—	4783	56.2	—

a BA, Basal area.
b IVI, Important Value Index [(% basal area + % density/2)].
c Understory Prestoea montana and Cyathea arborea were individuals with stems heights greater than 0.6 m but less than 1.3 m and were not included in the calculation of IVI.

Table 12.3. Structure and Composition of Three Mahogany Plantations in the Luquillo Experimental Forests, Puerto Rico, 1998

	El Verde	Sabana	Bisley	$F(Pr > F)$
Basal area ($m^2 ha^{-1}$)	43.4 ± 5.7 a[a]	59.4 ± 13.6 a	56.1 ± 22.6 a	0.29 (0.76)
Density (no. >2.5 cm ha^{-1})[b]	4250 ± 33 a	3600 ± 242 a	4783 ± 400 a	2.30 (0.18)
Species per plot	16 ± 2 a	1.17 ± 0.06 a	1.14 ± 0.08 a	1.84 (0.26)
Shannon diversity index	1.01 ± 0.05 a	1.17 ± 0.06 a	1.14 ± 0.08 a	1.84 (0.26)
Percentage (%) of forest floor with woody debris	2.13 ± 0.94 b	5.40 ± 1.86 ab	7.20 ± 0.66 a	4.15 (0.07)
IVI of mahogany	55.0 ± 14.5 a	35.0 ± 9.4 ab	12.3 ± 5.0 b	4.77 (0.06)
IVI of *Cecropia peltata*	3.1 ± 3.2 b	15.2 ± 4.3 ab	33.1 ± 15.9 a	3.61 (0.09)
PPFD with mahogany canopy ($mol s^{-1} m^{-2}$)	14.2 ± 1.0 a	15.8 ± 0.8 a	16.5 ± 1.1 a	1.53 (0.29)
PPFD without mahogany canopy ($mol s^{-1} m^{-2}$)[c]	68.7 ± 4.2 a	70.4 ± 5.3 a	29.8 ± 4.6 b	23.75 (0.00)
Ratio of PPFD with and without mahogany canopy[c]	4.8	4.5	1.8	

[a] Values with similar letters in a row are not significantly different.
[b] Values are means ± SD of three transects per plantation.
[c] PPFD, photsynthesis photon flux density.

Table 12.4. Population Structure of Mahogany in Three Plantations in the Luquillo Experimental Forest, Puerto Rico, 1997

	El Verde	Sabana	Bisley	$F(Pr > F)$
Seedlings/ha (height <50 cm)	168,667 ± 28,420 a	74,000 ± 5,039 b	1,333 ± 1,335 c	25.36 (0.00)
Young juveniles (height <50 cm, dbh <2.5 cm)	1,417 ± 696 a	267 ± 148 b	0.0 ± 0.0 b	3.36 (0.10)
Old juveniles (2.5 cm ≤ dbh <10 cm)	983 ± 249 a	200 ± 200 b	0.0 ± 0.0 b	7.95 (0.02)
Small trees/ha (10 cm ≤ dbh <20 cm)	83 ± 60 a	100 ± 100 a	0.0 ± 0.0 a	7.95 (0.02)
Adult trees/ha (dbh ≥ 20 cm)	183 ± 33 a	200 ± 50 a	66.7 ± 44.2 b	4.07 (0.08)

Means ± SD of three transect plots per plantation. Values with similar letters are not significantly different.

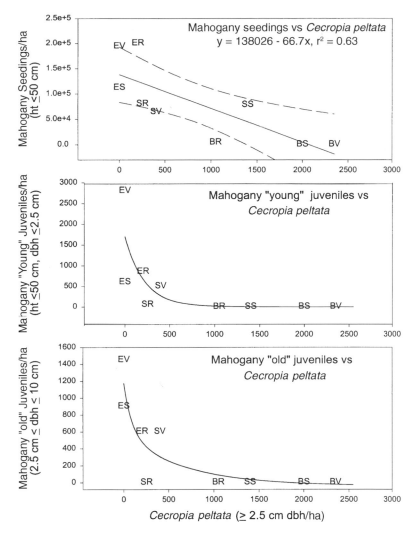

Figure 12.2. Relationships between the density of mahogany seedling and juveniles and the density of *Cecropia peltata* in mahogany plantations of the Luquillo Experimental Forest, Puerto Rico. Letters correspond to plantations (*E*, El Verde; *S*, Sabana; *B*, Bisley) and the topographic position of the transect (*R*, ridge; *S*, slope; *V*, valley).

size-classes (seedling, young juvenile, and old juvenile) were also negatively related to the amount of coarse woody debris on the forest floor (Fig. 12.3).

Phenology and Germination

The seasonal pattern of adult mahogany phenology was similar in all three sites (Fig. 12.4). In Bisley, however, the adult trees had suffered so much

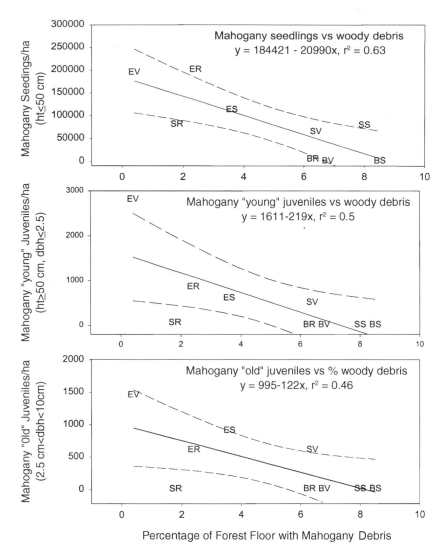

Figure 12.3. Relationships between the density of mahogany seedlings and juveniles and the percentage of forest floor covered by woody debris in mahogany plantations in the Luquillo Experimental Forest, Puerto Rico. Letters correspond to plantations (*E*, El Verde; *S*, Sabana; *B*, Bisley) and the topographic position of the transect (*R*, ridge; *S*, slope; *V*, valley).

crown damage during Hurricane Hugo that only one tree was producing seeds 8 years after the storm.

The mahoganies dropped most of their leaves after 2 relatively dry weeks near the end of March and continued to lose leaves through May. Observations made in the Bisley and Sabana area during the past 10 years indi-

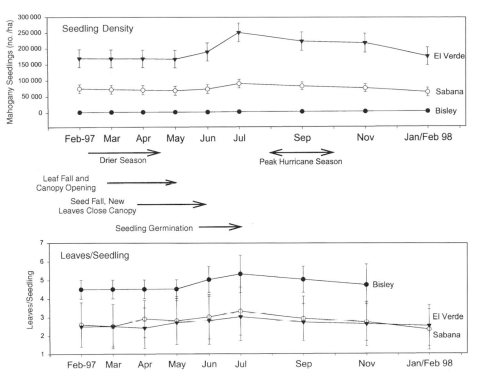

Figure 12.4. Seedling density, seasonal phenology of adult mahogany tree, and average number of leaves per seedling in mahogany plantations of the Luquillo Experimental Forest between February 1997 and February 1998.

cate that mahoganies commonly lose most of their leaves during similar periods of dry weather at this time of year (Scatena, personal observervation). The adult mahoganies began to sprout new leaves shortly after the old leaves fell, and the canopies were completely replaced by new leaves within 3 months. In all three sites, the light intensity 1 m above the forest floor increased after the mahogany trees had dropped their leaves (see Table 12.3). This temporary increase in light was similar at El Verde and Sabana (4.5 to 4.8 times closed-canopy conditions), but it was considerably less (1.8 times) at Bisley, where nondeciduous pioneer species dominated the posthurricane canopy.

After the initial leaf fall and the subsequent opening of the canopy, seed rain, new leaves, and flowers increased. The amount of annual seed rain varied considerably among sites (0–226,667 seeds ha^{-1}yr^{-1}) and was greatest in the least damaged site with the highest density of large trees (El Verde). Observations of seedlings and saplings 30 to 50 m from the edge of the plantations indicated that the seeds were transported at least that distance.

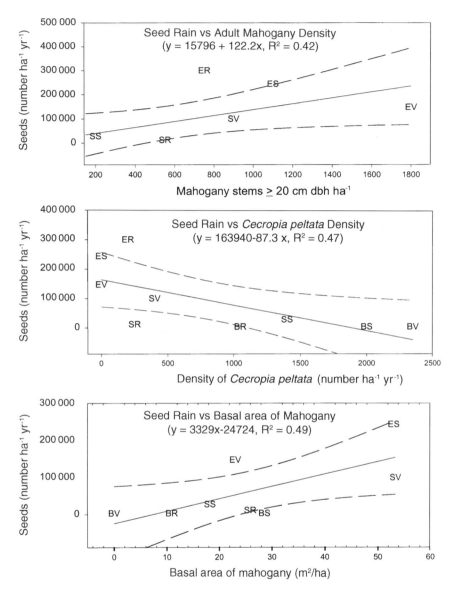

Figure 12.5. Annual rain of mahogany seeds versus stem density and basal area in mahogany plantations in the Luquillo Experimental Forest of Puerto Rico. Letters correspond to plantations (*E*, El Verde; *S*, Sabana; *B*, Bisley) and the topographic position of the transect (*R*, ridge; *S*, slope; *V*, valley).

The amount of annual seed rain was directly related to the basal area and density of mahogany and inversely related to the density of *C. petalta* and woody debris on the forest floor (Fig. 12.5). New seedlings were observed in the plots from June to July, after the canopy had releafed and closed.

Therefore, most of the annual crop of seedlings was not exposed to open-canopy conditions during most of their first year.

The percentage of viable seed was similar between the El Verde and Sabana plantations and averaged 68% (Table 12.5). Germination studies on other Forest mahogany seeds showed that nursery germination can range from 68.8% to 91.1%, depending on the age of seeds and how they are stored (Bauer 1987). In the field experiment reported here, germination began 6 weeks after seeds were spread on the forest floor (Fig. 12.6). Seeds sown under 4 cm of litter had a significantly higher germination rate (85%; $F = 7.4$, $P = 0.031$) compared to seeds sown on bare soil (58%) or placed above the litter layer (58%). For comparison, only 39% of mahogany seed germinated in a secondary dry forest in Costa Rica (Gerhardt 1996).

Seedling Dynamics

The densities of mahogany seedlings differed significantly among the three sites, with the greatest in El Verde, the least damaged plantation (Table 12.4). Over the course of this 1-year study, we found a net increase in seedling density in El Verde and Bisley and a net decrease in Sabana (Table 12.5, Fig. 12.4). All three plantations had more old seedlings (>1 year old) than new seedlings. The mortality of new seedlings was inversely related to seedling density; it decreased from Bisley (0%) to El Verde (47.5%; see Table 12.5). For comparison, the annual mortality for mahogany seedlings

Table 12.5. Mahogany Seed Rain, Seed Germination, and Seedling Dynamics in Three Mahogany Plantations in the Luquillo Experimental Forest, Puerto Rico in 1997

	El Verde	Sabana	Bisley
Seed rain and seed germination			
Seed rain (seeds ha^{-1}yr^{-1})	226,667 ± 44,435	42,667 ± 27,095	0 ± 0
Viable seeds (%)	67.5	68.0	—
Seedlings germinated before February 1997 ("old" seedlings)			
Density (seedlings/ha)	168,667 ± 28,420	74,000 ± 5,039	1,333 ± 1,335
Survival (%)	69.2 ± 13.8	56.8 ± 13.7	100 ± 0
Mortality (%)	30.8 ± 5.6	43.2 ± 7.3	0 ± 0
Resprouted "old" seedlings (seedlings ha^{-1}yr^{-1})	15,337 ± 3,960	4,000 ± 1,633	0 ± 0
Seedlings germinated between February 1997 and February 1998 (new seedlings)			
Density (seedlings ha^{-1})	106,667 ± 11,697	22,000 ± 4,995	667 ± 0
Survival (%)	52.5 ± 7.9	87.9 ± 21.0	100 ± 0
Mortality (%)	47.5 ± 5.6	12.1 ± 3.3	0 ± 0
Net variation in seedlings between February 1997 and February 1998			
(No. seedlings ha^{-1}yr^{-1})	+9,337 ± 4378	−8,667 ± 2546	+667 ± 367

Values are means ± SD of 15 sample plots per plantation.

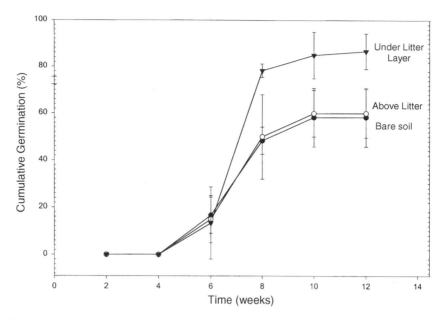

Figure 12.6. Germination of mahogany seeds in plantations under three litter treatments; seed sown under 4cm of litter (*triangles*); seeds sown on bare soil (*solid circles*); and seed sown beneath 4cm of litter (*empty circles*).

in a Forest nursery ranged from 10.4% to 17.6%, with no statistical difference between type of seed storage (refrigerator or room temperature) or species (big-leaf or the hybrid) (Bauer 1987).

Discussion

Eight years after Hurricane Hugo, stand structure and natural regeneration differed considerably in the three study plantations. The abundant seed rain and regeneration observed in these plantations before the hurricane suggested these differences are a consequence more of hurricane disturbance than of the age or management history of the plantation. Our results and those from other mahogany plantations in humid regions of Puerto Rico (Marrero 1947; Weaver and Bauer 1986; Bauer 1987; Lugo 1992; Rodríguez Pedraza 1993; Fu et al. 1996; Wadsworth et al., Chapter 17, this volume) suggest the following life history for these plantations.

After initial planting in sites with high light—abandoned pastures or young secondary forests—2 to 5 years are required before the plantations are well enough established not to need annual weeding or other maintenance. During this phase, diameter and height growth show considerable within- and between-stand variation. As the plantations develop, the canopy trees have annual, nearly synchronous leaf fall during the drier season of the year. Adult trees sprout new leaves shortly after leaf fall and completely replace their canopies within a few months.

After about 20 years, the dominant canopy trees begin to have annual seed rains. Seed production per tree increases with age and after stand thinning. The wind-dispersed seeds are spread throughout the plantation, commonly 30 to 50 m from its boundary. The age of seed production (Mayhew et al., Chapter 13, this volume) and the median and maximum distance of seed dispersal are similar to those observed elsewhere (Rodríguez et al. 1994; Gullison et al. 1996).

The percentage of germination is high in the weeks after seed fall, but the seeds do not remain viable for much longer than 1 year. Seeds typically sprout during the rainy summer months and shortly after the deciduous canopy releafs. During most of their first year, seedlings survive under closed canopies and partial to low light. They do not survive in areas where pioneer species or coarse woody debris is abundant. In favorable sites, more than 80% of the seedlings survive for at least 1 year. Nursery studies indicate that the light compensation point for big-leaf mahogany is $34\,\mu mol\,s^{-1}\,m^{-2}$ when it is grown in full sun (PPFD = $800\,\mu mol\,s^{-1}\,m^{-2}$) and $16\,\mu mol\,s^{-1}\,m^{-2}$ in shade (PPFD = $80\,\mu mol\,s^{-1}\,m^{-2}$; Ramos and Grace 1990). Given that the plantation seedlings germinate and live most of their first year under shade until the deciduous canopy opens to PPFDs of nearly $70\,\mu mol\,s^{-1}\,m^{-2}$ (Table 12.3), the seedlings apparently assimilate carbon and produce new leaves during these annual periods of open canopy. Although these seedlings can exist in a suppressed state for years, less than 1% eventually reaches the young juvenile stage.

As the plantation matures, considerable variation continues in the diameter and height growth of the planted trees. This uneven growth plus recruitment from natural regeneration combine to shift the frequency distribution of diameters from a normal, even-aged size distribution into an inverse J-shaped distribution. The gradual shift probably takes place within 30 to 40 years of establishment. During this period other, typically early-successional, species colonize the plantation, and a native understory develops. Nevertheless, mahogany remains the dominant species in the plantations.

Similar to the adjacent natural forest, these plantations are subject to effects associated with passing hurricanes. Hurricanes typically pass over the Forest during the late summer after the mahoganies have releafed and the annual seedling crop has established. When the hurricanes pass, they open the plantation canopy to a greater degree than in similar-aged natural forests. In heavily affected plantations, all the adults suffer some crown damage, and mortality can be well over 50%. Although some adult trees survive even if their upper trunks are completely snapped, they may not completely reestablish their crowns or produce seeds during the next decade. In contrast, trees that only suffer defoliation and branch damage will rapidly releaf and produce seeds within a few years of the hurricane.

In severely damaged areas in the plantations where large (>200-m²) canopy openings are created, light-demanding pioneer species invade, and mahogany regeneration is limited to nonexistent in the decade after the hurricane. Even in plantations where seed rain and advanced regeneration were relatively abundant, mahogany regeneration is limited in sites occupied by fast-growing pioneer species. In contrast, areas within the plantations that receive light to moderate stem damage and canopy opening regenerate abundantly. This finding is supported by nursery studies indicating that hybrid mahogany has better height and volume growth in medium-light environments (32% full daylight) than under low-light (6.2%) or high-light (79.8%) environments (Medina et al., Chapter 8, this volume).

These observations indicated that mahogany regenerates well in areas with intermediate to low canopy disturbance and not occupied by pioneers (see Figs. 12.2 and 12.3). Moreover, although mahogany seed production and regeneration are related to disturbance and mahogany trees grow well when planted and weeded in high-light environments, they do not occupy the same regeneration space (large forest openings) as pioneer species do. Given the large spatial variation in canopy damage caused by hurricanes, the ability to regenerate in areas with moderate canopy disturbance can be a very effective life history strategy. For example, when Hurricane Hugo passed over the Forest, areas like Bisley were classified as "very damaged" or "destroyed" and comprised 20% of a 5254-ha study area (Boose et al. 1994). Areas with "slight" to "moderate" damage, such as the plantations in El Verde and Sabana, comprised 75% of the area. Therefore, a species that can regenerate in these slight to moderately damaged areas had about 3.8 times the area for regeneration as a pioneer-type species that only regenerates in severely damaged areas with large canopy openings.

The ability to regenerate in areas with moderate canopy disturbance may explain why mahogany does well in inland areas subject to hurricanes in Central America. Because the destructive force of hurricanes rapidly decreases as they move inland, canopy damage in these inland areas is typically limited to defoliation and branch damage similar to what we observed in the Sabana and El Verde plantations. In contrast, trunk snapping and uprooting, similar to the Bisley site, is common in unprotected areas where hurricanes first cross land. Likewise, mahogany may also do well in parts of the Caribbean where rugged topography interacts with hurricanes and tropical storms to produce localized patches of moderately damaged canopies every few years and severe damage every 50 to 60 years (Scatena and Larsen 1991). Finally, this intermediate-amount-of-canopy-opening hypothesis may explain why mahogany does fairly well in areas with prolonged periods of lakelike flooding, as Gullison et al. (1996) described in Bolivia. Moreover, because this flooding typically produces standing dead trees and smaller canopy openings, it apparently creates the intermediate-light conditions suited to mahogany regeneration. In contrast, short-duration, high-velocity floods that modify floodplain morphology and completely denude large areas of vegetation create high-light environments suited to pioneer species (for examples, see Junk 1989; Foster 1990; Lamotte 1990; Kalliola et al. 1991; Thompson et al. 1992).

Implications for Managing Mahogany in Natural Stands

The three plantations studied here had slightly different ages and management histories. Nevertheless, except in the most severely damaged area, they had abundant natural regeneration and inverse J-shaped size distributions 8 years after the passage of a severe hurricane. The shade tolerance of the seedlings and the success of the natural regeneration documented here suggested that, if silvicultural treatments can create moderate canopy disturbance or if weeding can reduce competition during establishment, subsequent rotations can be obtained without additional planting (Mayhew et al., Chapter 13, this volume). In contrast, the lack of regeneration in highly disturbed plantations with large canopy openings and only a few seed trees indicates that harvesting practices that create large gaps and only leave a few seed trees will have very low natural regeneration and relatively long replacement times.

The success of these small plantations through numerous environmental events during the past decades indicates that establishing similar plantations may be an effective strategy for maintaining populations in human-caused disturbance areas. Given the tremendous amount of land disturbance associated with the selective harvesting of mahogany, ample areas of moderately disturbed canopy should be available for these plantings. Moreover, given that 0.5 km of road can be built per harvested mahogany tree (Watson 1996) and 0.11 to 2.8 ha of land can be disturbed per harvested tree (Veríssimo et al. 1995; Gullison et al. 1996) at a planting density similar to those studied here, hundreds to thousands of individuals could be planted per harvested tree. Regardless of what management strategies are used, the long periods needed for mahogany seed to reach commercial sizes (50–150 years) require a commitment to long-term management of mahogany-rich forests.

Acknowledgments. We acknowledge the support provided by the Taiwan Forestry Research Institute, the USDA Forest Service, the Department of Biology of the University of Puerto Rico, and the Luquillo Long-Term Ecological Research program. In addition, the help and comments of S. Moya, C. Estrada, A. García, S. Ward, F.H. Wadsworth, J.E. Mayhew, A.E. Lugo, and three anonymous reviewers are greatly appreciated.

Literature Cited

Basnet, K., Likens, G.E., Scatena, F.N., and Lugo, A.E. 1992. Hurricane Hugo: damage to a tropical rain forest in Puerto Rico. *Journal of Tropical Ecology* **8**:47–55.

Bauer, G.P. 1987. *Swietenia macrophylla* and *S. macrophylla* × *S. mahagoni* Development and Growth: The Nursery Phase and the Establishment Phase in Line Planting in the Caribbean National Forest, Puerto Rico. Masters thesis, College of Environmental Science and Forestry, Faculty of Forestry, State University of New York, Syracuse, NY.

Boose, E.R., Foster, D.R., and Flute, M. 1994. Hurricane impacts to tropical and temperate forest landscapes. *Ecological Monographs* **64**(4):396–400.

Ewel, J.J. 1963. Height growth of big-leaf mahogany. *Caribbean Forester* **24**(1):34–35.

Ewel, J.J., and Whitmore, J.L. 1973. *The Ecological Life Zones of Puerto Rico and the U.S. Virgin Islands*. ITF-8. U.S. Department of Agriculture, Forest Service, Río Piedras, PR.

Foster, R.B. 1990. Long-term changes in the successional forest community of the Río Manu floodplain. In *Four Neotropical Rainforests*, ed. A.H. Gentry, pp. 565–572. Yale University Press, New Haven, CT.

Fu, S., Rodríguez Pedraza, C., and Lugo, A.E. 1996. A twelve-year comparison of stand changes in a mahogany plantation and a paired natural forest of similar age. *Biotropica* **28**(4a):515–524.

García Martinó, A.R., Warner, G.S., Scatena, F.N., and Civco, D.L. 1996. Rainfall, runoff, and elevation relationships in the Luquillo Mountains of Puerto Rico. *Caribbean Journal of Science* **32**(4):413–424.

Gerhardt, K. 1996. Germination and development of sown mahogany (*Swietenia macrophylla* King) in secondary tropical dry forest habitats in Costa Rica. *Journal of Tropical Ecology* **12**:275–289.

Gullison, R.E., Panfil, S.N., Strouse, J.J., and Hubbell, S.P. 1996. Ecology and management of mahogany (*Swietenia macrophylla* King) in the Chimanes Forest, Beni, Bolivia. *Botanical Journal of the Linnean Society* **122**:9–34.

Johnson, M.S., and Chaffey, D.F. 1973. An inventory of the Chiquibel Forest Reserve, Belize. Land Resources Study 14, Foreign and Commonwealth Office, Overseas Development Administration, Land Resources Division, London, UK.

Junk, W.J. 1989. Flood tolerances and tree distributions in Central Amazonian floodplains. In *Tropical Forests: Botanical Dynamics, Speciation, and Diversity*, ed. A.H. Gentry, pp. 47–64. Academic Press, London.

Kalliola, R., Saol, J., Puhakka, M., and Rajasilta, M. 1991. New site formation and colonizing vegetation in primary succession on the western Amazon floodplains. *Journal of Ecology* **79**:877–901.

Keay, R.W.J. 1996. Introduction: the future for the genus *Swietenia* in its native forests. *Botanical Journal of the Linnean Society* **122**:3–7.

Lamb, F.B. 1966. *Mahogany in Tropical America. Its Ecology and Management*. University of Michigan Press, Ann Arbor, MI.

Lamotte, S. 1990. Fluvial dynamics and succession in the Lower Ucayali River Basin, Peruvian Amazonia. *Forest Ecology and Management* **33/34**:141–156.

Lamprecht, H. 1989. *Silviculture in the Tropics*. Deutsche Gesellschaft für Technische Zusammenarbeit (GTZ), Eschborn, Germany.

Liogier, H.A., and Martorell, L.F. 1982. *Flora of Puerto Rico and Adjacent islands: A Systematic Synopsis*. Editorial de la Universidad de Puerto Rico, Río Piedras, PR.

Lugo, A.E. 1992. Comparsions of tropical tree plantations with secondary forests of similar age. *Ecological Monographs* **62**:1–42.

Lugo, A.E., and Scatena, F.N. 1996. Background and catastrophic tree mortality in tropical moist, wet, and rain forests. *Biotropica* **28**(a):585–599.

Marrero, J. 1947. *A Survey of the Forest Plantations in the Caribbean National Forest*. Masters Thesis, School of Forestry and Conservation, University of Michigan, Ann Arbor, MI.

Ramos, J., and Grace, J. 1990. The effects of shade on the gas exchange of seedlings of four tropical trees from Mexico. *Functional Ecology* **4**:667–677.

Rodríguez, S., Chavelas, B.J., and García Cuevas, X. 1994. Dispersión de semillas y establecimiento de caoba después de un tratamiento mecanico del sitio. In *Madera, Chicle, Caza y Milpa: Contribuciones al Manejo Integral de las Selvas de Quintana Roo*, eds. L. Snook and A. Barrera de Jorgenson, pp. 81–91. INIFAP/PROAFT/AID/WWF-US, Mérida, México.

Rodríguez Pedraza, C.D. 1993. *Efectos del Huracán Hugo Sobre Plantaciones y Bosques Secundarios Pareados en el Bosque Experimental de Luquillo, Puerto Rico*. Master's thesis, University of Puerto Rico, Río Piedras, PR.

Scatena, F.N., and Larsen, M.C. 1991. Physical aspects of Hurricane Hugo in Puerto Rico. *Biotropica* **23**:317–323.

Scatena, F.N., Moya, S., Estrada, C., and Chinea, J.D. 1996. The first five years in the reorganization of aboveground biomass and nutrient use following Hurricane Hugo in the Bisley Experimental watersheds, Luquillo Experimental Forest, Puerto Rico. *Biotropica* **28**(4a):441–457.

Snook, L.K. 1996. Catastrophic disturbance, logging and the ecology of mahogany (*Swietenia macrophylla* King): grounds for listing a major tropical timber species in CITES. *Botanical Journal of the Linnean Society* **122**:35–46.

Swaine, M.D., and Whitmore, T.C. 1988. On the definition of ecological species groups in tropical rain forests. *Vegetatio* **75**:81–86.

Thompson, J., Proctor, J., Viana, V., Milliken, W., Ratter, J.A., and Scott, D.A. 1992. Ecological studies on a lowland evergreen rainforest on Maraca Island, Roraima, Brazil. 1. Physical environment, forest structure, and leaf chemistry. *Journal of Ecology* **80**:689–703.

Veríssimo A., Barreto, P., Tarifa, R., and Uhl, C. 1995. Extraction of a high-value natural resource from Amazonia: the case of mahogany. *Forest Ecology and Management* **72**:39–60.

Watson, F. 1996. A view from the forest floor: the impact of logging on indigenous peoples in Brazil. *Botanical Journal of the Linnean Society* **122**:75–81.

Weaver, P.W., and Bauer, J. 1986. Growth, survival and shoot borer damage in mahogany plantings in the Luquillo Forest in Puerto Rico. *Turrialba* **36**(4): 509–522.

3. Silviculture

13. Silvicultural Systems for Big-Leaf Mahogany Plantations

John E. Mayhew, M. Andrew, James H. Sandom,
S. Thayaparan, and Adrian C. Newton

Abstract. The different silvicultural techniques for establishing big-leaf mahogany plantations have been well documented, but these methods frequently take little account of the long-term development of stands or subsequent rotations and cannot properly be called silvicultural systems. As existing plantations begin to mature, the need to consider establishing the next rotation grows, and, therefore, the need to adopt a suitable silvicultural system. Where plantations of big-leaf mahogany have begun to regenerate naturally, the traditional system of clear-cutting and replanting may not be appropriate. Natural regeneration has the advantages of reducing nursery and establishment costs and possibly the extent of damage by shoot borer (*Hypsipyla* sp.). Early attempts to apply a single tree selection system to big-leaf mahogany plantations were unsuccessful, but recent plans to adopt a modified uniform shelterwood system in Sri Lanka and a group selection system in St. Lucia have been more carefully developed. The proposed systems are described in detail. Successful long-term management of big-leaf mahogany plantations depends on choosing a system

that accommodates both the light requirements of the
species and key local constraints under which plantation
managers are operating. Local constraints should be
carefully considered before a silvicultural system is
adopted.

Keywords: Mahogany, Natural regeneration, Silvicultural systems, Sri Lanka, St. Lucia

Introduction

The silvicultural techniques used for establishing and early tending of big-
leaf mahogany plantations have been well documented (Streets 1962; Lamb
1966; Chable 1967; Weaver 1987; Oliver 1992; Chaplin 1993). Applying these
"reproductive methods" (Smith 1986) does not amount to adopting a silvicultural system, however. A silvicultural system is "the process by which the
crops constituting the forest are tended, removed, and replaced by new
crops, resulting in the production of woods of a distinctive form" (Troup
1928). A silvicultural system therefore addresses establishing second and
subsequent rotations. Adopting a silvicultural system implies applying a
planned program of silvicultural treatment (Smith 1986).

The emphasis of existing literature on reproductive methods is to be
expected. The key stage in the life cycle of a big-leaf mahogany plantation
is the period of early growth, when shoot borer (*Hypsipyla* sp.) attack may
profoundly affect the form of the trees and thus the value of the plantation.
In many countries around the world, however, big-leaf mahogany trees have
grown well, and a valuable timber crop is already nearing maturity (Table
13.1). The threat of shoot borer damage to plantations should not be under-
estimated, but more careful consideration of long-term management, particularly in establishing the next rotation, is needed if managers are to make
the best use of the opportunities presented by maturing plantations. A long-
term perspective is also relevant to forest departments and other organizations committed to sustainable plantation management. Adopting a
suitable silvicultural system for big-leaf mahogany plantations makes both
financial and environmental sense.

In this chapter, we describe applying various silvicultural systems to big-
leaf mahogany plantations, evaluate the extent of natural regeneration
under existing plantations, and consider the advantages of systems that use
natural regeneration. Until recently, only one example of natural regeneration has been documented. In the last few years, however, a uniform shelterwood system has been adopted in Sri Lanka, and a group selection

Table 13.1. Countries with Mahogany Plantations Reaching or Having Reached Maturity

Country	Plantation Area Including Immature Stands (ha)	Age of Oldest Existing Stands Excluding Trials (years)	Source
Indonesia	116,000	—[a]	Fattah (1992)
Fiji	42,000	35	K. Singh (personal communication)[c]
Philippines	25,000	75[b]	E. Lapis (personal communication)[d]
Sri Lanka	4,500	95[b]	Sandom and Thayaparan (1995)
Guadeloupe	4,000	50	Soubieux (1983)
Martinique	1,500	50	Tillier (1995)
Puerto Rico	300	60	Bauer and Gillespie (1990)
Honduras	150	50	M. Cruz (personal communication)[e]
Belize	100	40	Ennion (1996)
St. Lucia	100	30	Andrew (1994)

[a] No details available.
[b] These selectively cut stands and may not contain trees from the original planting.
[c] Silvicultural Research Division, Fiji.
[d] Ecosystems Research and Development Bureau, Philippines.
[e] Lancetilla Botanic Gardens, Honduras.

system is being considered in St. Lucia (Lesser Antilles); both systems are described here.

Applying Silvicultural Systems to Big-Leaf Mahogany Plantations

Most silvicultural systems, which vary in complexity, were originally designed for species-poor temperate forests of western Europe. The simplest systems are little more than clear-cutting and replanting pure stands. The most complex systems include polycyclic cutting, liberation thinning, and natural regeneration in mixed stands. According to Smith (1986), four categories of silvicultural system, each with several variations (Table 13.2), are used. More detailed explanations can be found in Troup (1928), Smith (1986), or Matthews (1994).

Early foresters sought to apply temperate systems to natural forest in the tropics, but species-rich tropical forest proved far more difficult to manage and required several modifications. Lamprecht (1989) divides the adapted systems into two categories: systems for gradual transformation into uniform forest (including the Malayan uniform system and the tropical shelterwood system), and systems for gradual transformation into managed

Table 13.2. Categories of Silvicultural System

Silvicultural System	Variations
Clearcutting system	Replanting or reseeding with natural regeneration (including coppice)
Shelterwood system	Uniform
	Strip
	Group
	Irregular
Selection system	Single tree
	Strip
	Group
Coppice with standards	

Source: Smith (1986).

selection forest (including the Philippine selective logging system and the Queensland system). The systems provide practical guidelines for what would otherwise be an impossibly difficult task of managing a forest to suit the ecological requirements of many timber species.

Big-leaf mahogany has been established in pure stands, in high proportion in mixed stands (Marie 1949; Perera 1955), and in enrichment lines through natural forest (Weaver and Bauer 1986; Oliver 1992; Veríssimo et al. 1995). Pure or near-pure mahogany stands can be managed under a system designed for species-poor temperate forest without the need for much modification. Enrichment plantations in natural forest are species rich, however, and it could be argued that a silvicultural system designed for natural forest is more appropriate for managing them. In fact, the standard spacing of enrichment lines (9–11 m) allows big-leaf mahogany trees to close the canopy on reaching maturity. Some authors have referred to this kind of plantation establishment as "conversion line planting" (Dawkins 1965; Weaver 1987), indicating that the natural forest will ultimately be replaced by big-leaf mahogany (assuming reasonable growth and effective maintenance). We suggest that applying silvicultural systems designed for species-poor temperate forest to enrichment plantations is appropriate as well.

Natural Regeneration in Big-Leaf Mahogany Plantations

The silvicultural system most commonly applied to tropical plantations includes clear-cutting and replanting or coppicing to establish the next rotation. Clear-cutting and replanting have been applied to big-leaf mahogany plantations in Martinique with good results (D. Chabod, Office National des Forêts, personal communication), and managing many big-leaf mahogany plantations similarly elsewhere is likely. Clear-cutting is straightforward to implement but has several drawbacks. The cost of raising (or purchasing) nursery stock, transporting it, and planting are high, and, when mature trees are harvested, risks of causing soil erosion and of destroying any surviving or recolonizing native plant and animal populations are also high. In general, clear-cutting systems do not take advantage of any natural regeneration under mature trees, although the system can be adapted to make use of advance growth.

Big-leaf mahogany sets seed at an early age (Table 13.3). Well-spaced stands appear to develop fruits earlier than closely spaced stands where tree crowns are competing for light. Thinning has been found to stimulate fruiting in Sri Lanka (Table 13.3) and in Puerto Rico (F. Wadsworth, USDA Forest Service, personal communication). The observations of Busby (1967) in Fiji (Table 13.3), however, indicate that well-spaced trees may sometimes be slow to set seed and that other, possibly genetic factors may be important.

Big-leaf mahogany trees seed for several decades before they reach maturity and are ready for felling. Profuse natural regeneration has been noted under stands in Trinidad (Marshall 1939), the French Antilles (Anonymous 1959), Indonesia (Alrasjid and Mangsud 1973), the Solomon Islands (Chaplin 1993), the Philippines, Fiji, Puerto Rico, St. Lucia, and Sri Lanka (our observations). Big-leaf mahogany regeneration has

Table 13.3. Age of First Mahogany Seeding from Planted Trees

Age	Description of Seed Stand	Country	Source
8	Seed orchard (clonal) at 9- × 9-m spacing	Fiji	T. Kubuabola (personal communication)[b]
8–9	Isolated individuals	Perú	Tito and Rodríguez (1988)
9–10	Spacing initially 2 × 2 m	Philippines	Chinte (1952)
10+	Spacing 10 × 3 m in enrichment lines	Solomon Islands	Chaplin (1993)
12	"Favorable conditions"	Central America	Lamb (1966)
12	Spacing initially 2 × 4 m	México	C. Parraguirre (personal communication)[c]
12–13	—[a]	Indonesia	Noltee (1926)
12–13	—[a]	India	Troup (1932)
<15	Spacing initially 3 × 3 m	Trinidad	Marshall (1939)
15	Planted under *Tectona, Pericopsis, Artocarpus*	Sri Lanka	Perera (1955)
18–20	—[a]	Fiji	Streets (1962)
20	Recently thinned	Sri Lanka	Streets (1962)
23	Basal area 25 m^2 ha^{-1} at time of seeding	Puerto Rico	Tropical Forest Research Centre (1959)
30	Small stands and avenues	Fiji	Busby (1967)

[a] No details available.
[b] Silvicultural Research Division, Fiji.
[c] INIFAP, México.

often established under a closed canopy where light availability is low. The life expectancy of big-leaf mahogany seedlings in such conditions has not been properly investigated, but plantation-grown big-leaf mahogany trees clearly set seed often enough for ingrowth of young seedlings to match or exceed the mortality of older seedlings. With one of the systems listed in Table 13.2, this natural regeneration could be used to establish the next rotation.

The existence of mature big-leaf mahogany trees does not guarantee the presence of its natural regeneration in the understory, however; in both St. Lucia and Sri Lanka, some big-leaf mahogany stands are notably free of regeneration, even though a nearby stand may be regeneration rich. Unless natural regeneration can be induced, the next rotation will have to be established by planting, limiting the forest manager to a clear-cutting system (possibly using partial cuts staggered over several years if continuous cover is required).

No surveys of natural regeneration in big-leaf mahogany plantations have been documented, and suggested explanations of its absence have not been proved. Muttiah (1965) noted that in certain Sri Lankan plantations, roadsides and ridges were the only sites colonized, although older, pole-stage regeneration was present throughout. He suggested that the deep litter layer was preventing regeneration in undisturbed parts of the plantations and that the pole-stage trees were established after disturbance during an earlier selective felling. Parraguirre (Instituto Nacional de Investigaciones Forestales, Agrícolas y Pecuarias (INIFAP) personal communication) believes that deep litter is preventing natural regeneration under experimental plantations in México. Site factors may also be important. M. Bobb (Forestry Department, Castries, St. Lucia, West Indies, personal communication) suggested that the lack of regeneration on one of two similar sites in St. Lucia might be due to poor drainage and waterlogging after rains. Streets (1962) observed that big-leaf mahogany regenerates most profusely in Trinidad on calcareous clays and good soils, where moisture is adequate all year round. Wolffsohn (1961) convincingly demonstrated that insect predation of seeds was preventing regeneration of big-leaf mahogany in natural forests of Belize, which may be relevant to plantations. Low light intensity (Muttiah 1965) and root competition from mature big-leaf mahogany trees (Coster 1935) may also be factors. Low rates of seed production and poor seed viability are other possibilities.

Natural regeneration under closed plantation canopies is often remarkably free of shoot borer damage, although attack has been found at plantation edges. This observation is made based on our experience in big-leaf mahogany plantations of the Philippines, Sri Lanka, the Solomon Islands, St. Lucia, and Puerto Rico. Whitman (no date) also noted an absence of attack on natural regeneration in St. Lucia, suggesting that it may be due to the shade of the overstory (although the mechanism has yet to be identified). The lack of attack is surprising, given the amount of easily accessible shoot material on which the shoot borer could lay its eggs, and it is another reason forest managers may wish to consider using a silvicultural system other than clear-cutting and replanting.

Experience: Single-Tree Selection in Big-Leaf Mahogany Plantations of Sri Lanka

Big-leaf mahogany was first planted in Sri Lanka in 1889 (Tisseverasinghe and Satchithananthan 1957). The species was mixed with *Artocarpus inte-grifolia*, *Tectona grandis*, and local timber species, including *Chloroxylon swietenia*, *Mesua ferrea*, and *Pericopsis mooniana*, in what later became known as the Sundapola plantation. The successful growth of the mahogany led to the gradual establishment of many other plantations, which, between 1925 and 1960, were principally mixtures of big-leaf mahogany and *A. integrifolia* (Sandom and Thayaparan 1995). Subsequently, big-leaf mahogany was established in lines through degraded natural forest. The current area of big-leaf mahogany plantations is about 4500 ha, although several plantations have yet to be mapped. In all the pre-1960 plantations, big-leaf mahogany developed at the expense of other species (Muttiah 1965).

A single-tree selection system was adopted for managing the Sundapola plantation in 1954 because of its "historical and silvicultural importance," which would have been lost under a clear-cut system (Tissverasinghe and Satchithananthan 1957). The selection system was considered most suitable because of the presence, under the maturing stands, of natural regeneration that had already begun to develop into pole-stage trees. Diameter-class frequency histograms had begun to shift from a typical even-aged (or normal) distribution to what resembled a regenerating or inverse J-shaped distribution (Tisseverasinghe and Satchithananthan 1957; Muttiah 1965). Muttiah believes that the selection system is advantageous for several reasons:

- Site factors are exploited to the maximum by trees occupying the different strata in air and soil.
- The balance of young and old trees guarantees a steady income from saw timber.
- The system is highly flexible (individual trees can be retained as long so they are putting on valuable increment, and the timing of cut is varied accordingly).
- All seed years contribute to the seedling bank, which reduces establishment costs to a negligible amount.
- The growing stock and output may be higher than from more uniform systems, with an annual percentage increment in volume ranging from 5.6% to 9.2% (note that these figures were taken from measurements in stands in the process of conversion from an even-aged, clear-cut system to an uneven-aged selection system).

- The system produces a natural stand structure, making it less subject to pests and diseases.
- The plantation is esthetically attractive, and management activities use the professional skills of foresters.

Doubts over the suitability of a selection system were expressed at the time, however, and problems such as "expense," "lack of properly trained staff," and "the ease with which selection can degenerate into uncontrolled exploitation" were noted (Tisseverasinghe and Satchithananthan 1957).

The Sundapola plantation was carefully managed under single-tree selection for a decade. The system may have been applied to other big-leaf mahogany plantations, although because most plantation records were lost, it cannot be verified. At some stage after the mid-1960s, however, the fears of Tisseverasinghe and Satchithananthan were realized. The selection system degenerated into systematic (although low-intensity) high-grading, which subsequently depleted many of the big-leaf mahogany plantations of plus trees. Much of this work was apparently carried out in the mistaken belief that it constituted correct selection system silviculture. The selection system failed because it had assumed a detailed understanding of complex forestry principles and an intimate knowledge of forest structure by all (even the most junior) members of staff (Sandom and Thayaparan 1995). The training and continuity of service required were (and still are) unavailable to Forest Department staff.

One of the fundamental flaws of adopting the single-tree selection system is its unsuitability for light-demanding species such as big-leaf mahogany. In Sri Lanka, selection system management was not used long enough for this drawback to become evident. In uneven-aged, single-tree selection stands, however, natural regeneration must grow under relatively small openings in the canopy for long periods. Big-leaf mahogany does not grow well under such conditions, and the system is much better suited to shade-tolerant species. Although, in theory, a stand can be thinned to promote growth of a light-demanding species, in practice the amount of work (particularly for weed control) makes such treatments unrealistic. This realization led to adoption of an interim management plan in 1994 advocating a different silvicultural system (Sandom and Thayaparan 1994).

Use of a Modified Uniform Shelterwood System in Sri Lanka

Because of the shortcomings of the selection system, big-leaf mahogany plantations in Sri Lanka are now to be managed under a uniform shelterwood system (Sandom and Thayaparan 1995). The shelterwood system has not been applied to mahogany plantations previously, although the system has been considered in the French Antilles (Anonymous 1959; Tillier 1995) and Fiji (K. Singh, Silvicultural Research Division, Fiji, personal communication). The intention is to convert complex (mixed and multistoried) big-leaf mahogany stands into simpler, more productive and more easily managed stands (Fig. 13.1a). The system is thought to be suitable for a light-demanding species such as big-leaf mahogany because trees are exposed to conditions of full light through most of the rotation.

The first stage of the shelterwood system is regeneration felling (Fig. 13.1b), carried out in mature stands or those with some mature trees but otherwise relatively unproductive. Besides encouraging the growth of existing advance regeneration, the felling will aid in establishing new seedlings. Ideally, regeneration felling should coincide with good seed years, after the seed has ripened but before the onset of rains when the seed starts to germinate (Fig. 13.1c). Opening the canopy is likely to stimulate strong weed growth (Matthews 1994) and, if natural regeneration is not established quickly, foresters will be faced with the daunting task of cutting back secondary regrowth until the next good seed year.

In the absence of regeneration trials, determining how many seed trees should be retained is difficult. The shelterwood must be sufficiently dense to ensure that all parts of the forest floor are exposed to falling seeds. In the Sri Lankan management plan, a density of 35 stems ha^{-1} (representing a square spacing of about 17 m) is proposed. In contrast, forest managers in Fiji believe that 80 to 100 trees will be required to control weedy regrowth and to produce sufficient natural regeneration (K. Singh, Silvicultural Research Division, Fiji, personal communication), although mature tree crowns in Fiji are smaller than those in Sri Lanka because of the shorter (35-year) rotations. The main disadvantage of retaining a dense shelterwood is the damage to established natural regeneration at the time of the final or seed-tree felling (Fig. 13.1d). Alrasjid and Mangsud (1973) measured the amount of natural regeneration under a 28-year-old big-leaf mahogany stand in Indonesia before and after harvesting; 65% of the seedlings and 34% of the saplings were destroyed during felling operations. Although this survey was under a closed rather than shelterwood canopy, it illustrates the damaging effect of forest operations on natural regeneration. A low-density shelterwood complemented with suitable harvesting techniques (delimbing, directional felling) and extraction techniques (planned routes, use of animals instead of machinery), as outlined in the Sri Lankan management plan (Sandom and Thayaparan 1994), will reduce damage, however.

The stands will be thinned several times during the rotation. Initially, stands will be systematically thinned to space the natural regeneration (Fig. 13.1e); subsequently, stands will be selectively thinned to take out poorly formed stems and to expand the growing space of residual trees (Fig. 13.1f,g). The proposed thinning regime is based on predicted diameter growth of just under 1 cm yr^{-1}, but no growth and yield data are available from which to determine optimal thinning frequency and intensity. An important aspect of the shelterwood system is early selection of seed trees. Crown development takes time and cannot be achieved by a single,

Present stands are uneven-aged; mahogany stocking is variable, and other species are present in various proportions

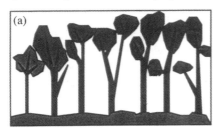

Year 0: **Regeneration felling**; selected seed trees are left, including other species such as *Artocarpus integrifolia*; 35 trees/ha

Years 0-5: **Maintenance**; undergrowth cleared and weeded to encourage natural density; regeneration

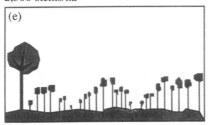

Year 5: **Seed-tree felling**; natural regeneration will have reached a high of 2,500 stems/ha

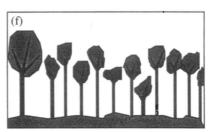

Year 8: **Singling and spacing** of natural stocking regeneration (transplantation); some nonmahogany species retained; 2,500 stems/ha

Years 20, 30, and 40: **Thinning**; stocking gradually reduced to 900, 600, and 270 stems/ha, respectively

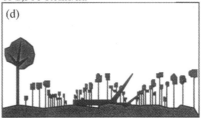

Year 50: **Final thinning**; final crop trees well spaced to maximize dbh increment; 100 stems/ha

Year 60: **Regeneration felling**; cycle repeated from years 0-5

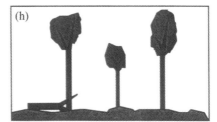

Figure 13.1a–h. Modified uniform shelterwood system in Sri Lanka.

Table 13.4. Constraints on Managing Mahogany in Sri Lanka[a]

Key Constraints	Advantages of the Modified Uniform Shelterwood System
Poorly trained workforce	Allows clear, straightforward guidelines for forest management to develop
Requirements of local communities	Makes provision for retaining *Artocarpus integrifolia* for jak fruit; maintains continuous tree cover, allaying fears of permanent forest clearance during fellings
Risk of soil erosion	Maintains continuous tree cover, which reduces runoff and erosion after harvesting and extraction
Conservation of local tree species	Provides for retaining certain local timber species (for example, *Chloroxylon swietenia* and *Vitex pinnata*) at low densities
Risk of shoot borer attack	Maintains a shelterwood for the first 5 years of the rotation, which may reduce attack on young trees

[a] Constraints are listed in decreasing order of importance.

last-minute thinning (Matthews 1994). Therefore, thinning sufficiently during the second half of the rotation is essential. In the design of an untried shelterwood system for the French Antilles, the recommendation (Anonymous 1959) was to make three to four selective cuts toward the end of the rotation, removing a third to a quarter of the timber volume each time. At least half the crop yield should be produced from thinnings (Matthews 1994). The thinning regime proposed (Fig. 13.1f,g) is in broad agreement with these recommendations. A good market exists for thinnings down to 15 cm for the local furniture industry. For some uses, branchwood is acceptable. Logs and poles less than 15 cm in diameter are also in demand for housebuilding, fencing, and farm implements.

Big-leaf mahogany plantations are managed under several constraints (Table 13.4), but the shelterwood system is still considered the most suitable, although compromises have had to be made. The need to accommodate certain nonmahogany species adversely affects the productivity of the plantations and complicates silvicultural operations.

Use of a Group Selection System in St. Lucia

St. Lucia (in the Lesser Antilles) has a small big-leaf mahogany resource of just over 100 ha (Andrew 1994). Big-leaf mahogany was introduced as an ornamental species in the 1930s, but plantations did not begin to be established until the 1960s. The species was in pure plantations, in a taungya system with bananas, or in a mixture with blue mahoe (*Hibiscus elatus*),

with big-leaf mahogany typically at an initial 3- × 3-m spacing. Thinning has
been erratic, with some stands heavily crown-thinned and others virtually
untreated (Whitman, no date). Recurrent hurricane damage has had a
noticeable effect, creating gaps in the canopy that have stimulated the
growth of preestablished natural regeneration.

Although no management plan has been designed, the plantations are
proposed to be managed under a variation of the selection system that corresponds roughly to the group selection system. A hypothetical sequence
of silvicultural treatments is outlined in Figure 13.2. Trees would be thinned
and felled on a 7-year cycle, with a rotation length of about 35 years, based
on existing growth data (Andrew 1994). Where blue mahoe is present, it
will be maintained at 15% of total stocking to encourage tall, straight, big-
leaf mahogany stems to develop. Many details, including tending and thinning natural regeneration and the desirable size of groups, have yet to be
worked out. To be suitable for a light-demanding species such as big-leaf
mahogany, felling groups may need to be large.

One of the difficulties with any selection system is harvesting and extracting trees without damaging the remaining trees. With group felling, selected
trees can be felled toward the middle of the group to avoid catching and
snagging surrounding trees. In St. Lucia, mechanized extraction is forbid-
den, and contractors use portable sawmills to saw logs at the harvest site.
Planks are carried out by foot, and the amount of damage caused to remaining trees is minimized.

Several constraints on managing big-leaf mahogany plantations in St.
Lucia have been identified (Table 13.5), and the need to protect watersheds
is considered paramount by the Forestry Department. Much of the big-leaf
mahogany has been planted on steep slopes where the threat of soil erosion
during forest operations is severe; adopting the selection system will
minimize the risks. The productivity of the system will be low, however, if
felling groups are small and natural regeneration is shaded by surrounding
trees. Forestry Department staff is well educated and enthusiastic enough
to put such a complex system into practice, but the demands on staff time
will be significant if the system is to work, despite the small size of the
resource.

Conclusions

Dense natural regeneration is found under many big-leaf mahogany plantations, offering possibilities for adopting silvicultural systems other than
traditional clear-cutting and replanting. The modified uniform shelterwood
and group selection systems appear to be suitable for managing big-leaf
mahogany plantations, although many years must pass before a quantitative assessment is possible. The systems discussed in this chapter are confined to those that have been or are to be implemented. Others (see Table

Present stands are becoming uneven-aged and multistoried; nonmahogany species (such as *Hibiscus elatus*) may be present

Year 0: First group felling; tending of natural regeneration in first group

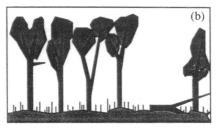

Year 7: Second group felling; tending of natural regeneration in second group; spacing of natural regeneration in first group

Year 14: Third group felling; regeneration in third group; spacing of regeneration in second group; thinning

Year 21: Group felling, tending, spacing and thinning in groups repeated

Year 28: Group felling; tending, thinning in groups repeated; natural regeneration from first group felling now nearing

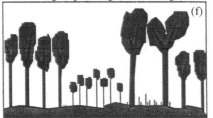

Year 35: Natural regeneration from first group felling mature and cut; stand now contains 5 different age groups; felling continued on a 7-year cycle

Figure 13.2a–g. Group selection system in St. Lucia.

Table 13.5. Constraints on Managing Mahogany in St. Lucia[a]

Key Constraints	Advantages of the Group Selection System
High risk of soil erosion	Maintains continuous tree cover, reducing runoff and erosion after harvesting and extraction; reduces effects of operations at any one time to a few scattered groups
Risk of hurricane damage	Reduces the proportion of susceptible (tall) trees in the stand; allows gaps in the canopy created by hurricane damage to be incorporated into the system like any other felling group
Risk of shoot borer attack	Maintains high side shade around groups, which may reduce attacks on young trees, encourage strong vertical growth (hence recovery), or both
Conservation of local bird species	Creates an uneven-aged forest that may contain more habitat for rare local birds, such as the St. Lucian parrot (especially in plantations around surviving pockets of natural forest)

[a] Constraints listed in decreasing order of importance.

13.2) may also be appropriate, if they are managed in such a way that big-leaf mahogany trees can receive sufficient light. For some systems, such as the single-tree selection system, the high intensity of management required to achieve suitable light conditions may not be practical.

The successful adoption of a silvicultural system does not depend simply on finding a system that meets the light requirements of big-leaf mahogany. Forest managers responsible for the recent design of silvicultural systems in both Sri Lanka and St. Lucia have also taken careful note of the key local constraints on plantation management practices. Failure to take local constraints into account is likely to result in the failure of the adopted system, as was demonstrated by the selection system in Sri Lanka. Based on the experiences described in this chapter, some general guidelines have been produced for those interested in adopting a silvicultural system for managing their mahogany plantations (Table 13.6).

The least complex (clear-cutting) systems produce even-aged, structurally uniform plantations unable to accommodate many ecological, environmental, and social constraints. The institutional constraints described, however, can be accommodated because the clear-cutting system is easy to manage and puts few demands on the workforce. For a light-demanding species such as big-leaf mahogany, uniform plantations are probably more productive—and therefore more profitable—than uneven-aged plantations, but nursery costs will be high (unless natural regeneration can be used), and incidences of shoot borer attack associated with single-canopied plantations in some countries may reduce profitability considerably.

Table 13.6. The Suitability of Silvicultural Systems for Managing Plantations Under Different Constraints[a]

Types of Constraint	Examples of Constraint	Clear-cut System	Shelterwood System	Selection System
Ecological	Shoot borer attack is a high risk; conservation of native species found in plantations is a high priority	Not suitable	Suitable with modification	Suitable
Environmental	Soil erosion and hurricane damage are high risks	Not suitable	Suitable	Suitable
Social	Producing benefits for local communities is a high priority	Not suitable	Suitable with modification	Suitable
Institutional	Silvicultural malpractice (because of inadequate, poorly managed, untrained workforce) is a high risk	Suitable	Suitable	Not suitable
Economic	Maximizing stand profitability is a high priority	Suitable?	Suitable	Not suitable

[a] These suggestions are based on work in Sri Lanka and St. Lucia.

The more complex (selection and shelterwood) systems produce structurally diverse plantations that can accommodate many of the ecological, environmental, and social constraints on management. Selection systems cannot easily accommodate institutional or economic constraints, however, because of the need for frequent and highly involved silvicultural treatments, which put heavy demands on the workforce and greatly increase management costs. If plantation managers respond to all the constraints described, the shelterwood system appears to offer the best compromise solution for managing mahogany over the long term.

Acknowledgments. We acknowledge the helpful assistance of employees of the Forest Department of Sri Lanka; the Forestry Department of St. Lucia (Brian James, Lindon John, Michael Bobb, Peter Vidal); the Silvicultural Research Division of Fiji (Kuldeep Singh, Tevita Evo, Tom Kubuabola); the Ecological Research and Development Bureau of the Philippines (Celso Díaz, Eraneo Lapis, Norma Pablo); the USDA Forest Service of Puerto Rico (Frank Wadsworth, Peter Weaver, John Francis); and INIFAP of México (Conrado Parraguirre). The ideas, comments, and suggestions of all these foresters and ecologists have been invaluable in developing this paper. This publication resulted from a research project funded by the Department for International Development of the United Kingdom. The Department, however, can accept no responsibility for any information provided or views expressed. Project code R6351, Forestry Research Programme.

Literature Cited

Alrasjid, H., and Mangsud. 1973. Natural regeneration trials with mahogany (*Swietenia* spp.) in the Ngraho and Tobo Forest Circles, E. Java. Lembaga Penelitian Hutan, Laporan 165, Bogor, Indonesia.

Andrew, M. 1994. *Growth and Yield of Mahogany* (Swietenia macrophylla *King*) *Plantations in St. Lucia*. B.S. thesis, University of New Brunswick, New Brunswick, Canada.

Anonymous. 1959. *Swietenia macrophylla* King: Caractères sylvicoles et méthodes de plantation. *Bois et Forêts des Tropiques* **65**(3):37–42.

Bauer, G.P., and Gillespie, A.J.R. 1990. *Volume Tables for Young Plantation-Grown Hybrid mahogany* (Swietenia macrophylla × S. mahagoni) *in the Luquillo Experimental Forest of Puerto Rico*. Research paper SO-257. U.S. Department of Agriculture, Forest Service, Southern Forest Experiment Station, New Orleans, LA.

Busby, R.J.N. 1967. Reforestation in Fiji with large-leaf mahogany. Paper presented at the Ninth Commonwealth Forestry Conference, India. Department of Forestry, Suva, Fiji.

Chable, A.C. 1967. Reforestation in the Republic of Honduras, Central America. *Ceiba* **13**(2):1–56.

Chaplin, G. 1993. *Silvicultural manual for the Solomon Islands. ODA Forestry Series 1.* Overseas Development Administration, London, UK.

Chinte, F.O. 1952. Trial planting of large-leaf mahogany (*Swietenia macrophylla* King). *Caribbean Forester* **13**(2):75–84.

Coster, C. 1935. Licht, ondergroei en wortelconcurrentie. Abstract 386 in *Indonesian Forestry Abstracts: Dutch Literature Until About 1960*. Centre for Agricultural Publishing and Documentation, Wageningen, The Netherlands.

Dawkins, H.C. 1965. Problems of natural regeneration, plantations and research. Unpublished report for the Forestry Division of the Ministry of Natural Resources, Sierra Leone.

Ennion, R.C. 1996. Evaluation of four taungya mahogany increment plots age 36–38 years, Columbia River Forest Reserve. Unpublished report for the Belize Forest Department, Ministry of Natural Resources.

Fattah, H.A. 1992. Mahogany forestry in Indonesia. In *Mahogany Workshop: Review and Implications of CITES*. Tropical Forest Foundation, Washington, DC.

Lamb, F.B. 1966. *Mahogany of Tropical America: Its Ecology and Management*. University of Michigan Press, Ann Arbor, MI.

Lamprecht, H. 1989. *Silviculture in the Tropics*. Deutsche Gesellschaft für Technische Zusammenarbeit (GTZ), Eschborn, Germany.

Marie, E. 1949. Notes on reforestation with *Swietenia macrophylla* King in Martinique. *Caribbean Forester* **10**(3):206–222.

Marshall, R.C. 1939. *Silviculture of the Trees of Trinidad and Tobago*. Oxford University Press, Oxford.

Matthews, J.D. 1994. *Silvicultural Systems*. Clarendon Press, Oxford.

Muttiah, S. 1965. A comparison of three repeated inventories of Sundapola mixed selection working circle and future management. *Ceylon Forester* **7**(1):3–35.

Noltee, A.C. 1926. *Swietenia mahagoni* and *Swietenia macrophylla*. Abstract 735 in *Indonesian Forestry Abstracts: Dutch Literature Until About 1960*. Centre for Agricultural Publishing and Documentation, Wageningen, The Netherlands.

Oliver, W.W. 1992. *Plantation Forestry in the South Pacific: A Compilation and Assessment of Practices*. Field Document 8, June 1992. U.S. Department of Agriculture, Forest Service, Pacific Southwest Research Station, Redding, CA.

Perera, S.P. 1955. *Swietenia macrophylla* and its propagation by striplings in Ceylon. *Ceylon Forester* **2**(2):75–79.

Sandom, J.H., and Thayaparan, S. 1994. An interim management plan for the mahogany forests of Sri Lanka. Unpublished report for the Forest Department, Colombo, Sri Lanka.

Sandom, J.H., and Thayaparan, S. 1995. A revision of the interim management plan for the mixed mahogany forests of Sri Lanka. Unpublished report for the Forest Department, Colombo, Sri Lanka.

Smith, D.M. 1986. *The Practice of Silviculture*, eighth edition. Wiley, New York.

Soubieux, J.M. 1983. Croissance et production du mahogany (*Swietenia macrophylla* King) en peuplements artificiels en Guadeloupe. Ecole Nationale des Ingenieurs des Travaux des Eaux et Forêts, Direction Régionale pour la Guadeloupe, Office National des Forêts (unpublished).

Streets, R.J. 1962. *Exotic Trees in the British Commonwealth*. Clarendon Press, Oxford.

Tillier, S. 1995. Le mahogany grandes feuilles en Martinique. *Bois et Forêts de Tropiques* **244**(2):55–66.

Tisseverasinghe, A.E.K., and Satchithananthan, S. 1957. The management of Sundapola plantation. *Ceylon Forester* **3**(1):82–93.

Tito, B.V., and Rodríguez, M. 1988. Estudio silvicultural preliminar de la caoba en la zona de Tingo María. Documentos de Trabajo 11. Avances de la Silvicultura en la Amazonia Peruana. República del Perú Instituto Nacional de Desarrollo, Perú.

Tropical Forest Research Centre. 1959. Nineteenth annual report. *Caribbean Forester* **20**(1):1–10.

Troup, R.S. 1928. *Silvicultural Systems*. Oxford Science Publications, Clarendon Press, Oxford.

Troup, R.S. 1932. *Exotic Forest Trees in the British Empire*. Clarendon Press, Oxford.

Veríssimo, A., Barreto, P., Tarifa, R., and Uhl, C. 1995. Extraction of a high-value natural resource in Amazonia: the case of mahogany. *Forest Ecology and Management* **72**:39–60.

Weaver, P.L. 1987. Enrichment plantings in tropical America. In *Management of the Forests of Tropical America: Prospects and Technologies*, eds. J.C. Figueroa, F.H. Wadsworth, and S. Branham, pp. 259–278. U.S. Department of Agriculture, Forest Service, Institute of Tropical Forestry, Río Piedras, PR.

Weaver, P.L., and Bauer, G.P. 1986. Growth, survival and shoot borer damage in mahogany plantings in the Luquillo Forest in Puerto Rico. *Turrialba* **36**(4):509–522.

Whitman, D. (no date). Provisional yield tables for Honduras mahogany, *Swietenia macrophylla*, in St. Lucia. Unpublished report for the Forestry Department, St. Lucia.

Wolffsohn, A.L. 1961. An experiment concerning mahogany germination. *Empire Forestry Review* **40**(1):71–72.

14. Enrichment Planting of Big-Leaf Mahogany and Spanish Cedar in Quintana Roo, México

Patricia Negreros-Castillo and Carl W. Mize

Abstract. Enrichment planting, a tool for regenerating tropical forests, has been used but not evaluated for many years in community forests in Quintana Roo, México. This study was designed to determine if enrichment planting would assure regeneration of big-leaf mahogany and Spanish cedar in these forests. Survival and growth of seedlings of both species planted in two communities were evaluated. Seedlings had been planted from 1986 to 1993 along main and log-hauling roads, in log landings, and in narrow strips opened under forest canopies. Survival was 18%; it varied considerably from year to year but was similar for both species and among the four types of areas planted. Available light appeared to be the principal limiting factor for survival in the enrichment plantings.

Keywords: Big-leaf mahogany, Spanish cedar, Enrichment planting, *Swietenia macrophylla, Cedrela odorata*

Introduction

Satisfactory natural regeneration of commercial species in tropical forests is often difficult to obtain, and it requires silvicultural expertise and intensive effort (Evans 1992). Failure or limited success in establishing regeneration results from administrative mistakes (such as leaving inadequate seed trees, felling before seed release, inadequate weeding, lack of cleaning and release cuttings), insufficient knowledge of tropical ecology, or both (Weaver and Bauer 1986).

Three common approaches are used to obtain regeneration in tropical forests: encouraging existing regeneration, establishing forest plantations (replacement planting), and enrichment plantings (Wadsworth 1966; Lamb 1969; Evans 1992). Enrichment planting uses a suite of artificial forest regeneration systems that supplement a stand's natural regeneration by planting commercial species (Dawkins 1961; Negreros-Castillo 1994). The ultimate objective of enrichment planting is to improve the composition and growth of degraded forests and render the forests more commercially valuable (Weaver and Bauer 1986).

Enrichment planting is generally chosen over other regeneration methods when harvesting would destroy existing regeneration or seed trees, when advanced regeneration is inadequate (Negreros-Castillo 1991; Evans 1992; Snook 1993), or when no regeneration method has been developed (Negreros-Castillo 1994). At least 20 different techniques are used for enrichment planting, and at least 163 species have been established through enrichment plantings in at least 12 countries (Pierront 1995). Most plantings have been experimental, however (Lamb 1969; Palmer 1981; Weaver and Bauer 1986; Bauer 1987; Ramos and Del Amo 1992; Adjers et al. 1995).

Enrichment planting on a management scale has been practiced in México. During the early 1940s, big-leaf mahogany and Spanish cedar were planted on an estimated 20,000 ha of abandoned logging roads and adjacent areas (Cuevas 1947). Efforts to regenerate the forests of Quintana Roo, México, by enrichment planting began around 1943. Between 1945 and 1949, about 90,000 big-leaf mahogany and Spanish cedar seedlings were planted (Cuevas 1947). Between 1960 and 1980, a local timber company also carried out enrichment plantings by planting seedlings in lines under the forest canopy (Galletti 1993). Although 40 million seedlings were planted between 1945 and 1994 (Cuevas 1947; Alejandro Osorio 1993, personal communication; Rafael Vaena 1995, personal communication; Javier Chavelas 1995; personal communication), none of the plantings were monitored over the long term.

In 1994, we began to evaluate the enrichment plantings installed by a community forestry organization in Quintana Roo. The major question we addressed was whether enrichment planting could assure regeneration of big-leaf mahogany and Spanish cedar.

The Study Area

Regional Characteristics

The state of Quintana Roo is along the eastern portion of the Yucatán Península (Fig. 14.1). The peninsula, projecting northward into the Gulf of México, is a large limestone platform joined at its base to the Sierras of northern Central America. Quintana Roo has a total area of 50,500 km² of which 3400 km² is forest.

The predominant soil types are Rendolls (USA soil taxonomy) or Rendzinas (USA old classification) (Brady 1984). The annual precipitation is between 1000 and 1200 mm, distributed unevenly, with a dry period of 3 to 4 months (February to May). The mean annual temperature is 26°C, with a minimum of 8°C (INEGI 1994). Rzedowski and Rzedowski (1989) classify the vegetation of this region as tropical semievergreen forest. Forest composition is a result of the interaction of the precipitation regime with soil types and disturbances such as hurricanes, fires, and agricultural activities (Barrerra et al. 1977). The most abundant tall tree species are ramon, *Brosimum alicastrum* Sw., and *Manilkara zapota* (Pennington and Sarukhán 1968).

The forests in the state of Quintana Roo represent one of the last large, continuous, seasonal tropical forests in México (Toledo and Ordoñez 1993); big-leaf mahogany and Spanish cedar reach their optimum natural growth and development in these forests (Lamb 1966). Although more than 100 tree species can be found growing on a single hectare (Pennington and Sarukhán 1968; Negreros-Castillo and Mize 1993), big-leaf mahogany and Spanish cedar are by far the most commercially important.

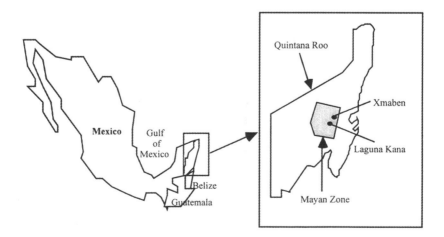

Figure 14.1. Quintana Roo in the Yucatán Peninsula of México.

Community Forest Management

In Quintana Roo, 60% of the land is owned by communities, called ejidos, which own and manage 80% of the forest area in the state (INEGI 1994). In 1984, after centuries of commercial extraction by outside loggers (Carreón et al. 1987a; Galletti 1993), ejidos were granted the legal right to profit from their own forests. Unfortunately, the forests at that time were seriously understocked with commercial species, and communities lacked the technical knowledge and economic resources to manage them (Negreros-Castillo 1991; Galletti 1993; Snook 1993). With the support of the Mexican government and international aid programs, however, some of the ejidos—those with the highest timber potential—organized themselves into forestry community organizations and hired foresters to help them meet their new challenge. Five such community forestry organizations are now operating in Quintana Roo.

One such group is the "Organización de ejidos productores forestales de la Zona Maya" (Mayan zone ejidos), formed in 1985. It is composed of 19 ejidos with about 500,000 ha of land, about half of which is forested. Inventories conducted by the organization's foresters indicated little or no regeneration of young growing stock of big-leaf mahogany and Spanish cedar (Carreón et al. 1986, 1987b). To increase the forest's future value, the foresters initiated enrichment plantings in 1986. Although not all the ejidos participated in the beginning, enrichment plantings had been established on all the ejidos by 1990 (Carreón 1990). Although many plantings have been established, none of them had been evaluated by 1993.

Enrichment Planting by the Mayan Zone Ejidos

Nursery-grown seedlings, between 4 and 5 months old and generally not more than 30 cm tall, were used in all enrichment plantings (Alejandro Osorio 1993, personal communication). Enrichment seedlings were planted in four settings: along both sides of main roads, on log landings, along both sides of log-hauling roads, and in narrow strips that the ejiditarios intentionally opened under the canopy of recently harvested forests. We hereafter refer to these four settings as "planting conditions."

Seedlings were planted after timber harvest and at the beginning of the rainy season, usually in June and July. In some years, however, because of insufficient resources and other administrative problems, plantings were as late as September and October. The foresters trained the ejidatarios to do the planting and suggested where and how to plant seedlings. Big-leaf mahogany and Spanish cedar were the only species planted.

Methods

The study was conducted in two of the Mayan Zone ejidos, Laguna Kana and Xmaben, which were selected because one of them had planted seedlings for many years and both were in areas representative of the zone (see Fig. 14.1). In both, ejidatarios who had done some of the enrichment plantings assisted in locating the planted sites and taking measurements.

To carry out the evaluation, we separated planted areas by year of planting and by planting condition. Seedlings were not planted in all four planting conditions every year (Table 14.1). To estimate the success of the plantings, we developed detailed maps for both ejidos. On a base map of each ejido, we measured the length of all planted main and log-hauling forest roads and noted the locations of the plantings. Log landings were also mapped. The narrow strips that had been opened and planted were located, their lengths measured, and their locations added to the map.

We developed a sampling scheme for each planting condition. For plantings along main roads, sampling was systematic. Generally, seedlings had been planted on both sides of the roads; the two sides were sampled separately, so the total length of the planting was about twice the length of the road along which the seedlings were planted. A 20-m-long sample unit was established about every 300 m, and the number of seedlings still alive was recorded. Because the average distance between seedlings planted along main roads for both ejidos had been about 4 m, we expected to find five planted seedlings per 20-m sample unit. An estimate of the number of seedlings planted along the road was calculated by dividing the total length of the plantings by four. The number of surviving seedlings was estimated by multiplying the length of the plantings by the average number of

Table 14.1. Number of Areas Planted, by Planting Condition and Year of Planting, for the Ejidos of Laguna Kana and Xmaben

Planting Condition	Year Planted	Total Areas
Laguna Kana		
Main road	1986	2
Main road	1992	1
Main road	1993	1
Landing	1989	1
Landing	1992	1
Landing	1993	1
Under canopy	1989	1
Under canopy	1991	2
Under canopy	1992	2
Haul road	1993	1
Xmaben		
Main road	1990	1
Under canopy	1991	1
Under canopy	1992	2
Under canopy	1993	2

survivors per meter. Percentage survival was estimated by dividing the estimated number of survivors by the estimated number planted and multiplying by 100.

For seedlings planted in narrow strips opened under the forest canopy, 50% to 10% of the strips planted in a particular year were randomly selected for sampling. Sample units, 50 m long, were located in each selected strip, and the number of seedlings still alive in each sample unit was recorded. Initial spacing of seedlings was 3 m. The number of seedlings planted in each narrow strip was estimated by dividing the length of the strip by three. The number of surviving seedlings for all strips was estimated by multiplying the total length of all strips by the average number of survivors per meter for the strips sampled.

For log landings, the size of the landing was estimated, and three 2-m-wide transects that traversed the landing were established. The number of seedlings that appeared to have been planted in each transect was recorded. The original spacing of the seedlings in landings was about 3 by 3 m. The number of seedlings planted in a landing was estimated by dividing the area of the landing by nine because of an average spacing of one seedling per 9 m². The number of seedlings still alive in a landing was estimated by multiplying the estimate of the number of seedlings planted by the average number of seedlings per 9 m² found in the transects.

Log-hauling roads were planted infrequently; when they were, seedlings were planted on both sides of the road, as on the main roads. For those that appeared to have been planted, two 20-m-long sample units were established. The number of surviving seedlings was recorded for both sample units. Estimates of the number of seedlings planted, the number of surviving seedlings, and the percentage survival were calculated as they were for main roads.

Each surviving seedling was identified and recorded by species, and its height was measured and recorded. When a seedling was taller than 1.3 m, diameter at breast height (dbh) was also measured and recorded.

Results

The total survival of planted seedlings averaged 15% in Laguna Kana (Table 14.2) and 23% in Xmaben (Table 14.3). In Laguna Kana, percentage survival varied considerably, from 0% to 52% (Table 14.2); in Xmaben, survival ranged between 11% and 29% (Table 14.3). For all the sample units combined, average survival was 18%, representing an estimated 4300 surviving seedlings from an estimated 24,200 planted seedlings.

Only two sites had seedlings tall enough for measuring their diameters. The sites were the oldest, planted in 1986 in Laguna Kana, and only big-leaf mahogany had been planted. Those trees averaged 25 mm dbh, and the annual diameter increment was 4 mm yr⁻¹.

Table 14.2. Status in 1994 of Big-Leaf Mahogany and Spanish Cedar in Enrichment Plantings Established Between 1986 and 1993 in Laguna Kana

Planting Condition	Year Planted	Species	Total Sample Units Measured	Estimated Seedlings Planted	Survival Percentage	Average Height (m)
Main road	1986	M	38	2850	15	4.3
Main road	1986	M	29	2700	5	3.4
Main road	1992	M	24	2500	12	0.66
Main road	1993	M	21	8750	6	0.59
Landing	1989	M	2	218	0	—
Landing	1992	SC	3	21	52	0.31
Landing	1993	M	2	308	12	0.47
Under canopy	1989	SC	3	750	8	0.24
Under canopy	1991	M	53	2800	35	0.72
Under canopy	1991		4	200	31	0.53
Under canopy	1992	M	20	2040	3	0.50
Under canopy	1992	SC	18	1360	2	1.2
Haul road	1993	M	2	40	35	0.44

M, big-leaf mahogany; SC, Spanish cedar.

Survival, although variable from year to year, appeared to be similar among the four planting conditions. The highest survival (52%) was on a landing but so was the lowest (see Table 14.2). The landing with no survival was covered with lilies and probably too wet for good seedling survival. Another landing in Laguna Kana had 75 big-leaf mahogany seedlings that appeared to be natural regeneration because the seedlings were in groups and of very different heights. A mature big-leaf mahogany growing just 20m from the landing was probably the seed source.

Comparisons of survival among species and years were difficult because of the unbalanced planting conditions and years sampled. When big-leaf

Table 14.3. Status in 1994 of Big-Leaf Mahogany and Spanish Cedar in Enrichment Plantings Established Between 1990 and 1993 in Xmaben

Planting	Year Planted	Species	Total Sample Units Measured	Estimated Seedlings Planted (number)	Survival (percentage)	Average Height (m)
Main road	1990	SC	21	800	29	1.48
Under canopy	1991	SC	5	4500	24	0.99
Under canopy	1992	M	8	660	22	0.56
Under canopy	1992	SC	6	420	22	0.41
Under canopy	1993	M	13	160	13	0.80
Under canopy	1993	SC	8	630	11	0.57

M, big-leaf mahogany; SC, Spanish cedar.

mahogany and Spanish cedar were planted in the same year and in similar conditions, however, survival for the two species was similar. Yet, plantings in some years appeared to have survived better than in others.

Discussion

When data on both ejidos and both species were combined, about 18% of all seedlings survived; such low survival would be unacceptable in most planting operations. Thus, enrichment planting, as it has been practiced, is apparently not an acceptable way to establish big-leaf mahogany and Spanish cedar in the forests of the Mayan zone ejidos.

The highest survival (52%) was on a landing, and another landing had 75 seedlings, representing 1730 seedlings ha^{-1}, that seemed to be a result of natural regeneration. Survival on landings was evaluated in another forestry community organization in Quintana Roo. Survival averaged 70% on 11 landings planted between 1988 and 1993 (Sánchez 1994). Sánchez did not evaluate survival in other planting conditions, however. Snook (1993) and Lamb (1966) have also reported successful mahogany natural regeneration on landings. Negreros-Castillo and Mize (1993) found successful natural regeneration on stands where canopies were opened by at least 50%.

Thus, light was very likely a limiting factor for survival in the enrichment plantings. In addition, much of Quintana Roo has very rocky, shallow soils; very few spots have soils deeper than 15 cm. How the seedlings in the enrichment plantings were planted could also have decreased survival. Some of the ejidatarios who helped plant the measured seedlings said that sometimes the holes dug for them were not deep enough, and occasionally the plastic bags the seedlings were grown in were not removed. During the evaluation, the ejidatarios recognized some of the potential causes for the low survival.

Because of this evaluation, the foresters and forestry committee of the Mayan zone ejidos decided in 1995 not to do enrichment planting along main roads and haul roads under forest canopies. They continued enrichment planting on landings, and they have added another planting condition they believe will produce higher survival: areas of open secondary vegetation (sites with vegetation not older than 12 years, such as abandoned crop fields or degraded pasture land).

Plantings since 1995 will be evaluated soon, and survival is expected to be higher than was observed in our study. Landings in the forests of the Mayan zone ejidos, however, represent only 0.5% of the harvested area (Negreros-Castillo and Mize 1993), and secondary vegetation is not generally available because the land is used for agriculture. Therefore, even if survival increases dramatically, the area that will be regenerated with big-leaf mahogany and Spanish cedar will be insignificant.

Thus, the future harvest of big-leaf mahogany and Spanish cedar by the Mayan zone ejidos is not assured, and the question of how to improve the stocking of big-leaf mahogany and Spanish cedar in natural stands remains unanswered. A possible solution might arise from a recently initiated study of the survival and growth of big-leaf mahogany and Spanish cedar planted into openings created during the harvest for railroad ties. Because these harvests are across large areas of forest, they create widely scattered openings. Beyond this possibility, silvicultural treatments for regeneration need to be developed to assure the long-term "commercial" existence of these two extremely valuable tropical timber species.

Acknowledgments. The study was partially funded by the Directorate on Tropical Ecosystems of the United States Man and the Biosphere Program. This contribution is paper 17921 of the Iowa Agriculture and Home Economics Experiment Station, Iowa State University, Ames, IA, USA, Project 3209. We thank the ejidatarios who helped with the field-work: Jacinto Cob, Silverio Petch Chuc, Juan de Dios Varela Bad, and Francisco Cob. Also, Idelfonso Yam Pool and Angelica Navarro, two college students studying forest biology, were research assistants and their participation was of great value. Special thanks to the foresters of the Mayan zone ejidos, especially Ingenieras Rosa Ledesma Santos and Victoria Santos Jiménez.

Literature Cited

Adjers, G., Hadengganan, S., Kuusipalo, J., and Nuryanto, K. 1995. Enrichment planting of dipterocarps in logged-over secondary forests: effect of width, direction and maintenance method of planting line on selected *Shorea* species. *Forest Ecology and Management* **3**:259–270.

Barrerra, A., Gómez-Pompa, A., and Yanes, C. 1977. El manejo de las selvas por los Mayas: sus implicaciones silvícolas y agrícolas. *Biótica* **22**:47–61.

Bauer, G.P. 1987. Swietenia macrophylla *and* S. macrophylla x S. mahagoni *Development and Growth: The Nursery Phase and the Establishment Phase in Line Planting, in the Caribbean National Forest, Puerto Rico.* Masters thesis, State University of New York, Syracuse, NY.

Brady, N.C. 1984. *The Nature and Properties of Soils*, ninth edition. Collier Macmillan, New York.

Carreón, M.M. 1990. Seis años de actividad forestal en la Zona Maya de Quintana Roo, OEPF-Zona Maya. Reporte anual Felipe Carrillo Puerto, Quintana Roo, México.

Carreón, M.M., Galletty, H., and Santos, V. 1986. Plan de manejo de X-Hazil y F. Carrillo Puerto. OEPFZM, Felipe Carrillo Puerto, Quintana Roo, México.

Carreón, M.M., Santos, V., and Ledezma, R. 1987a. Plan de manejo de Chunhuas. OEPFZM, Felipe Carrillo Puerto, Quintana Roo, México.

Carreón, M.M., Santos, V.J., and Martin, P.E. 1987b. Logros en la actividad forestal en la Zona Maya de Quintana Roo. OEPF-Zona Maya. Felipe Carrillo Puerto, Quintana Roo, México.

Cuevas, L.A. 1947. *Explotación de Tres Especies Forestales y Propagación Artificial de Caoba en Quintana Roo.* Tesis, Escuela Nacional de Agricultura, Chapingo, México.

Dawkins, H.C. 1961. New methods of improving stand composition in tropical forests. *Caribbean Forester* **22**(1–2):12–20.

Evans, J. 1992. *Plantation Forestry in the Tropics: Tree Planting for Industrial, Social, Environmental, and Agroforestry Purposes*, second edition. Clarendon Press, Oxford.

Galletti, H.A. 1993. Actividades forestales y su desarrollo histórico. In *Estudio Integral de la Frontera México-Belice: Análisis Socioeconómico. CIQROO*, pp. 131–198. Chetumal, Quintana Roo, México.

INEGI. 1994. *Anuario Estadístico del Estado de Quintana Roo*. Instituto Nacional de Estadística Geografía e Informática, Chetumal, Quintana Roo, México.

Lamb, A.F.A. 1969. *Enrichment Planting in English Speaking Countries in the Tropics*. FAO committee on forest development in the tropics. Second session. Special paper FDT-69/4. FAO, Rome, Italy.

Lamb, F.B. 1966. *Mahogany of Tropical America: Its Ecology and Management*. University of Michigan, Ann Arbor, MI.

Negreros-Castillo, P. 1991. *Ecology and Management of Mahogany* (Swietenia macrophylla *King*) *Regeneration in Quintana Roo, México*. Doctoral dissertation, Iowa State University, Ames, IA.

Negreros-Castillo, P. 1994. Evaluation of enrichment planting and postharvesting natural regeneration in Quintana Roo, México. MAB/Tropical Ecosystems Directorate Report.

Negreros-Castillo, P., and Mize, C. 1993. Effects of partial overstory removal on the natural regeneration of a tropical forest in Quintana Roo, México. *Forest Ecology and Management* **58**:259–272.

Palmer, J.R. 1981. Enrichment line planting in the Neotropics. Special publication. CATIE, Turrialba, Costa Rica, Central America.

Pennington, T.D., and Sarukhán, J. 1968. *Árboles Tropicales de México*. INIF/FAO, México.

Pierront, M.K. 1995. Enrichment planting: a literature review and brief proposal. Thesis proposal. University of Florida, Department of Forestry. Gainesville, FL.

Ramos, J.M., and Del Amo, S. 1992. Enrichment planting in a tropical secondary forest in Veracruz, México. *Forest Ecology and Management* **54**:289–304.

Rzedowski, J., and Rzedowski, G.C. 1989. Transisthmic México (Campeche, Chiapas, Quintana Roo, Tabasco and Yucatán). In *Floristic Inventory of Tropical Countries*, eds. D.G. Campbell and H.D. Hammond, pp. 270–280. The New York Botanical Garden, New York.

Sánchez, R.L.A. 1994. *Evaluación de las Reforestaciones con Caoba* (Swietenia macrophylla *King*) *en Bacadillas de los Ejidos Noh-Bec y Petcacab, Quintana Roo*. Tesis profesional, Universidad Autónoma de Chapingo, Chapingo, México.

Snook, L.K. 1993. *Stand Dynamics of Mahogany* (Swietenia macrophylla *King*) *and Associated Species After Fire and Hurricane in the Tropical Forests of the Yucatán Península, México*. Doctoral dissertation, Yale School of Forestry and Environmental Studies, Yale University, New Haven, CT.

Toledo, V.M., and Ordoñez, M. de J. 1993. The biodiversity scenario of México: a review of terrestrial habitats. In *The Biological Diversity of México: Origins and Distribution*, eds. R.B. Ramammorthy, A. Lot, and J. Fa, pp. 757–777. Oxford University Press, New York.

Wadsworth, F.H. 1966. La orientación de las investigaciones de silvicultura para Lationoamérica. *Turrialba* **16**:390–95.

Weaver, P.L., and Bauer, G.P. 1986. Growth, survival and shoot borer damage in mahogany plantings in the Luquillo Forest in Puerto Rico. *Turrialba* **36**:509–522.

15. Structure and Dynamics of Mahogany Plantations in Puerto Rico

Ariel E. Lugo and Shenglei Fu

Abstract.   We review research on the species composition, structure, functioning, and response to disturbance of mahogany plantations in Puerto Rico. The review includes nine plantations at seven sites from dry to wet life zones, two soil parent materials (limestone and volcanic), three plantation designs (block, line, and taungya), and plantation ages from 18 to 64 years. Mahogany grew well regardless of plantation design or soil parent material but the species differed. Small-leaf mahogany grew better than big-leaf and the big-leaf × small-leaf hybrid mahogany in the dry life zone, on degraded sites, and on shallow limestone soils. Small-leaf mahogany failed to grow in moist to wet life zones. Big-leaf and hybrid mahogany had the fastest rates of growth and yield in moist life zones, regardless of soil parent material. All plantations were invaded by native tree species and increased in species richness and vertical stratification with age. All plantation species exhibited abundant regeneration under mature plantation canopies. More nutrients accumulated in vegetation and plantation litter than in vegetation and fine litter of paired secondary forests. Nutrient return to the forest floor was pulsed because of dry season peaks of leaf fall and mast fruit production by mahogany trees. Fruit production in both plantations and paired secondary forests influenced nutrient cycles by increasing nutrient return and accumulation on the forest floor, thereby decreasing within-stand nutrient use efficiency. Small-leaf mahogany had half the litterfall but more litter standing stock than did big-leaf or hybrid mahogany. Hurricane Hugo reduced tree biomass, opened the canopy, and created a pulse of mass and nutrient fall in paired plots of hybrid mahogany and a secondary forest stand. Six years after the hurricane, the plantation returned more nutrients to the forest floor than did the paired secondary forest but had a lower nutrient use efficiency and a similar nutrient turnover rate.

Keywords: Mahogany, *Swietenia*, *Swietenia macrophylla*, *Swietenia mahagoni*, Tropical tree plantations, Secondary forests, Hurricane disturbance

The ecological amplitude of S. macrophylla *over its wide geographic range places it in that group of organisms, often referred to as pioneers or successional species, which can move into disturbed areas or "tension zones" in greater abundance than in situations where the vegetation has achieved ... [climax conditions]. This combination of characteristics indicates the potential of this species as plant material for productive land management in the tropics.*

—F. Bruce Lamb, 1966, p. 91.

Introduction

Mahogany is one of the most valuable tropical timbers (Lamb 1966). Its wood has been traded internationally for centuries, and current trade to the United States averages about $100,000 \, m^3 yr^{-1}$ (Fig. 15.1). An annual trade between 70,000 and $140,000 \, m^3$ has been sustained for almost 200 years. The rate of importation in the 1990s was similar to that in the 1950s. The high economic value of mahogany wood and concerns about the fate of its natural commercial stocks provide impetus to establishing mahogany plantations.

The global area of mahogany plantations has been increasing rapidly in the past 15 years and is now estimated at some 150,000 ha (Pandey, in press).

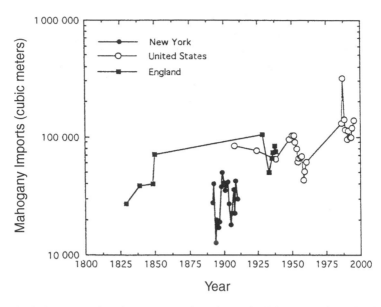

Figure 15.1. Imports of mahogany wood to the United States and England. Data for the United States were obtained from the Bureau of the Census, Foreign Trade Division, Trade Data Inquiries and Controls Section in Washington D.C. Data for England were obtained from Watson (1938, 1939, 1940).

The growth, yield, and survival of mahogany in plantations are poorly understood, however, and how plantations contribute to the conservation of natural stocks of the species is unclear. Moreover, some people believe that plantation forestry is not a land-use alternative to secondary forests because the ecological functions of plantation forests, such as nutrient cycling and biodiversity, are somehow inferior to those of secondary forests (cf. Lugo 1997).

A comprehensive conservation effort for mahogany will benefit from quantitative information about the ecological characteristics and life history of mahogany growing under plantation conditions. For example, how is species richness of native plants affected by the establishment of mahogany plantations? How productive are mahogany plantations under different environmental conditions? What effects do mahogany plantations have on the nutrient cycling and soil fertility of the sites where they grow? How do structural and functional parameters of mahogany plantations compare with those of paired or adjacent secondary forests? How do mahogany plantations respond to disturbances? The purpose of this chapter is to address these questions with information gathered in Puerto Rico.

Plantations of big-leaf, Pacific coast, and small-leaf mahogany—and the hybrid of big- and small-leaf mahogany—were established throughout Puerto Rico, beginning in the 1920s, under different soil and climate conditions (Francis 1995). Some of these mahogany plantations are now among the oldest in the world and provide an opportunity to study the long-term ecological functioning of these systems. Information about the growth rate of trees in these plantations is reported elsewhere (Wadsworth et al., Chapter 17, this volume; Ward and Lugo, Chapter 3, this volume). Here, we review the results of ecological studies in plantations of small-leaf, big-leaf, and the hybrid between the two mahoganies. Comparative information for plantations of Pacific coast mahogany was not available. Medina and Cuevas (Chapter 7, this volume) and Ward and Lugo (Chapter 3, this volume) provide information on plantations of this species.

Approach

Our review is based on ecological measurements at nine plantations in seven sites representing three climatic life zones (dry, moist, and wet), two soil parent materials (limestone and volcanic), three types of plantings (blocks, lines, and taungya), and plantations between 18 and 64 years old (Table 15.1). The site identification in Table 15.1, including the life zone and the species planted, is used throughout the chapter to describe each plantation. All plantations were established in sites previously deforested for agricultural uses. Trees were usually planted from nursery seedling stock at a spacing of 3 by 3 m when planted in blocks, and in 2-m-wide lines spaced at 10-m intervals when planted in lines. At the arboretum site (wb-3), big-leaf mahogany was 1 of 68 tree species planted in blocks (Francis 1989).

Taungya plantations are planted in lines among crops (Weaver 1989). Plantations were usually weeded during the establishment phase but were left unmanaged after the first few years. Details of the planting and managing of these mahogany plantations are in papers listed in the literature cited and in Marrero (1950), Weaver and Bauer (1983, 1986), and Lugo (1992).

We base our comparisons among plantation sites on species composition, plantation structure, primary production, litter dynamics, nutrient cycling, and response to disturbances. For nutrient cycling, we report comparative data on N, P, and K, and for other elements when available. We compare mahogany plantations to paired or adjacent secondary forests and to native big-leaf mahogany forests in Central and South America. The reason for these comparisons is to attain a better understanding of the ecological trade-offs in establishing and managing mahogany plantations in the tropics. We report new data on the response to hurricanes of mahogany plantations and paired secondary forests.

We report new data on plantation structure, litter dynamics (fall, accumulation, and decomposition), and nutrient cycling for plantations of big-leaf (site wb-1), small-leaf (site ds-2), and hybrid mahogany (site wh-2).

Table 15.1. Description of Mahogany Plantation Sites Reported in This Study: Hybrid Mahogany Refers to a Hybrid Between Big-Leaf and Small-Leaf Mahogany

Site[a]	Location	Type	Elevation (m)	Year of Study	Age (years)	Source
Small-leaf mahogany						
ds-1[b]	Mona Island	Block	Sea level	1974	35	Cintrón and Rogers (1991)
ds-2[b]	Guánica	Block	135–185	1974–1998	40–64	Lugo et al. (1978); Molina Colón (1986); Chinea (1990); Quigley (1994)
Big-leaf mahogany						
mb-1[b]	Río Abajo[c]	Block	350	1977, 1982	27, 32	Álvarez-López et al. (1983); Dugger et al. (1979)
mb-2[b]	Río Abajo	Taungya	400	1984	48	Weaver (1989)
wb-1	Río Chiquito, LEF[d]	Block	170	1981–1994	18–31	Lugo (1992); Rodríguez Pedraza (1993)
wb-2	LEF	Taungya	250	1984	45	Weaver (1989)
wb-3	Arboretum, LEF[d]	Block	650	1986	26	Lugo et al. (1990)
Hybrid mahogany (big-leaf × small-leaf)						
wh-1	Río Chiquito, LEF[d]	Line	180–200	1981, 1983	18–20	Weaver and Bauer (1983, 1986)
wh-2	Harvey, LEF[d]	Block	200	1981–1995	50–64	Lugo (1992); Rodríguez Pedraza (1993); Fu et al. (1996)

[a] The first letter of the site code denotes the life zone (d = dry, m = wet, and w = wet), and the second letter denotes the mahogany species (s = small-leaf, b = big-leaf, and h = hybrid). All life zones are subtropical *sensu* Holdridge (1967).
[b] Soil parent material was limestone for sites ds-1 to mb-2.
[c] Two valley sites, near each other.
[d] LEF, Luquillo Experimental Forest.

These data were collected in 1982 for site wb-1, 1994 to 1996 in site wh-2 (including its paired secondary forest), and 1998 in site ds-2. Data for 1994 to 1996 at site wh-2 were collected after Hurricane Hugo. For all sites, we used the methods reported by Lugo (1992).

Results

Species Composition

Tree species counts at sites mb-1, wb-1, and wb-3 varied between 9 species per hectare in the overstory and 26 species/0.01 ha in the understory (Table 15.2). Hybrid mahogany plantations (sites wh-1 and wh-2) had between 9 species/0.2 ha in the overstory and 31 species/0.01 ha in the understory. With two exceptions, all the tree species that invaded big-leaf, small-leaf, and hybrid mahogany plantations were native. For example, a small-leaf mahogany plantation studied in 1974 at site ds-2 had saplings of 8 species: *Schaefferia frutescens, Bucida buceras, Leucaena leucocephala, Capparis flexuosa, Krugiodendron ferreum, Erythroxylon areolatum, Eugenia foetida,* and *Amyris elemifera* (Cintrón et al. 1975). All were native species with the exception of *Leucaena,* an introduced pioneer, and they all grew in the surrounding mature forest.

In the 1980s, Molina Colón (1986) found 25 tree species in a 0.1-ha plot at site ds-2 and showed that the epiphyte *Tillandsia recurvata* was selective toward small-leaf mahogany trees. Chinea (1990) studied the same plantation and found 42 tree species associated with small-leaf mahogany; all but 2 were native. In two separate enumerations of woody species more than 50 cm tall, Chinea found 25 species/0.02 ha and 30 species/0.04 ha. These values are at the asymptote of the species–area curve between 0.04 and 0.2 ha. The highest tree species density reported in Table 15.2 was for a small-leaf mahogany plantation at site ds-1 (27 species/0.004 ha).

Quigley (1994) found that the species–area curve for the small-leaf mahogany plantation at site ds-2 saturated at about 90 tree species over a 1-ha sample. He identified 103 plant species in four strata: understory (<2.5 m tall) with 77 species, 8 of which were exclusive to that stratum; subcanopy (>2.5 m tall but crown without access to full light) with 81 species, 5 of which were exclusive to that stratum; canopy (crown exposed to full sunlight) with 52 species, none exclusive to that stratum; and lianas with 11 species.

The results of studies on species richness of mahogany plantations show that monocultures of mahogany trees are invaded by other tree species when the understory is allowed to develop (Lugo 1992). The overstory of mahogany plantations was diversified with the growth of the invading trees. Species counts at all sites except wb-2 and mb-1, which were not assessed, showed that species invaded widely, although the measured

Table 15.2. Structural Data for Mahogany Plantations in Puerto Rico

Site[a]	Diameter Limit (cm)	Stratum	Species/Area (number ha^{-1})	Density (stems ha^{-1})	Basal Area[b] (m^{2}ha^{-1})	Height (m)	Source
Small-leaf mahogany							
ds-1	>1.5 m height	Overstory	2/0.004	4,800		6	Cintrón and Rogers (1991)
	>3.8	Understory	13–27/0.004	18,500–78,000			Cintrón and Rogers (1991)
ds-2		Seedlings	8/0.0016	68,125		0.15–0.6	Álvarez (unpublished)
		All plants	0/0.04; 25/0.02				Chinea (1990)
	>50 cm height	All plants	103/1	6,644	17.7		Quigley (1994)
	>50 cm height	All plants	25/0.1	14,540			Molina Colón (1986)
	>1 m height	Trees		4,570	16.3		This study
	>1	All stems		8,490	21.6		This study
	>10 cm height						
Big-leaf mahogany							
mb-1	>2.5	Trees	22/0.1	790	27.8	25.7	Álvarez López et al. (1983)
	>10	Trees	10/0.1	580	27.3	25.7	Álvarez López et al. (1983)
	>10	Trees	9/0.1	490	31.6	32.7	Dugger et al. (1979)
wb-1	>4	Overstory	15/0.2	974	29.5	25	Lugo (1992)
	<4	Understory	25/0.01	62,400	3.6		Lugo (1992)
wb-3	>4	Overstory		400–816	21.9–39	15.5	Rodríguez Pedraza (1993)
	0.5–4.0	Understory	26/0.01	28,300	3.1	2.7	International Institute of Tropical Forestry (unpublished)
Hybrid mahogany							
wh-1	>10	Trees	16/0.2	373	28.2	19.1	Weaver and Bauer (1983)
wh-2	>4	Overstory	31/0.01	1,242	31.9	27	Lugo (1992)
	>4	Understory		106,800	2.9		Lugo (1992)
	>4	Overstory[c]	9–11/0.2	531–800	21.1–28.5	14	Rodríguez Pedraza (1993)
	>4	Overstory[c]	15/0.2	1,650	34.5		Fu et al. (1996)

[a] Table 15.1 identifies and describes the sites.
[b] Empty cells, no data.
[c] Posthurricane.

Table 15.3. Importance Value[a] of Tree Species with the Highest Value in Each Plantation, and Density of Seedlings of Small-Leaf, Big-Leaf, and Hybrid Mahogany Grown in Plantations

Site[b]	Importance Value[a] (%)	Mahogany Seedling Density (stems ha⁻¹)	Source
Small-leaf mahogany			
ds-1	12[c]	3,300	Cintrón and Rogers (1991)
ds-2	[d]	7,500	Álvarez (unpublished)
	9[c]		Molina Colón (1986)
	17[c]		Quigley (1994)
	52	3,190	This study
Big-leaf mahogany			
mb-1	30		Álvarez López et al. (1983)
	57		Dugger et al. (1979)
wb-1	60	3,500	Lugo (1992)
Hybrid mahogany			
wh-2	49–60		Fu et al. (1996)
	(12-yr period)	71,700	Lugo (1992)

[a] Importance value is based on tree density and basal area.
[b] See Table 15.1 for site descriptions.
[c] Based on all stems.
[d] Empty cell, no data.

species density varied with the minimum diameter or height used (see Table 15.2). More nonplantation species were found in small-leaf mahogany plantations (sites ds-1 and ds-2) than in either big-leaf or hybrid mahogany plantations.

The invasion of tree species into mahogany plantations changes the dominance relations (*sensu* Whittaker 1965) of the community. A greater species density results in lowering the importance value of the dominant tree species (Table 15.3). Mahogany usually retained a high importance value, however, and more dominance by mahogany was observed in plantations from the wet and moist sites than in plantations in the dry life zone. The plantations in the dry life zone had the lowest mahogany dominance of all those studied (Table 15.3).

Plantation Structure

Vertical Stratification

Quigley (1994) described four vegetation strata in a small-leaf mahogany plantation at site ds-2 (understory, subcanopy, canopy, and lianas). He found that all tree sizes greater than 2 cm in diameter contributed equally to the basal area of the forest because of the high density of small-diameter trees relative to the low density of large-diameter trees. Emergent trees were about 15 m tall. A similar stratification was reported in big-leaf and hybrid mahogany plantations (Lugo 1992). Cruz (1987) found that these vegetation strata were used differentially for foraging by 14 avian species. The vegetation strata are a result of both the invasion and growth of nonplantation tree species and the development of mahogany seedlings into saplings, poles, and mature trees (Fig. 15.2; see Wang and Scatena, Chapter 12, this volume).

The tallest mahogany plantations in Puerto Rico are the big-leaf plantations at site mb-1 (Table 15.2), where individual trees could exceed 40m. Individual hybrid mahogany plantation trees at site wh-2 also exceed 40m, but the plantation as a whole was not as tall as the mb-1 plantation, perhaps because of its frequent exposure to winds.

Mature small-leaf mahogany trees were smallest. At site ds-1, for example, small-leaf mahogany trees averaged 6m in height and had a mean dbh of 4.1 cm at age 35 years (Cintrón and Rogers 1991). At site ds-2, small-leaf mahogany trees averaged 3.1 m in height at age 49 years (Molina Colón 1986). Stands at this site have a high density of small trees with heights up to 2.5 m and dbh of 4 cm and a lower density of larger canopy trees that exceed 3 m in height and 10 cm in diameter (Fig. 15.2).

Seedlings and Saplings

Seedlings of various species, including mahogany, can grow under the canopy of the plantations so far studied (Tables 15.2 and 15.3). The highest densities of mahogany seedlings were found in site wh-2 and the lowest were observed in sites sd-1, sd-2, and wb-1 (Table 15.3). Tree size distribution in big-leaf (Wang and Scatena, Chapter 12, this volume; Wadsworth et al., Chapter 17, this volume), small-leaf (Fig. 15.2; Molina Colón 1986; Quigley 1994), and hybrid (Lugo 1992; Wang and Scatena, Chapter 12, this volume) mahogany plantations all show representation of all size-classes from seedlings to poles to canopy trees. Mahogany regenerates in Puerto Rico under its own shade when it grows in plantations.

Stand Basal Area and Tree Density

Basal areas are greater in big-leaf and hybrid mahogany plantations from the moist and wet life zones (sites mb-1 to wh-2) than in small-leaf plantations from the dry life zones (sites ds-1 and ds-2; Table 15.2). This generalization holds in spite of age differences (plantations from the dry life zone were older; Table 15.1). Trees greater than 1 cm dbh at site ds-2 had a basal area of $16.3 \pm 2.1\,m^2\,ha^{-1}$ ($n = 10$), and, even when all stems were counted, the basal area was less than those measured in block plantations in the moist and wet sites.

The basal area of line-planted and taungya big-leaf and hybrid mahogany plantations in moist (mb-2) and wet (wh-1) sites was similar to the basal area of small-leaf mahogany block plantations in dry sites (sites ds-1 and ds-2). This similarity was the result of the lower density of the

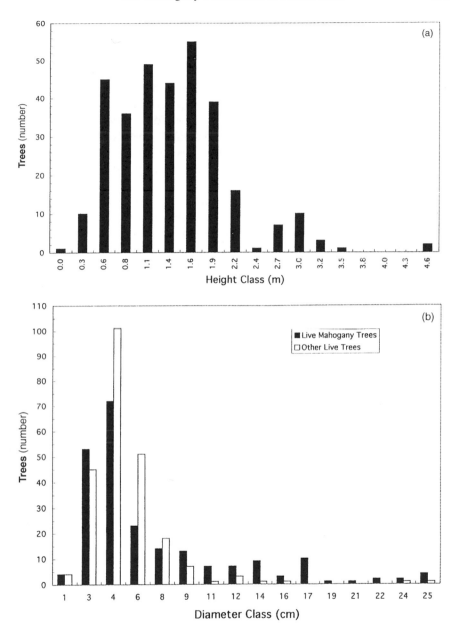

Figure 15.2a,b. Height (**a**) and diameter (**b**) size-class distribution of small-leaf mahogany saplings (dbh <1 cm) and trees (dbh >1 cm) in a 64-year-old plantation at the Guánica Forest (site ds-2) in Puerto Rico.

mahoganies in the line and taungya plantations. Trees in moist and wet conditions grew to larger basal areas than those in dry life zones, regardless of the plantation scheme.

Tree density in small-leaf mahogany plantations in dry sites (ds-1 and ds-2) was higher than those of big-leaf and hybrid mahogany plantations in moist and wet sites (mb-1 to wh-2; Table 15.2). The larger stature of big-leaf and hybrid mahogany trees and their faster growth rate results in complete canopy closure, but the canopy remains fairly open for longer periods in small-leaf mahogany plantations. As a result, more stems and more species can grow under the small-leaf mahogany canopy than in plantations of big-leaf and hybrid mahogany.

The data suggest that mahogany plantations develop complex vertical structure. Seedlings and saplings grow at high densities under closed or open canopies, including those of invading species from surrounding forested areas and those of the planted mahogany. Basal areas and tree densities change with age, disturbances, and stand development.

Biomass

Most of the biomass in big-leaf and hybrid mahogany plantations is above ground, particularly in wood (Table 15.4). Roots comprise a small fraction (5%–7%) of the total biomass in the plantation. Leaf, root, understory, and fine-litter biomass are all at about the same order of magnitude (3–12 Mg ha^{-1}). Leaf biomass ranged from 7.8 to 12.7 Mg ha^{-1}. Fine litter ranged from 8 to 12 Mg ha^{-1} (Table 15.5), and understory biomass of big-leaf (7.7 Mg ha^{-1}) and hybrid (3.7 Mg ha^{-1}) mahogany plantations was a small fraction of the total plantation biomass (Lugo 1992). Small-leaf mahogany had more fine-litter biomass than big-leaf or hybrid mahogany (Table 15.5).

Variations in biomass distribution among mahogany plantations (Table 15.4) are difficult to interpret because of the effects of species age and the management history of each plantation. For example, line plantings of hybrid mahogany (site wh-1) had a lower mahogany biomass than did block-planted mahogany because the lines of trees are scattered in the secondary forest.

Big-leaf and hybrid mahogany plantations had most of their root biomass in diameter sizes greater than 1 mm (Lugo 1992). Smaller-diameter roots comprised a low fraction of the total root biomass in the plantations relative to that of paired secondary forests. The density of roots, measured as roots per meter squared along the face of a soil pit, ranged from 4400 in hybrid plantations to 4700 in big-leaf mahogany plantations. The hybrid mahogany plantation had a lower root biomass than the big-leaf mahogany plantation (Fig. 15.3). In both plantations, 47% to 66% of the roots were in the top 10 cm of soil, and no roots were found below 0.7-m depth. We did not sample directly below trees, however.

Table 15.4. Biomass Above- and Belowground in Mahogany Plantations in Puerto Rico

Site[a]	Leaves (Mg ha^{-1})	Wood		Aboveground (Mg ha^{-1})	Roots[b] (Mg ha^{-1})	Total mass (Mg ha^{-1})	Source
		Volume (m^3 ha^{-1})	Boles (Mg ha^{-1})				
Big-leaf mahogany							
IVa	7.8[c]	145	94	102	8.1	110	Lugo (1992)
	12.7[d]	[e]	125.7	261.3[f]			Rodríguez Pedraza (1993)
Hybrid mahogany							
IVb	9.0	247	51	135.8[f]			Weaver and Bauer (1983)
		292	161	171[f]			Weaver and Bauer (1986)
VII	8.5[c]	419	116	125	5.7	131	Lugo (1992)
	10.0[d]		111.5	236.8[f]			Rodríguez Pedraza (1993)

[a] Table 15.1 identifies and describes the sites.
[b] Root depth is to 1 m.
[c] 1982.
[d] 1989.
[e] Empty cells, no data.
[f] Includes branches.

Table 15.5. Fine-Litter Standing Stock in Four Mahogany Plantations in Puerto Rico

Site[a]	Fine Litter (Mg ha^{-1})			Source
	Leaves[e]	Wood	Total	
Small-leaf mahogany				
ds-2[b]	9.9	2.5	12.4	Lugo et al. (1978)
ds-2[c]	16.0	2.7	18/8	This study
Big-leaf mahogany				
wb-1	8.4	3.0	11.4	Lugo (1992)
wb-3	7.3	4.7	12.0	Lugo et al. (1990)
Hybrid mahogany				
wh-2[d]	5.5	2.8	8.3	Lugo (1992)

[a] Table 15.1 identifies and describes sites.
[b] 1973.
[c] 1997.
[d] 1982.
[e] Includes miscellaneous.

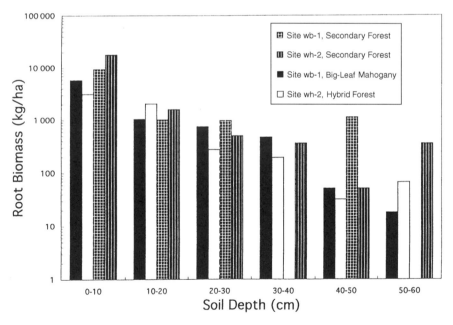

Figure 15.3. Root distribution with depth in big-leaf (*site wb-1*) and hybrid (*site wh-2*) mahogany plantations and paired secondary forests in the Luquillo Experimental Forest, Puerto Rico.

Primary Productivity and Tree Growth

The net aboveground primary productivity (NPP) of big-leaf (17 Mg ha^{-1} yr^{-1}, site wb-1) and hybrid mahogany (19 Mg ha^{-1} yr^{-1}, wh-2) plantations in wet life zones was high (Table 15.6). Wood volume increment ranged from 4 to 15 m^3 ha^{-1} yr^{-1}. Line plantings had the highest and taungya plantings had the lowest volume increment values (Table 15.6).

The fraction of NPP allocated to tree biomass was between 36% and 39% with the balance resulting from litterfall (Table 15.6). When grown in line plantings through native secondary forest in areas previously farmed at site wb-2, 20-year-old big-leaf mahogany yielded up to 15 m^3 ha^{-1} yr^{-1} or 8.5 Mg ha^{-1} yr^{-1}. Under block plantation conditions, big-leaf mahogany yielded 5.9 Mg ha^{-1} yr^{-1} of biomass, and hybrid mahogany yielded 7.5 Mg ha^{-1} yr^{-1}.

Average tree growth, expressed in diameter or basal area, was fairly similar in big-leaf and hybrid mahogany growing either in blocks or line plantings (Table 15.6). Line-planted trees had different growth rates, depending on site topography and canopy position (Weaver and Bauer 1983). They attained greater diameter and height in bottomland and lower slopes compared to midslopes and ridges. Trees growing on ridges had the lowest diameter and height. Trees with dominant crowns were significantly larger than trees with codominant crowns, which in turn were significantly larger than trees with intermediate crowns (Weaver and Bauer 1983, 1986). Trees with suppressed crowns were not encountered in line plantings.

The results of primary productivity measurements in Puerto Rico were exclusive of thinnings, which would add to the productivity of the

Table 15.6. Parameters of Tree Growth and Net Primary Productivity (NPP) for Mahogany Plantations in Puerto Rico

| | | Growth Rate | | Biomass[b] NPP[c] | | |
Site[a]	Diameter (cm yr^{-1})	Basal Area (m^2 ha^{-1} yr^{-1})	Volume (m^3 ha^{-1} yr^{-1})	(Mg ha^{-1} yr^{-1})		Source
Big-leaf mahogany						
wb-2		[d]	7.6–12.1			Weaver (1989)
			3.8–7.8			Weaver (1989)
		1.47	7.63	5.9	16.6	Lugo (1992)
Hybrid mahogany						
wh-1	1.5	1.40	12.4	6.8		Weaver and Bauer (1983)
	1.4	1.36	14.6	8.5		Weaver and Bauer (1986)
wh-2			8.2	7.5	19.1	Lugo (1992)

[a] See Table 15.1 for site descriptions.
[b] Biomass increment of surviving trees.
[c] Includes leaves and branches.
[d] Empty cells, no data.

plantations. Annual rates of wood volume production by mahogany plantations in Puerto Rico compared favorably with the 8 to $18\,m^3\,ha^{-1}$ reported by Pandey (1983) for big-leaf mahogany plantations in Indonesia. In addition, the variation in tree growth with topographic position underscores the importance of site variation in regulating growth and yield of mahogany.

Fine-Litter Dynamics

Litterfall

Annual rates of leaf fall and total litterfall were remarkably similar in hybrid and big-leaf mahogany plantations (Table 15.7). Annual rates of leaf and total litterfall in the small-leaf mahogany plantation at site ds-2 were half as high as those in big-leaf and hybrid mahogany plantations. Fine wood fall and miscellaneous fall varied by month or annually in all the plantations studied. In short-term studies of one to several years, miscellaneous litterfall parameters could be similar in big- and small-leaf and hybrid mahogany plantations. The reason is that the pattern of fruit production by mahogany appears to fluctuate year to year and includes mast years. Therefore, a short-term study may include a mast year for small-leaf mahogany and a low year for big-leaf mahogany (or vice versa), making comparisons difficult to interpret.

Fine litterfall in mahogany plantations is seasonal (Lugo et al. 1978; Lugo 1992). The seasonal pulse in litterfall is associated with the seasonal shedding of leaves by mahogany trees. Cuevas and Lugo (1998) explained 41% of the monthly leaf fall of big-leaf mahogany at site wb-3 based on the number of dry days per month. For sites wb-1, wb-3, and wh-2, the pulse in leaf fall tended to be between February and March. During the rest of the

Table 15.7. Litterfall in Four Mahogany Plantations in Puerto Rico

Site[a]	Leaf	Wood	Flower and Fruit	Miscellaneous	Total	Source
Small-leaf mahogany						
ds-2	4.2[b]	1.2	[c]		5.4	Lugo et al. (1978)
Big-leaf mahogany						
wb-1	8.5	1.2	0.1	0.2	10.0	Lugo (1992)
wb-3	7.0	2.2	0.5	0.2	9.9	Cuevas and Lugo (1998)
Hybrid mahogany						
wh-2	9.0	1.0	0.4	0.4	10.7	Lugo (1992)

Litterfall column group heading: Litterfall $(Mg\,ha^{-1}\,yr^{-1})$

[a] Table 15.1 identifies and describes sites.
[b] Includes miscellaneous.
[c] Empty cells, no data.

year, litterfall was low in both types of plantations. Leaf fall was between 70% and 85% of total litterfall.

In the small-leaf mahogany plantation at site ds-2, fine litterfall peaked during April to June (Lugo et al. 1978). Mahogany parts also dominated the composition of litterfall. Leaf fall was about 70% of the total litterfall but reached more than 80% during peak fall. The fall of fine wood and miscellaneous plant parts were about 15% each of the total litterfall. Three years of observation suggested that the pulse of leaf fall can be at different times each year. During the rest of the year, litterfall was fairly constant at a low rate.

Fine-Litter Decomposition

Lugo (1992) found a fine-litter decomposition rate of about $7\,Mg\,ha^{-1}\,yr^{-1}$ at sites wb-1 and wh-2, a value lower than the annual rate of litter production. When divided by the fine-litter stock in those plantations, the turnover rate was $0.62\,yr^{-1}$ for wb-1 and $0.82\,yr^{-1}$ for wh-2. The proportion of leaf and fine wood in forest floor litter was 85% for wb-1 and 15% for wh-2. This proportion is similar to the proportion of these compartments in litterfall, suggesting similar rates of decomposition for both plant parts.

Leaf litter decomposition at site wh-2 was fast, with a decomposition constant (k; *sensu* Olson 1963) of $1.35\,yr^{-1}$ and a half-life of 0.52 years. Simultaneous measurements of leaf decomposition rates for a paired secondary forest resulted in a k value of $1.11\,yr^{-1}$ and a half-life of 0.63 years.

Fine-litter decomposition studies showed fast rates for both leaves and fine-wood compartments. Leaf decomposition in decomposition bags resulted in a turnover rate of $1.3\,yr^{-1}$, similar to estimates based on the ratio of litterfall to fine-litter standing stock in both the plantation (1.3) and the paired secondary forest (1.4).

Changes in Fine-Litter Stocks

The standing stock of fine litter of mahogany plantations reflects the seasonal production of litter and its subsequent decomposition. Seasonality of fine-litter standing stock has been observed in small-leaf (ds-2; Lugo et al. 1978) and big-leaf (wb-1) and hybrid (wh-2; Lugo 1992) mahogany plantations. The 1981 to 1983 fine-litter standing stock for a hybrid mahogany plantation (Fig. 15.4) illustrated the near steady-state pattern of increase and decrease as a result of peaks of fine-litter production and decomposition. Litter was produced during the dry months when decomposition was slow. The decomposition rate accelerated during the wet months, when litter production was low. Because droughts are more prevalent in the dry life zones, fine litter at site ds-2 (see Table 15.5) accumulated more than in wet sites in spite of the lower rate of litterfall.

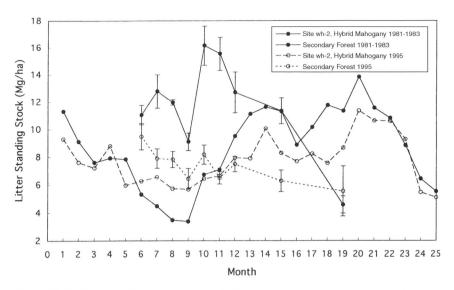

Figure 15.4. Total fine-litter mass in a hybrid mahogany plantation and a paired secondary forest (*site wh-2*) in the Luquillo Experimental Forest, Puerto Rico.

Nutrient Cycling

Nutrient Concentrations

Medina and Cuevas (Chapter 7, this volume) discuss various nutritional aspects of mahogany plantations throughout Puerto Rico, including how differences in the nutritional acquisition among mahogany species influence their growth and survival under a variety of conditions. Their study includes data on the variation of nutrient concentration in tissues of mahogany trees. Additional nutrient concentration data for mahogany tissue is reported in Sánchez et al. (1997), Lugo (1992), and Cuevas and Lugo (1998).

Lugo (1992) described changes in the N, P, and K concentrations of bigleaf and hybrid mahogany leaves in this progression: mature leaves, leaves in leaf fall, and leaves on ground litter at different depths in the litter layer. As mature leaves senesce, they undergo a reduction in nutrient concentration from retranslocation before leaf fall. Leaves on the forest floor, however, increased in nutrient concentration as they were buried in the litter layer. This increase results from biomass decomposition and mineralization and from immobilization of nutrients in nutrient-rich microbial biomass (Lodge 1996). We found a similar pattern in a small-leaf mahogany plantation (site ds-2; Table 15.8).

The pattern of change in nutrient concentration in decomposing leaves from site wh-2 was different from the pattern of change observed for mass. While leaf mass steadily decreased at a rate of 0.4%/day, N, P, Fe, and Al increased in both concentration and absolute amount; K, Mn, and Ca

Table 15.8. Concentration of Elements (mg g^{-1}) in Mature Small-Leaf Mahogany Leaves and Three Layers of Fine Litter at a Plantation in Guánica, Puerto Rico (site ds-2)

Element (%)	Mature Leaves[a]	Fine-Litter Layers[b]		
		Upper	Middle	Lower
N	1.32	1.35	1.61	1.50
P	0.69	0.36	0.44	0.48
K	7.20	3.55	4.80	6.21
Ca	10.86	38.01	61.11	81.23
Mg	1.10	2.05	2.84	3.54
Mn	0.01	0.04	0.08	0.12
Fe	[c]	0.88	2.42	4.69
Al	0.06	1.05	2.80	5.11
S		0.38	0.36	0.24

[a] From Sánchez et al. (1997).
[b] Fine litter layers identified by appearance and depth.
[c] Empty cell, no data.

decreased rapidly; S increased slightly; and Mg changed little (Fig. 15.5). The increases in concentration are due mostly to immobilization and accumulation by decomposing microbial populations. Potassium leached rapidly from litter while Ca initially decreased faster than mass, but then stabilized and showed no further change.

Amounts of Nutrients in Vegetation, Fine Litter, and Soil

The accumulation and movement of nutrients in vegetation varies according to the nutrient. For N, P, and K, the highest amounts are in the soil (Lugo 1992). The distribution of N, P, and K (kg ha^{-1}) in the biotic compartments of big-leaf (site wb-1) and hybrid (site wh-2) mahogany plantations followed the rank order of wood > leaves ≥ litter > roots, but roots had more K than did fine litter, and fine litter had the same or more N and P as the leaves did.

Nutrient (kg ha^{-1})	Biotic Compartment			
	Wood	Leaves	Litter	Roots
Big-leaf (site wb-1)				
Nitrogen	409	112	136	44
Phosphorus	21	5	5	3
Potassium	293	47	11	24
Hybrid (site wh-2)				
Nitrogen	835	121	91	44
Phosphorus	29	5	4	3
Potassium	333	57	8	24

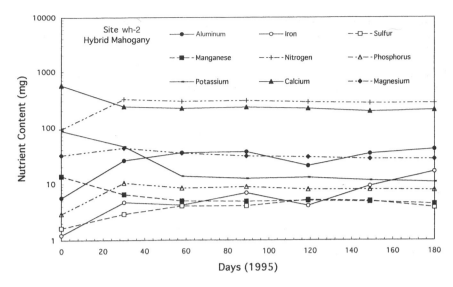

Figure 15.5. Change in N, P, K, Ca, Mg, Mn, Al, Fe, and S content in decomposing leaves in a hybrid mahogany plantation at site wh-2.

Big-leaf mahogany fine litter at site wb-3 accumulated large amounts of Ca and N. Among mahogany and other plantation species at the site, total litter mass and fine-wood litter mass had an inverse curvilinear relation with N concentration, and species leaf litter mass had a similar relation with Ca concentration (Lugo et al. 1990).

Fine litter of the hybrid mahogany plantation at site wh-2 had less N, P, and K than did fine litter in two big-leaf mahogany plantations at sites wb-1 and wb-3 (Table 15.9). The difference in fine-litter stock was most significant for N and not as dramatic for P. The big-leaf mahogany plantation at site wb-3 had particularly high K in fine litter because the amount of woody litter contributing to the total amount was high. The nutrient stocks in fine litter at site ds-2 were the highest reported in this study (see Table 15.9). These high quantities were mostly caused by the large standing crop of fine litter (see Table 15.5). This generalization holds for all fine-litter compartments and most nutrients.

Temporal Patterns of Nutrients in Fine Litter

The nutrient content of fine litter of mahogany plantations had a temporal pattern that followed that of fine-litter standing stock; that is, nutrient stocks rose and fell according to the rise and fall of fine-litter stocks (Lugo 1992).

Nutrient Fluxes

The largest nutrient fluxes in mahogany plantations were those associated with leaf retranslocation, followed by nutrient uptake by roots (Lugo 1992).

Table 15.9. Nutrient Storage (kg ha^{-1}) in Loose Litter of Mahogany Plantations in Puerto Rico

Site[a]	All Litter					Leaves[b]					Wood					Reproductive Parts					Source
	N	P	K	Ca	Mg	N	P	K	Ca	Mg	N	P	K	Ca	Mg	N	P	K	Ca	Mg	
Small-leaf mahogany																					
ds-2[c]	276	8.0	79	1020	45	248	7.2	70	914	42	25	0.7	8	96	3	3	0.1	0.1	10	0.5	This study
Big-leaf mahogany																					
wb-1	140	4.7	11	[d]	113	113	3.8	9	23	23	0.8	2									Lugo (1992)
wb-3	140	4.1	34	201	35	101	2.9	23	133	22	29	1.0	9	59	12	2	0.1	0.4	1	0.3	Lugo et al. (1990)
Hybrid mahogany																					
wh-2	94	3.7	8			66	2.6	6			25	1.4	2								Lugo (1992)

[a] See Table 15.1 for site descriptions.
[b] Includes miscellaneous and/or reproductive parts when any of these values are not reported.
[c] 1997.
[d] Empty cells, no data.

Nutrient uptake and retranslocation were similar in the hybrid mahogany plantation at site wh-2.

Nutrient	Nutrient Uptake (kg ha^{-1} yr)	Nutrient Retranslocation (kg ha^{-1} yr)
Nitrogen	109	109
Phosphorus	2.9	6.2
Potassium	36	61

Leaf fall and wood fall comprise the next highest nutrient fluxes in mahogany plantations (Lugo 1992).

The fraction of nutrients in flux relative to nutrients in stocks was highest in leaves, mostly because of retranslocation (N, 115%; P, 116%–132%; and K, 115%), followed by litter (N, 24%–47%; P, 26%–38%; and K, 48%–62%). Stemwood and the plantation as a whole had a small fraction of nutrients in flux relative to the amount of nutrients in stock. For example, the amount of aboveground nutrients in circulation (nutrient uptake plus nutrients in litterfall) divided by their respective mass in all the aboveground compartments (leaves, stemwood, and fine litter) was equivalent to N, 15%; P, 11% to 13%; and K, 10%.

The nutrient return in litterfall varied with the compartment. As expected, leaf fall produces the largest amounts of nutrients returned to the forest floor, followed by the fall of reproductive parts (flowers and fruits), which can return more nutrients to the forest floor than fine woody litter does. For example, Cuevas and Lugo (1998) found that a big-leaf mahogany plantation annually returned.

Nutrient	Woodfall (kg ha^{-1})	Fall of Reproductive Parts (kg ha^{-1})
Nitrogen	4.4	14.6
Phosphorus	0.2	0.8
Potassium	5.2	8.2

This high nutrient return rate is due to the high concentrations in flowers and fruits and the periodic high mass fluxes associated with fruit fall.

Lugo and Frangi (1993) showed that fruit fall in mahogany plantations is highly seasonal with peak fall rates similar to leaf fall rates ($2 \, \mathrm{g} \, \mathrm{m}^{-2} \, \mathrm{d}^{-1}$). For this reason, the return of nutrients to the forest floor by reproductive parts is high but limited to particular periods. Apparently, when the mahoganies are fruiting, they invest a large fraction of the nutrient capital in producing reproductive parts.

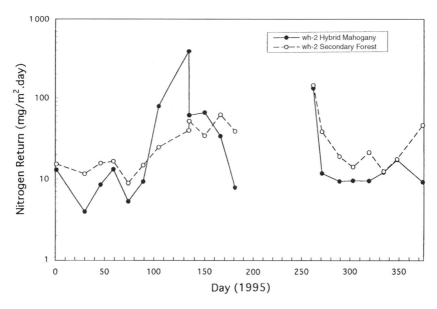

Figure 15.6. Pattern of N return in total fine litterfall in hybrid mahogany plantation and paired secondary forest (site wh-2) in the Luquillo Experimental Forest, Puerto Rico.

Patterns of Nutrient Return

The temporal pattern of nutrient return by mahogany plantations follows the pattern of mass fall. The pattern is illustrated with N in a hybrid mahogany plantation in Figure 15.6. Other nutrients had the same pattern of nutrient return but with different rates. Except during the peak of mass fall, nutrient return to the forest floor in mahogany plantations was low.

Nutrient Use Efficiency

Lugo (1992) found a similar efficiency of nutrient use as a result of leaf fall in big-leaf and hybrid mahogany (sites wb-1 and wh-2). Both had higher within-stand nutrient use efficiency (*sensu* Vitousek 1984) for P than for N (Table 15.10). Cuevas and Lugo (1998) found variability in the month-to-month within-stand nutrient use efficiency of leaf fall in a big-leaf mahogany plantation at site wb-3. Just before the pulse of leaf fall, retranslocation of P and N increased, as did their respective within-stand use efficiency. For example, as leaf fall increased in March and April, the N concentration of leaves decreased to its lowest annual value (from about 12.5 to about 6.5 mg g^{-1}). Higher leaf N concentrations were observed when leaf fall was at its lowest rates (as high as about 16 mg g^{-1}). We found a similar behavior in hybrid mahogany at site wh-2.

Table 15.10. Indices of Nutrient Use Efficiency and Turnover Rates of Various Compartments of Mahogany Plantations[a,b]

Efficiency Parameter	Site wh-2				Site wb-1				Site wb-3[d]	
	N	P	K	Mass	N	P	K	Mass	N	P
Within-stand[c]										
All leaves	88	2464	554	na	76	2485	519	na	70–120[a]	2400–4100
Litterfall	93	2403	588	na	90	2780	633	na		
NPP/uptake	175	6586	531	na	259	5929	553	na		
Root biomass/uptake	52	1971	159	na	126	2884	269	na		
Annual turnover										
Canopy leaf mass	na	na	na	1.15	na	na	na	1.16		
Leaf litter	0.46	0.38	0.65	1.60	1.56	0.18	0.30	1.02		
Wood litter	0.51	0.36	0.52	0.68	0.43	0.63	1.32	0.47		
All litter (1982, prehurricane)	0.46	0.41	0.61	1.29	0.24	0.23	0.46	0.88	0.79	

[a] Empty cells, no data.
[b] na, not applicable.
[c] *Sensu* Vitousek (1984).
[d] Data for 1986–1987.

The rank order of retranslocation rate of N and P is reversed in relation to the within-stand nutrient use efficiency of these elements. More N is retranslocated than P, when measured in absolute amounts; that is, 106 versus $5.2\,kg\,ha^{-1}\,yr^{-1}$ at wb-2 and 109 versus $6.2\,kg\,ha^{-1}\,yr^{-1}$ at wh-2. Measured as the fraction of stored nutrient, however, the retranslocation of P accounts for 100% and that of N accounts for 95% of the nutrient content in the leaf compartment of the big-leaf mahogany plantation at wb-1. A larger fraction of the P than of the N standing stock is retranslocated annually. Cuevas and Lugo (1998) found lower rates of N and P retranslocation (78% for N and 68% for P) at wb-1.

The efficiency of nutrient use based on nutrient uptake (measured as the ratio of NPP/uptake or root biomass/nutrient uptake) showed that the highest efficiency is for P, followed by K and N (Table 15.10). These results are consistent with Vitousek's (1984) within-stand nutrient use efficiency and also showed the high multiplier value of P uptake relative to NPP and root biomass.

The annual turnover rates of most nutrients in fine litter in these mahogany plantations were less than 1.0 (Table 15.10), which suggested a tendency to conserve nutrients through slow turnover. This tendency contrasts with the higher turnover rates for mass relative to nutrients in that mass. The turnover rates for P are particularly low; those of N and K are somewhat higher. When total plantation nutrient stocks are considered in relation to fluxes, the annual turnover rates become much lower (0.10–0.15), in terms of both mass and nutrients, and the system turns out to be a sink because it is accruing biomass (NPP >1) and nutrients (net uptake was measured).

Response to Disturbances

Mahogany plantations in Puerto Rico are susceptible to three principal types of disturbance: the shoot borer *Hypsipyla grandella* Zeller, drought, and hurricanes. The experience in the Caribbean with the shoot borer is that trees can recover from attacks (Weaver and Bauer 1986; Frederick 1997), and its effects can be mitigated both in the nursery (Bauer 1987) and the field (Weaver and Bauer 1983, 1986) by following strict guidelines of control and planting strategies. The growth rate of trees exposed to stressful conditions of the lower montane wet forest life zone, that is, excessive soil moisture and low air temperatures, was significantly reduced by the shoot borer (Ward and Lugo, Chapter 3, this volume). They found that the growth rate of mahogany was not affected by the shoot borer at lower elevations in dry, moist, and wet life zones.

Drought

The effects of drought on mahogany plantations have not been sufficiently studied. Medina and Cuevas (Chapter 7, this volume) point out that small-

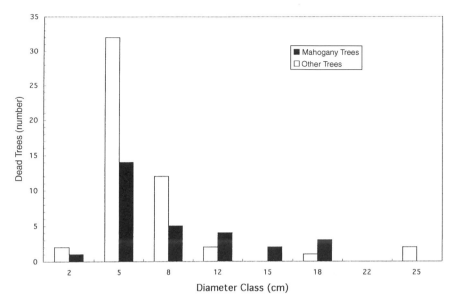

Figure 15.7. Diameter size-class distribution of dead trees (dbh >1 cm) in a 64-year-old small-leaf mahogany plantation (site ds-2) at the Guánica Forest (site ds-2) in Puerto Rico.

leaf mahogany is drought tolerant and big-leaf mahogany is drought sensitive. They show this behavior to be associated with nutritional differences, such as in tissue concentration. The N : P ratio of mahogany leaves increased with site humidity, and the Ca : Mg ratio reflected the higher concentration of soluble Ca in drier environments. In spite of their drought tolerance, small-leaf mahogany suffers increased mortality during extensive droughts—that is, those that last for more than a year—with below-average rainfall in dry life zones. We have observed mortality of clusters of large small-leaf mahogany trees at site ds-2 as a result of extended drought.

The 1998 basal area of dead trees at site ds-2 after prolonged droughts during a 5-year period was $5.2 \, m^2 ha^{-1}$ or 25% of the total basal area of the stand, corresponding to 800 ± 141 ($n = 10$) stems/ha, of which at least 510 ± 174 ($n = 10$) were small-leaf mahogany. Mahogany trees tended to die in the larger-diameter size-classes than other tree species invading the plantation (Fig. 15.7). The top branches of emergent trees died first and eventually the whole tree was dead, forming groups of dead trees equivalent to a dozen or more trees in plots of $100 \, m^2$.

Hurricanes

After the passage of Hurricane Betsy in 1956, Wadsworth and Englerth (1959) classified big-leaf mahogany as prone to windthrow and small-leaf mahogany as resistant. Big-leaf and hybrid mahogany trees are susceptible to wind damage because of their shallow and sparse root system (see Fig. 15.3) and large spreading crowns. Trees tend to tip over, lose large branches, or both when exposed to high-velocity wind.

Hurricane Hugo passed over the Luquillo Experimental Forest in 1989 and exposed mahogany plantations to hurricane-strength winds. Fifty-two percent of the trees in a big-leaf mahogany plantation on the windward side (site wb-1) suffered severe damage from Hurricane Hugo, and 73% of those with severe damage were windthrown (Rodríguez Pedraza 1993). At site wh-2, wind damage was less severe because the hybrid plantation was in the lee of the hurricane. Sixty percent of the trees in the hybrid mahogany plantation suffered moderate damage from winds and 39% suffered severe damage. In terms of biomass transfer to the forest floor, however, the big-leaf plantation at site wb-1 transferred 41% of its aboveground biomass and the hybrid mahogany plantation at site wh-2 transferred 55%. Larger trees at site wh-2 than at site wb-1 accounted for the higher fraction of biomass transfer to the forest floor, even though site wh-2 was in the lee of the hurricane's path.

The rate of tree mortality in affected mahogany plantations changed from about 50 trees $ha^{-1}yr^{-1}$ before Hurricane Hugo to about 250 trees $ha^{-1}yr^{-1}$ immediately after the hurricane (Rodríguez Pedraza 1993). Five years later, the tree mortality rate was below the rate observed before the hurricane (Fu et al. 1996). Expressed as a percentage of live trees per year, the rates of tree mortality increased from 0% to 51% at site wb-1 and from 5% to 34% at site wh-2 immediately after the hurricane.

Hurricane Hugo reduced tree density of hybrid mahogany by more than 1700 trees ha^{-1} and basal area at site wh-2 to $30 m^2 ha^{-1}$ (Rodríguez Pedraza 1993; Fu et al. 1996). Plantations at sites wb-1 and wh-2 had reduced basal area right after Hurricane Hugo (see Table 15.2). Five years after the hurricane, however, these parameters had a positive rate of change. The plantation at site wh-2 suffered a net reduction of eight tree species in plots of 0.2 ha.

Tree species counts in the oldest hybrid mahogany plantation (site wh-2) first decreased and then increased after Hurricane Hugo in 1989. After the hurricane, regeneration in the hybrid mahogany plantation (site wh-2) accelerated, as did tree growth and species invasions (Fu et al. 1996). Five years after the hurricane, the plantation had recovered and exceeded its prehurricane tree density and basal area. Ingrowth had increased from 32 trees $ha^{-1}yr^{-1}$ before the hurricane to 135 trees $ha^{-1}yr^{-1}$ afterward. Plots of 0.2 ha in the plantation gained six new tree species 5 years after the hurricane. Expressed in species per thousand individuals, however, the plantation changed from 20 in 1984, to 17 in 1989, and 11 in 1994. Mahogany maintained its dominance and importance value throughout the 12 years of observation of the plantation at site wh-2 (Table 15.3; Fu et al. 1996).

When a hurricane passed 100 km south of a small-leaf mahogany plantation at site ds-2, Cintrón et al. (1975) and Cintrón and Lugo (1990) measured a pulse in litterfall. Woodfall was particularly affected by the increased winds, and the pulse lasted for several months after the event. The same observation, but at a more severe scale, was made after Hurricane Hugo passed over sites wb-1 and wh-2. Six years after Hurricane Hugo, the hybrid mahogany plantation at site wh-2 was still producing much fine-wood litter (2.5 Mg ha^{-1} yr^{-1}), high leaf litter (9.4 Mg ha^{-1} yr^{-1}), and a large peak of flowers and fruits (3.6 Mg ha^{-1} yr^{-1}). Total litterfall was 15.8 Mg ha^{-1} yr^{-1}, higher than the prehurricane amount, 10.7 Mg ha^{-1} yr^{-1} (see Table 15.7). The fraction of leaves in total litterfall was reduced from between 70% and 85% to 59% because of a peak production of flowers, fruits, and fine-wood litter and not because of a reduction in leaf fall. In fact, leaf fall in 1995 was higher than normal at site wh-2.

The pattern of change in litter standing stock in 1995 at site wh-2 (see Fig. 15.4) was different from that observed between 1981 and 1983. More litter biomass (in Mg ha^{-1}) accrued on the forest floor during the first 6 months of 1995 (6.8 of leaves, 2.3 of wood; total, 11.7) than in 1981 to 1983 (see Table 15.5). Despite the effect of the hurricane, however, fine-litter biomass increased during the period of peak litterfall and rapidly decreased afterward (Fig. 15.5).

Nutrient content in fine litter (Table 15.11) was also higher than previously observed at this site (see Table 15.9). In part because of the higher contribution by leaf litter, which enriched the nutrient pool of fine litter, this enrichment was in turn due to Hurricane Hugo, which caused massive leaf fall before leaves senesced and could retranslocate nutrients in the plant. Also, successional species that invaded the plantation after the hurricane produced litter of higher nutrient content (Scatena et al. 1996) than that normally produced by the plantation. The result of these two trends was that, 6 years after the hurricane, the litter compartment had not returned to prehurricane conditions and was enriched in terms of both mass and nutrients.

Table 15.11. Nutrient Storage (kg/ha) in Loose Litter and Nutrient Return (kg ha^{-1} yr^{-1}) in Litterfall of a Hybrid Mahogany Plantation (wh-2)[a] in Puerto Rico, 1995

Parameter	All Litter					Leaves				
	N	P	K	Ca	Mg	N	P	K	Ca	Mg
Storage	130	4.7	32	235	29	101	3	20	193	22
Return	157	7.3	91	257	45	87	2.6	43	223	26
Return/storage[b]	1.2	1.6	2.9	1.1	1.6	0.9	0.8	2.1	1.2	1.2

Parameter	Wood					Reproductive Parts				
	N	P	K	Ca	Mg	N	P	K	Ca	Mg
Storage	15	0.6	4	31	5	12	0.7	7	10	3
Return	14	0.8	14	18	3	51	3.7	33	12	13
Return/storage[b]	0.9	1.3	3.1	0.6	0.5	4.2	5.3	4.7	1.2	48

[a] See Table 15.1 for site descriptions.
[b] The corresponding value for biomass was 1.35.

The seasonal patterns of nutrient content in fine litter that we found in 1995 at site wh-2 contrasted with those published by Lugo (1992) for 1981 to 1983 at the same location. We found that all values declined over the year of study (Fig. 15.8), except for the brief period when mahogany trees shed their leaves and caused a temporary increase in the nutrient pool of litter. We attribute this finding to a transitional state resulting from Hurricane Hugo. Six years later, the flux of nutrients from fine litterfall (Table 15.11) was higher than any nutrient flux measured before the hurricane.

After Hurricane Hugo, the nutrient return by reproductive parts was sufficient to influence the annual return of nutrients in the whole stand (see Table 15.11). At site wh-2, reproductive parts comprised the following percentages of nutrients in the hybrid mahogany plantation.

Nutrient	Percentage
Nitrogen	32
Phosphorus	53
Potassium	36
Calcium	5
Magnesium	29

This flux was also seasonal (Fig. 15.9). After they fell to the forest floor, reproductive parts in fine litter continued to account for important quantities of stored nutrients (Fig. 15.10).

Our results also showed that the pulses of P returned as fruit fall (Fig. 15.9, Table 15.11) influenced the within-stand nutrient use efficiency of hybrid mahogany at site wh-2. For example, the within-stand P use efficiency for the collection of April 30, 1995, was 2156 with fruit fall and 4242 without fruit fall. In that collection, fruit fall accounted for 36% of the stand's P return.

Discussion

Mahogany plantations in subtropical moist and wet life zones are characterized by high rates of aboveground net primary productivity, which includes fast rates of tree growth and litter production, net accumulation of mass and nutrients, and variable efficiency in the cycling of nutrients. Other tree species (as well as big-leaf, small-leaf, and hybrid mahogany) can regenerate under the canopy of mahogany plantations (see Table 15.3). Birds (Cruz 1987), earthworms (González et al. 1996), epiphytic plants and climbers (Molina Colón 1986; Quigley 1994), and other types of organisms grow well in mahogany plantations.

Results from planting mahogany in Puerto Rico showed that it could grow under a wide diversity of conditions, but the rate of growth can vary significantly both within and among species (Ward and Lugo, Chapter 3, this volume; Wadsworth et al., Chapter 17, this volume). The wide range of variation in the structural and functional parameters of mahogany planta-

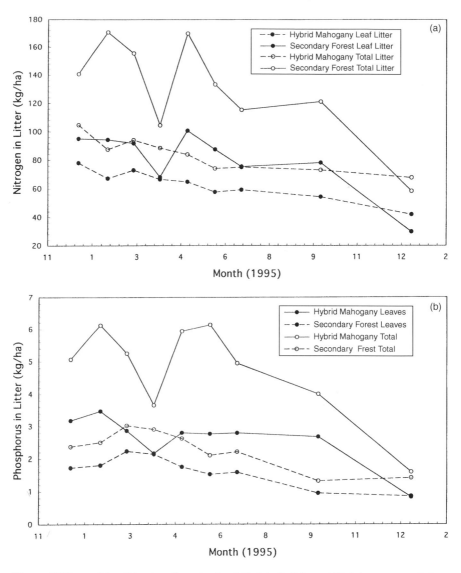

Figure 15.8a–c. Monthly standing stock of N (**a**), P (**b**), and K (**c**) content of fine litter in a hybrid mahogany plantation and a paired secondary forest (site wh-2) in the Luquillo Experimental Forest, Puerto Rico.

tions can be explained by species or climate differences, age of the plantations, differences in soils or topographic position, and disturbance regime. We discuss each of these in turn and conclude with comparisons of mahogany plantations with adjacent native forests and of mahogany plantations with natural mahogany stands.

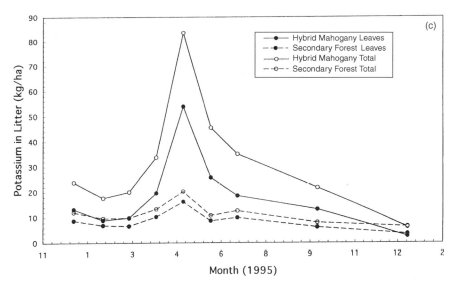

Figure 15.8a–c. *Continued.*

Species Differences and Similarities

Big-leaf mahogany, small-leaf mahogany, and their hybrid exhibit seasonal leaf fall. Small-leaf mahogany appears to be more finely tuned to drought, and big-leaf mahogany maintains its seasonal leaf fall even where the rain-

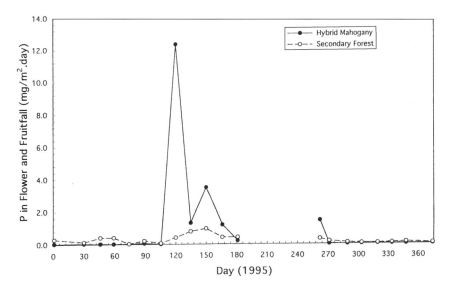

Figure 15.9. Monthly return of P by flower fall and fruit fall in a hybrid mahogany plantation and paired secondary forest (site wh-2) in the Luquillo Experimental Forest, Puerto Rico.

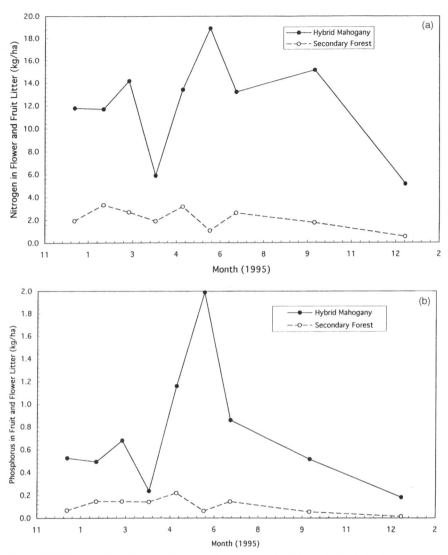

Figure 15.10a–c. Monthly standing stock of N (**a**), P (**b**), and K (**c**) in flowers and fruits in a hybrid mahogany plantation and paired secondary forest (site wh-2) in the Luquillo Experimental Forest, Puerto Rico.

fall is abundant or not seasonal. The monthly variation in retranslocation particularly reveals the natural tendency of the species to conserve nutrients before massive leaf fall (Cuevas and Lugo 1998). Other nutritional differences among these species are discussed in Medina and Cuevas (Chapter 7, this volume). For example, they found that small-leaf mahogany accumulated significantly lower amounts of Ca and Mg and big-leaf mahogany accumulated less Al.

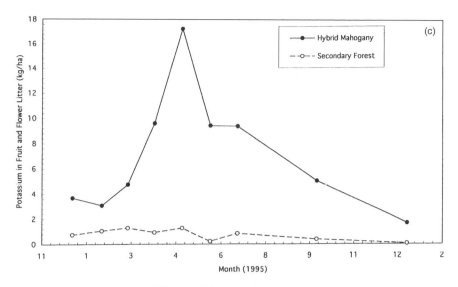

Figure 15.10a–c. *Continued.*

Differences between small-leaf and big-leaf mahogany reflect the contrasting conditions under which they grow. Small-leaf mahogany planted in moist to wet life zones in Puerto Rico did not grow well (Marrero 1950; Francis 1995). Medina and Cuevas (Chapter 7, this volume) attribute this trait to the species sensitivity to low pH and high Al mobility. Similarly, big-leaf mahogany failed to grow in the dry life zone. Associated with this intolerance to low rainfall and drought, Medina and Cuevas found an apparent low tolerance by big-leaf mahogany to high concentrations of soluble Ca.

Ward and Lugo (Chapter 3, this volume) combined tree growth data by species and showed that big-leaf mahogany outgrows both small-leaf and Pacific coast mahogany in Puerto Rico. Hybrid mahogany maintained the growth capacity of big-leaf mahogany and grew best in the moist and wet conditions of the island. The shallow root system and spreading large crowns of big-leaf mahogany explain its susceptibility to wind disturbance and provide another contrast with small-leaf mahogany. In addition, Whitmore and Hinojosa (1977) suggested that small-leaf mahogany is less susceptible to the shoot borer than is big-leaf mahogany.

Climate Effects

The highest biomass and highest primary productivity and litterfall were observed in the moist and wet environments (see Tables 15.4, 15.6, and 15.7). In the dry environments, values for structural and functional variables were lower than in moist and wet environments. A critical climatic factor that separates these species is drought. Low rainfall reduces the growth and yield of plantations and, beyond an unknown threshold, determines the success and distribution of the species.

Age-Dependent Effects

Plantation processes are age dependent. For example, the function of plantations as foster ecosystems for native species has been proposed as a technique to restore native forests in degraded lands (Lugo 1988, 1997; Parrotta 1992, 1993, 1995; Parrotta and Turnbull 1997). This process is age dependent because time is needed for trees to invade and develop in the plantation. Time is also needed to allow the accumulation of basal area, volume, biomass, and nutrient capital in these forests. All these parameters tend to increase with the age of the plantation. Apparently, the fraction of nutrient need that is retranslocated and recycled in the plantation also increases with tree age.

Soil and Topographic Effects

Weaver and Bauer (1983, 1986) showed how, within a given site, topographic differences and canopy position affect tree growth, so that the size of trees (diameter and height) can vary significantly from one place to another. They pointed out that topography was a good indicator of the performance of mahogany because edaphic factors, such as soil depth and aeration, are associated with particular topographic conditions.

Mahogany performs well in Puerto Rico on a variety of soils including those derived from volcanic, limestone, serpentine, and alluvial geologic parent materials (Marrero 1950; Weaver and Bauer 1986; Ward and Lugo, Chapter 3, this volume; and Wadsworth et al., Chapter 17, this volume). When soil becomes too dry or if the pH is high, small-leaf mahogany tends to prevail over big-leaf mahogany. For example, small-leaf mahogany can grow on serpentine soils, degraded sites, and limestone slopes, but big-leaf mahogany cannot. Big-leaf mahogany grows best on volcanic soils, limestone valleys, and alluvial deposits and does not grow well in poorly aerated or highly degraded soils. Relative to the other mahogany species, however, big-leaf mahogany tolerates the wettest conditions.

Disturbance Regime

The period of observation of the mahogany plantations in sites ds-2, wb-1, and wh-2 included drought and hurricane disturbances. Besides noting the obvious effects of such disturbances to increase tree mortality, reduce tree density and basal area, and damage crowns and stems (cf. Rodríguez Pedraza 1993; Fu et al. 1996), we were able to observe some of the functional consequences of these events on the plantations. These functional effects can be summarized in three types of ecosystem attributes: species composition, primary productivity, and nutrient cycling.

The main effect of disturbances on the species composition of mahogany plantations is to reduce the numbers of species as a result of differential mortality from the disturbance force. We observed this mortality in the hybrid mahogany plantation at site wh-2 after Hurricane Hugo (Fu et al. 1996) and in the small-leaf mahogany plantation at site ds-2 after several years of drought. During the recovery period, stands increased in tree density and basal area and also in species richness. However, the number of species per number of individuals decreased until the stand was naturally thinned through normal succession.

Primary productivity of mahogany plantations increased after disturbance events. This increase was shown by the fast recovery of stem density, basal area, and tree height after Hurricane Hugo passed over a hybrid mahogany plantation at site wh-2 (Fu et al. 1996). We have no information on the effects of drought recovery on primary productivity of small-leaf mahogany plantations. Our field observations showed rapid understory development in the canopy gaps formed by the mortality of dominant small-leaf mahogany trees. These gaps supported rapid growth of both small-leaf mahogany seedlings and saplings and trees of other dry-forest species.

Both hurricane and drought disturbances result in a legacy of large amounts of standing dead or downed coarse woody debris. In addition, hurricanes are associated with large transfers of living biomass to the forest floor (Rodríguez Pedraza 1993). This instantaneous transfer of organic debris has significant implications to the nutrient and mass fluxes of these ecosystems. In general, they are larger than nondisturbance fluxes and create large accumulations of nutrients and biomass on the forest floor. The system is thus pulsed by the massive transfer of mass and nutrients and the simultaneous burst in new vegetation regrowth. It may take decades for the forest floor and vegetation to return to predisturbance states.

Six years after Hurricane Hugo, fine-litter production rates, nutrient return to the forest floor, fine-litter biomass, and nutrients stored in fine litter were all higher than predisturbance values (compare posthurricane data in Tables 15.11 and 15.12 and in the results section with prehurricane data in Tables 15.5, 15.7, and 15.9). Part of the reason for the high litter production was the mortality of fast-growing species that invaded the site after the hurricane. The falling tissues of these invading species in turn enrich the litter because they have higher nutrient concentrations than those of trees in the prehurricane forest (Scatena et al. 1996). As a consequence of the disturbance, the forest floor of a plantation receives more nutrients than it would normally receive without the disturbance. Higher nutrient input is beneficial to litter decomposition; thus, the mass of litter decreases rapidly despite the overwhelming input from the transfer of part of the canopy to the forest floor.

Comparison with Adjacent Secondary Forests

Mahogany plantations, particularly mature ones, approach secondary forests in structure and functioning. Lugo (1992) compared 73 parameters of hybrid and big-leaf mahogany plantations with paired secondary forests

Table 15.12. Comparison of Measures of Litter Dynamics Between a Hybrid Mahogany Plantation and a Paired Secondary Forest 7 Years Before and 6 Years After Hurricane Hugo[a]

Parameter	Plantation		Secondary Forest	
	1982	1995	1982	1995
Fine litter ($Mg\,ha^{-1}$)	8.3	11.7	7.4	7.3
Litterfall ($Mg\,ha^{-1}\,yr^{-1}$)	10.7	15.8	9.7	10.9
Nutrients in fine litter ($kg\,ha^{-1}$)				
N	93.5	130	87.3	83
P	3.7	4.7	2.7	2.3
K	8.2	31.8	8	11.3
Nutrient return in litterfall ($kg\,ha^{-1}\,yr^{-1}$)				
N	43	157	117	111
P	1.5	7.3	3.1	3.5
K	5	91	16.3	41.6
Within-stand nutrient use efficiency[b]				
N	93	84	92	99
P	2403	2182	2999	2964
K	588	256	565	311

[a] Data from 1982 are from Lugo (1992).
[b] *Sensu* Vitousek (1984).

of similar age. He found no consistent pattern of difference between the paired forests, other than the expected ones in plant species richness and composition. Other differences had to do with magnitudes of fluxes or structural parameters. Plantations outproduced the paired secondary forest in wood and fine litterfall as well as in tree growth. Significant differences were found in the distribution of biomass. For example, secondary forests have more belowground biomass (see Fig. 15.3), more roots, and more roots in diameter classes below 1 mm than do plantations. Differences were also found in the dynamics of litter: more litter was accumulated and more litter produced in plantations than in secondary forests (see Fig. 15.4, Table 15.12).

As mahogany plantations in subtropical moist and wet life zones age, the standing stock of fine litter decreases, perhaps because of the effects of increasing species richness in the plantation. The larger number of species introduced a greater variety of litter quality to decomposing microbial populations. Chances are that the invading species have leaf litter with faster turnover than the plantation species because their litters had higher nutrient quality. For example, in 1982 before Hurricane Hugo, the paired secondary forest at site wh-2 returned more nutrients in litterfall than did the hybrid mahogany plantation (see Table 15.12). Also, understory tissues and litter are of higher nutrient concentration than plantation species tissues and litter (Lugo 1992). These differences will eventually result in faster litter turnover and lower litter biomass as plantations age.

The amount of plantation nutrient input to litter is low and pulsed, and nutrient input to litter in secondary forests is higher and more evenly distributed year round (Lugo 1992). Relative to secondary forests, plantations tended to behave more like carbon and nutrient sinks in that they had a higher rate of aboveground biomass accumulation and higher accumulation of nutrients in litter. Before Hurricane Hugo, nutrients in secondary forests had higher turnover rates (ratio of nutrient return rate to nutrient stocks in fine litter) than they did in plantations (Lugo 1992). Thus, standing stocks were smaller and fluxes were larger for a given nutrient in paired secondary forests, relative to plantations. Plantation nutrients remained locked in litter longer than they did in paired secondary forests. Mahogany plantations accumulated more litter on the forest floor (see Table 15.5), had more wood biomass, and more nutrients in vegetation and litter (see Table 15.12) than did paired secondary forests (Lugo 1992).

After Hurricane Hugo, the pattern of nutrient return to the forest floor changed, as did the nutrient turnover rate (see Fig. 15.6; Tables 15.11 and 15.12). The mahogany plantation had more nutrients in flux (Fig. 15.6) and in standing stocks (Fig. 15.8) than it did before the hurricane, and nutrients were turning over at the same rate as those in the fine litter of the paired secondary forest (Tables 15.12 and 15.13). The within-stand nutrient use efficiency of the plantation was also lower after the hurricane than before. In contrast, the litter dynamics of the secondary forest and its within-stand nutrient use efficiency was very similar to prehurricane measurements. Potassium was the exception. The data showed that the low use efficiency for K was due to large fruit fall in secondary forests. For example, in the May 30, 1995, collection, fruit fall accounted for 70% of the total K return of the paired secondary forest stand.

Fruit fall and flower fall after the hurricane were particularly high in the plantation and contributed more nutrients to the forest floor than did fruit and flower fall in the secondary forest (see Fig. 15.9). The large input was reflected in the accumulation of nutrients in fine-litter fruits and flowers (see Fig. 15.10). In both forest types, nutrient use efficiency can be influenced by mast fruit production, which returns large nutrient amounts but small biomass.

Fu et al. (1996) compared a hybrid mahogany plantation at site wh-2 with a paired secondary forest in terms of response to hurricane disturbance and found that the plantation exhibits faster rates and wider fluctuations in structural and functional parameters than does the paired secondary forest. All patterns of response to the hurricane disturbance (tree mortality, basal area, tree density, and species composition) were identical in both plantation and paired secondary forest. One important difference was that the importance value of mahogany was unchanged over the record of observation, but the importance value of other tree species changed in both plantation and paired secondary forest. Apparently, plantation structure—and some management actions such as thinning and simplification of

Table 15.13. Nutrients (kg ha^{-1}; $n = 9$) in Litter of Paired Stands at Site wh-2

Nutrient and Compartment[a]	Plantation		Secondary Forest	
	Mean	SE	Mean	SE
Nitrogen				
Leaves*	10.7	1.1	7.0	0.4
Wood*	1.5	0.1	1.1	0.2
F&F*	1.2	0.1	0.2	0.03
Total*	13.4	1.1	8.3	0.4
Phosphorus				
Leaves*	3.4	0.4	1.9	0.2
Wood*	0.6	0.05	0.3	0.03
F&F*	0.7	0.2	0.1	0.02
Total*	4.7	0.5	2.4	0.2
Potassium				
Leaves*	20.4	5.5	9.1	1.2
Wood*	4.4	0.7	1.5	0.2
F&F*	7	1.6	0.7	0.1
Total*	31.8	7.5	11.3	1.3
Calcium				
Leaves*	193	33.4	63.9	1.5
Wood*	31.4	4.8	13.3	1.7
F&F*	10.4	1.2	0.9	0.2
Total*	235	34	78	7.8
Magnesium				
Leaves*	22	3.4	11.2	1.0
Wood*	4.8	0.6	2	0.2
F&F*	2.7	0.3	0.3	0.06
Total*	29.5	4.0	13.5	1.0
Manganese				
Leaves*	2.8	0.4	6.0	0.6
Wood*	0.5	0.1	0.7	0.1
F&F*	0.1	0.02	0.03	0.005
Total*	3.4	0.4	6.7	0.6
Iron				
Leaves*	39.8	7.3	24.9	3.0
Wood*	4.3	1.8	1.9	0.9
F&F*	1.3	0.6	0.2	0.04
Total	45.4	8.1	27.0	3.7
Aluminum				
Leaves*	39.4	7.2	30.9	3.7
Wood*	3.6	1.1	2.5	1.1
F&F*	1.4	0.6	0.3	0.04
Total*	44.4	7.6	33.7	4.4
Sulfur				
Leaves*	3.0	0.6	1.8	0.2
Wood*	0.3	0.03	0.2	0.03
F&F*	0.3	0.04	0.04	0.01
Total*	3.6	0.6	2.1	0.1

[a] Means with an asterisk (*) are significantly different at $p < 0.01$.
SE, standard error; F&F, flower and fruit fall.

structure—make mahogany plantations vulnerable to disturbances such as hurricanes and shoot borers. Small-leaf mahogany plantations may become vulnerable to drought when sites are overstocked.

Medina (1995) compared the big-leaf mahogany plantation at site wb-3 with plantations of other species and secondary forests in Venezuela and found that the mahogany behaved more like a secondary forest. The old small-leaf mahogany plantations at sites ds-1 and ds-2 also resembled adjacent mature forests in terms of species richness, species composition, and physiognomy (Cintrón and Rogers 1991; Quigley 1994). The same can be said about mahogany plantations on limestone substrates (site mb-1). The Holdridge (1967) complexity index (the product of basal area, tree density, number of species, and tree height on a tenth-hectare basis times 0.001) for big-leaf mahogany plantations at that site was 40 to 45. These values were as high as those of the adjacent native vegetation (Álvarez López et al. 1983).

Comparison with Native Mahogany Forests

Comparing mahogany plantations with native big-leaf mahogany forests is difficult because data on native forests are not abundant and because the growth characteristics of mahogany in its natural range are so diverse (Lamb 1966) that any comparison would have to be site specific. The examples we cite correspond to commercial natural stands. Lamb (1966) described nearly a dozen different vegetation types where mahogany grows and reproduces successfully but with statures and stem densities different from those observed in commercial stands.

Gullison et al. (1996) described native mahogany stands in the Chimanes Forest of Bolivia and found low stem densities of seedlings and saplings ($<200\,ha^{-1}$) and trees ($<1\,ha^{-1}$). Mature trees reached 50 m in height and 2 m in diameter, with crown radii of 15 m and buttresses 3 m tall. These trees shed leaves in June and seeds in July, the driest month. Leaf and flower production peaked in September with the onset of the rainy season. More seed capsules were produced as trees attained diameters greater than 80 cm. As trees increased in size, diameter growth declined, but basal area, volume, and biomass growth increased, as did the reproductive output of the trees.

The enormous size of these trees is attributed to the low frequency of disturbance. In this Bolivian floodplain forest, flooding, sediment deposition, and erosion are the main disturbance forces. Where disturbances are more frequent, trees are limited to smaller-diameter sizes, as in Belize (Weaver and Sabido 1997) with hurricanes and the Yucatán of México (Snook 1996) with hurricanes and fires. Snook (1996), and Lamb before her (1966), pointed out that the tendency of mahogany to grow as even-aged cohorts is a result of reproduction related to particular disturbance effects.

The behavior of mahogany in plantations is consistent with its behavior in native forests, with two important caveats. Under plantation conditions, trees grow at higher density and with competition from other species; the range of environmental conditions, including disturbance regimes under which both plantations and native forests grow, make comparisons difficult at best because equal conditions must be compared by using data derived with the same methods. As the quotation from Lamb at the beginning of this chapter indicates and our results confirm, however, the broad ecological plasticity of this species is what makes it attractive for management.

Acknowledgments. This work was done in cooperation with the University of Puerto Rico and with Academia Sinica of the Peoples Republic of China. It is part of the USDA Forest Service contribution to the National Science Foundation Long-Term Ecological Research Program at the Luquillo Experimental Forest (Grant BSR-8811902 to the Institute of Tropical Ecosystems, University of Puerto Rico, and the International Institute of Tropical Forestry, USDA Forest Service). We thank M. Aide, M. Alayón, S. Brown, C. Domínguez Cristóbal, P.G. Murphy, J. Francis, J. Parrotta, N. Popper, G. Reyes, M. Rivera, A. Rodríguez, B. Ruiz, I. Ruiz, M. Salgado, M.J. Sánchez, M. Santiago, T.C. Whitmore, and G. Miller and his University of North Carolina at Asheville class (particularly S. Wright, C. McCready, E. Clarke, and K. Foxworthy) for their contributions to this manuscript.

Literature Cited

Álvarez López, M., Acevedo Rodríguez, P., and Vázquez Otero, M. 1983. Quantitative description of the structure and diversity of the vegetation in the limestone forest of Río Abajo, Arecibo-Utuado, Puerto Rico. Progress Report Project W-10. Department of Natural Resources, San Juan, PR.

Bauer, G.P. 1987. Swietenia macrophylla *and* Swietenia macrophylla *x* mahagoni *Development and Growth: The Nursery Phase and the Establishment Phase in Line Planting in the Caribbean National Forest, Puerto Rico.* Thesis, College of Environmental Sciences and Forestry, State University of New York, Syracuse, NY.

Chinea, J.D. 1990. Árboles introducidos a la reserva de Guánica, Puerto Rico. *Acta Científica* **4**:51–59.

Cintrón, B.B., and Lugo, A.E. 1990. Litter fall in a subtropical dry forest: Guánica, Puerto Rico. *Acta Científica* **4**:37–49.

Cintrón, B., and Rogers, L. 1991. Plant communities in Mona Island. *Acta Científica* **5**:10–64.

Cintrón, B., González Liboy, J.A., and Lugo, A.E. 1975. Floristic composition of Guánica Forest with respect to litter production. In *Segundo Simposio del Departamento de Recursos Naturales*, pp. 191–216. Department of Natural Resources, San Juan, PR.

Cruz, A. 1987. Avian community organization in a mahogany plantation on a Neotropical island. *Caribbean Journal of Science* **23**:286–296.

Cuevas, E., and Lugo, A.E. 1998. Dynamics of organic matter and nutrient return from litterfall of ten tropical tree plantation species. *Forest Ecology and Management* **112**:263–279.

Dugger, K.R., Cardona, J., González Liboy, J., and Pool, D. 1979. Habitat evaluations in the wet limestone forest of Río Abajo. Final report. Project W-8, U.S. Fish and Wildlife Service and Puerto Rico Department of Natural Resources, San Juan, PR.

Francis, J.K. 1989. *The Luquillo Experimental Forest Arboretum*. Research note SO-358. U.S. Department of Agriculture, Forest Service, Southern Forest Experiment Station. New Orleans, LA.

Francis, J.K. 1995. Forest plantations in Puerto Rico. In *Tropical Forests: Management and Ecology*, eds. A.E. Lugo and C. Lowe, pp. 210–223. Springer Verlag, New York.

Frederick, R. 1997. An interim report on the growth of hybrid mahogany (*Swietenia mahagoni × S. macrophylla*) using line planting methods in secondary forest in Grenada. In *Protected Areas Management*, eds. C. Yocum and A.E. Lugo, pp. 111–117. Proceedings of the eighth meeting of Caribbean Foresters at Grenada. U.S. Department of Agriculture, Forest Service, International Institute of Tropical Forestry, Río Piedras, PR.

Fu, S., Rodríguez Pedraza, C., and Lugo, A.E. 1996. A twelve year comparison of stand changes in a mahogany plantation and a paired natural forest of similar age. *Biotropica* **28**:515–524.

González, G., Zou, X., and Borges, S. 1996. Earthworm abundance and species composition in tropical croplands: comparisons of tree plantations and secondary forests. *Pedobiologia* **40**:385–391.

Gullison, R.E., Panfil, S.N., Strouse, J.J., and Hubbell, S.P. 1996. Ecology and management of mahogany (*Swietenia macrophylla* King) in the Chimales Forest, Beni, Bolivia. *Botanical Journal of the Linnean Society* **122**:9–34.

Holdridge, L.R. 1967. *Life Zone Ecology*: Tropical Science Center, San Jose, Costa Rica.

Lamb, F.B. 1966. *Mahogany of Tropical America. Its Ecology and Management*. University of Michigan Press, Ann Arbor, MI.

Lodge, D.J. 1996. Microorganisms. In *The Food Web of a Tropical Rain Forest*, eds. D.P. Reagan and R.B. Waide, pp. 53–108. University of Chicago Press, Chicago.

Lugo, A.E. 1988. The future of the forest: ecosystem rehabilitation in the tropics. *Environment* **30**(7):16–20, 41–45.

Lugo, A.E. 1992. Comparison of tropical tree plantations with secondary forests of similar age. *Ecological Monographs* **62**:1–41.

Lugo, A.E. 1997. The apparent paradox of re-establishing species richness on degraded lands with tree monocultures. *Forest Ecology and Management* **99**:9–19.

Lugo, A.E., and Frangi, J.L. 1993. Fruit fall in the Luquillo Experimental Forest, Puerto Rico. *Biotropica* **25**:73–84.

Lugo, A.E., González Liboy, J., Cintrón, B., and Dugger, K. 1978. Structure, productivity, and transpiration of a subtropical dry forest in Puerto Rico. *Biotropica* **10**:278–291.

Lugo, A.E., Cuevas, E., and Sánchez, M.J. 1990. Nutrients and mass in litter and top soil of ten tropical tree plantations. *Plant and Soil* **125**:263–280.

Marrero, J. 1950. Results of forest planting in the insular forests of Puerto Rico. *Caribbean Forester* **11**:107–147.

Medina, E. 1995. Physiological ecology of trees and application to forest management. In *Tropical Forests: Management and Ecology*, eds. A.E. Lugo and C. Lowe, pp. 289–307. Springer-Verlag, New York.

Molina Colón, S. 1986. *Estudio de la Relación Entre Tillandsia recurvata L. (Bromeliaceae) y sus Árboles Hospederos*. Thesis, University of Puerto Rico at Mayagüez, PR.

Olson, J.S. 1963. Energy storage and the balance of producers and decomposers in ecological systems. *Ecology* **44**:322–333.

Pandey, D. 1983. *Growth and Yield of Plantation Species in the Tropics*. Food and Agriculture Organization of the United Nations, Rome, Italy.

Pandey, D. 1997. Tropical forest plantation areas 1995. FAO, Rome. Report to Food and Agriculture Organization of the United Nations GCP/INT/628/UK.

Parrotta, J.A. 1992. The role of plantation forests in rehabilitating degraded ecosystems. *Agriculture Ecosystems and Environment* **41**:115–133.

Parrotta, J.A. 1993. Secondary forest regeneration on degraded tropical lands. The role of plantations as "foster ecosystems." In *Restoration of Tropical Forest Ecosystems*, eds. H. Lieth and M. Lohmann, pp. 63–73. Kluwer, Dordrecht.

Parrotta, J.A. 1995. Influence of overstory composition on understory colonization by native species in plantations on a degraded tropical site. *Journal of Vegetation Science* **6**:627–636.

Parrotta, J.A., and Turnbull, J.W., eds. 1997. Catalizing native forest regeneration on degraded tropical lands. *Forest Ecology and Management* **99**:1–290.

Quigley, M.F. 1994. *Latitudinal Gradients in Temperate and Tropical Seasonal Forests*. Dissertation, Louisiana State University, Baton Rouge, LA.

Rodríguez Pedraza, C.D. 1993. *Efectos del Huracán Hugo Sobre Plantaciones y Bosques Secundarios Pareados en el Bosque Experimental de Luquillo, Puerto Rico*. Thesis, University of Puerto Rico, Río Piedras, PR.

Sánchez, M.J., López, E., and Lugo, A.E. 1997. *Chemical and Physical Analyses of Selected Plants and Soils from Puerto Rico (1981–1990)*. Research Note IITF-RN-1. USDA Forest Service, International Institute of Tropical Forestry, Río Piedras, PR.

Scatena, F.N., Moya, S., Estrada, C., and Chinea, J.D. 1996. The first five years in the reorganization of aboveground biomass and nutrient use following Hurricane Hugo in the Bisley Experimental Watersheds, Luquillo Experimental Forest, Puerto Rico. *Biotropica* **28**:424–440.

Snook, L.K. 1996. Catastrophic disturbance, logging and the ecology of mahogany (*Swietenia macrophylla* King): grounds for listing a major tropical timber species in CITES. *Botanical Journal of the Linnean Society* **122**:35–46.

Vitousek, P.M. 1984. Litterfall, nutrient cycling, and nutrient limitations in tropical forests. *Ecology* **65**:285–298.

Wadsworth, F.H., and Englerth, G.H. 1959. Effects of the 1956 hurricane on forests in Puerto. *Caribbean Forester* **20**:38–51.

Watson, H. 1938. Imports of timber to the United Kingdom, 1937. *Empire Forestry Journal* **17**:75–79.

Watson, H. 1939. Imports of timber to the United Kingdom, 1938. *Empire Forestry Journal* **18**:77–82.

Watson, H. 1940. Imports of timber to the United Kingdom. *Empire Forestry Journal* **19**:74–76.

Weaver, P.L. 1989. Taungya plantings in Puerto Rico. *Journal of Forestry* **87**(3):37–41.

Weaver, P.L., and Bauer, G.P. 1983. Crecimiento de caoba sembrada en líneas en la Sierra de Luquillo de Puerto Rico. In *Décimo Simposio de Recursos Naturales*, pp. 31–39. Department of Natural Resources, San Juan, PR.

Weaver, P.L., and Bauer, G.P. 1986. Growth, survival and shoot borer damage in mahogany plantings in the Luquillo Forest in Puerto Rico. *Turrialba* **36**:509–522.

Weaver, P.L., and Sabido, O.A. 1997. *Mahogany in Belize: A Historical Perspective*. General Technical Report IITF-2. U.S. Department of Agriculture, Southern Forest Experiment Station, Asheville, NC.

Whittaker, R.H. 1965. Dominance and diversity in land plant communities. *Science* **147**:250–260.

Whitmore, J.L., and Hinojosa, G. 1977. *Mahogany* (Swietenia) *hybrids*. Research Paper ITF-23. U.S. Department of Agriculture, Forest Service, Institute of Tropical Forestry, Río Piedras, PR.

16. Mahogany Planting and Research in Puerto Rico and the U.S. Virgin Islands

John K. Francis

Abstract. The mahoganies are not native to Puerto Rico and the U.S. Virgin Islands but were introduced as timber and shade trees. The first known plantings of small-leaf mahogany on St. Croix, Virgin Islands, were in about 1790. Big-leaf mahogany was apparently introduced in Puerto Rico in 1904. The Pacific coast mahogany, arriving much later, has not performed well and grows only in research plots. The hybrid between big-leaf and small-leaf mahogany was first noted in 1935 in Puerto Rico. Regular nursery production of mahoganies began in the 1920s in Puerto Rico, and planting for reforestation and timber production became intense between 1936 and 1942, when about 5 million mahogany seedlings were planted. Unfortunately, only about 1200 ha of plantations remain today. A few tens of thousands of seedlings per year are now being planted, mostly as ornamentals. The Forest Service's International Institute of Tropical Forestry (the Institute) has been a principal player in mahogany research from the 1930s onward. Information gained through studies on seed handling, nursery production, planting techniques, stand establishment, agroforestry applications, thinnings, characteristics of hybrid mahogany, species site adaptability, mahogany shoot borer management, wood properties, growth and yield, nutrient and carbon cycles, damage to pavement, hurricane damage and recovery, and molecular genetics are briefly summarized. Studies sponsored by the Institute continue in provenance tests, molecular genetics, and reproduction problems.

Keywords: Mahogany, Tree planting, Research, Puerto Rico, U.S. Virgin Islands, *Swietenia macrophylla*, *Swietenia mahagoni*, *Swietenia macrophylla × mahagoni*, *Swietenia humilis*

Introduction

Through recent geologic history, the 110-km-wide Mona Channel has separated Puerto Rico from the mahogany range in the island of Hispaniola. That barrier was eventually overridden through human intervention, but most people in Puerto Rico and the U.S. Virgin Islands now think of the mahoganies as native species. The mahoganies have become part of the local heritage, if not through harvesting and wood production, at least from their familiar presence as ornamental and shade trees and from their intensive use in reforestation. Although much present interest is focused on big-leaf mahogany, the reality of the continuing scientific and conservation history demands that I include the other three New World mahoganies: the small-leaf, the Pacific coast, and the hybrid of small- and big-leaf mahogany.

History

The cultivation of mahogany began in St. Croix about 1790 with the planting of small-leaf mahogany, along Mahogany Road (Weaver and Francis 1988). The seed source of these early plantings is presumed to be Jamaica. Many of these trees are prospering today despite severe hurricanes. These giants range from 1 to 2 m in diameter and from 17 to 25 m in height. Assuming an age of 200 years, this indicates an average diameter growth of 0.63 cm yr^{-1}. These trees are the parents of thousands of ornamental shade trees in St. Croix and possibly other of the Virgin Islands, and their progeny have naturalized widely in those islands. More than 100 ha of forested hills west of Christiansted, St. Croix, have seeded from fencerow trees and now support a dominant overstory of small-leaf mahogany. Surplus and storm-felled mahogany trees supply a small industry of furniture, arts, and crafts in St. Croix.

Small-leaf mahogany was introduced and planted on sugar estates and in southern Puerto Rican towns during the Spanish colonial era. Evidence suggests that the species was already being planted during the 1860s (Renato de Grosourdy 1864; Fernández Ledón 1869). Its cultivation gradually became more widespread: at least two small groves were established before the turn of the century. A traditionally held belief is that big-leaf mahogany was introduced from Belize in 1904 by the U.S. Department of Agriculture, as part of a useful plant introduction program; it was first planted in Mayagüez, Puerto Rico. In a subsequent planting trial reported by the Department, the best trees had reached 3.3 m tall by 2 years old (May 1914). For many years, mahoganies of South American origin were thought to represent a different species (*S. candollei*) (Holdridge and Marrero 1941), but they have since been recognized as big-leaf mahogany, and they are lumped in current statistics. The Pacific coast mahogany was introduced in 1964 during provenance testing of mahoganies by personnel of the Institute of Tropical Forestry (now IITF), U.S. Department of Agriculture (Geary et al. 1973). Pacific coast mahogany has not grown well in Puerto Rico (Institute of Tropical Forestry 1963) or St. Croix, and it has not been planted outside of forestry research plots.

Hybrid mahogany trees were discovered in Puerto Rico in 1935 (Whitmore and Hinojosa 1977). The hybrid was once described as a new species in Central America (Stehle 1958), but it is now recognized as the same hybrid described in Puerto Rico. The mahogany species hybridize freely wherever they grow close to each other. As a result, this hybrid grows naturally in several places in Puerto Rico and has been used to establish plantations in Puerto Rico and St. Croix. The seed has been widely exported.

During the 1920s, Puerto Rico's Insular Forest Service produced trees for distribution to landowners; among them were at least a few mahoganies (Brush 1925b). References from the period refer to plantations of small-leaf and big-leaf mahoganies in the Susúa and Maricao Commonwealth Forests of Puerto Rico, established about 1923 (Brush 1925a). Another small block known as the Harvey Plantation, apparently of hybrids, was established in an old field in the Caribbean National Forest in 1931. Between 1936 and 1942, former agricultural lands in federal and insular forests were reforested by the Civilian Conservation Corps and the Puerto Rico Reconstruction Administration. Several species were planted, including small-leaf and big-leaf mahoganies. The small-leaf mahogany seed was either collected locally or imported from Haiti; the big-leaf mahogany seed was obtained from Panamá and, later, from Venezuela. By the end of the program, 5 million big-leaf mahogany trees had been planted on 4760 ha of land (Wadsworth et al., Chapter 17, this volume), and smaller quantities of small-leaf mahogany seedlings had also been planted. Unfortunately, because of poor planting practices and inadequate maintenance after planting, most of the seedlings of both species eventually died. By 1988, 1170 ha of mahogany plantations from all planting periods were known to exist in Puerto Rico (Francis 1995).

After World War II, mahogany plantations continued to be established, but on a much smaller scale. Sometimes, the postplanting care was effective, increasing the proportion of successful plantations. Small-leaf mahogany seedlings were planted only on dry sites, and the hybrid mahogany began to be planted more frequently both on moist and dry sites. Plantations of hybrids in St. Croix were a particular success (Weaver and Francis 1988). Many of the plantations established during the 1960s for research are mentioned later. The most recent successful development has been in the use of line planting of big-leaf and hybrid mahoganies. Just a few tens of thousands of mahogany seedlings of all types are being produced annually now. Because Puerto Rico and the Virgin Islands are now largely urban, these trees are used mostly for conservation plantings and as ornamental or shade trees.

Because little sawmilling is done in Puerto Rico (800 m^3 of all species in 1988; Francis 1995), few mahogany trees are harvested. The small amount of big-leaf, hybrid, and small-leaf volume harvested is used in the local furniture and craft industry. Many times the locally harvested volume is imported annually, principally from Brazil, for use in furniture, cabinetry, and interior trim.

Meanwhile, the mahogany trees established by all those years of reforestation continue to grow, self-thin, and sometimes die out. We are beginning to have well-dispersed dominants and superdominants with scattered groups of small-sawlog- and pole-sized trees. Mahoganies reproduce to some degree in most areas where they have been planted in the Caribbean. In the dry, southern part of Puerto Rico, in St. Thomas, and especially in St. Croix, small-leaf mahogany is slowly spreading (Francis and Liogier 1991) and becoming the most common dominant in some places. Big-leaf mahogany reproduces sparingly to abundantly in the moist and wet forests of Puerto Rico. It usually requires disturbance to reach a dominant-codominant canopy position and so succeeds as scattered trees and small groups. Hybridization continues where the big-leaf and small-leaf mahoganies grow in proximity. To obtain seed of the pure species, seed must be collected from isolated stands. The hybrids seed freely and produce progeny (F_2 and F_3 generations) much like their parents, with a mixture of traits from both parent species.

Research on Mahoganies

The Institute has conducted low-key but steady research on mahogany from its first activities to the present. The research has sometimes been pioneer and original and sometimes corroborative. Considerable effort has been made through informal trials, reported briefly in such sources as annual reports. Most of the early mahogany effort was utilitarian, on how to grow nursery seedlings and establish plantations.

Seed Production, Germination, and Seedling Establishment

The weights of mahogany seeds, numbers per capsule, and the rates and periods of germination for small-leaf, big-leaf, and hybrid mahoganies were determined and reported (Marrero 1949; Bauer 1987; Francis and Rodríguez 1993). The Venezuelan source of big-leaf mahogany was reported to have 13,000, seeds kg^{-1}, and the Panamanian source was reported to yield 20,000 seeds kg^{-1} (Holdridge and Marrero 1941). Marrero (1943) reported that big-leaf mahogany seeds lost viability abruptly after 2 months of unprotected storage at room temperature. Germination was slightly higher at room temperature in sealed containers, but viability was not retained longer. Refrigeration extended germinability to more than 8 months; refrigeration in sealed containers gave the best results.

Planting techniques have been investigated in several studies. The best depth of covering sown seeds was found to be 1.3 cm for sand and 1.9 cm for composts; seed should be sown in nursery beds at densities to achieve 65 to 85 seedlings m^{-2} (Holdridge and Marrero 1941). Although mahogany seedlings of widely varying sizes have been outplanted, no consensus has been reached as to which size is best and most successful in establishing plantations. Seedlings 0.6 to 1.5 m tall seem to be the least demanding of site conditions (Holdridge and Marrero 1941; Marrero 1950a; Bauer 1987). Spacing in solid-block plantings has not been formally tested. Most solid-block plantings were spaced somewhere between 1.8 by 1.8 m and 3 by 3 m. All plantings had to be precommercially thinned, although natural mortality by competition would do the thinning but at an unknown cost in lost growth.

Examples of some of the wisdom gained through trial and error are that small-leaf mahogany grows poorly and suffers high mortality in high-rainfall areas (Marrero 1950a) and that it requires an excessive amount of cleaning to get it above the weeds and vines in moist, fertile sites. Big-leaf mahoganies from the Venezuelan seed source grew better, and its leaves were a deeper green than those of the Panamanian source in wet forests. Mahogany seedlings have been direct-seeded and planted as bare-rooted seedlings, stumps, striplings, earth-balled seedlings, and potted in various containers. Although several successes were recorded in wet and moist forests (Marrero 1950a; Tropical Forest Research Center 1955), direct seeding did not give the new plants enough time to develop a deep root system to survive seasonal droughts and weed competition in several areas of Puerto Rico, including Guánica (Tropical Forest Experiment Station 1953), Salinas, and Vieques (Francis 1993), areas that receive 760 to 1200 mm yr^{-1} of precipitation. Bare-rooted seedlings have generally not given good survival in moist and dry forests. Striplings and stumps were better, but some form of containerized seedling was recommended (Marrero 1950a; Bauer 1987; Francis 1991).

Establishing Plantations

Plantations have been established in solid blocks in old fields and cleared secondary forests, in shelterwoods, and in cleared lines in forests. Solid-block planting in cleared areas is a perfectly sound way of establishing plantations, although attacks by the mahogany shoot borer have been somewhat more intense than with other planting systems (Holdridge and Marrero 1941). Mahogany can also be established by underplanting light shelterwood stands (Marrero 1950a); however, the overstory must be removed or severely thinned within a year or so to avoid growth loss and mortality (Holdridge and Marrero 1941). The taungya system, in which timber trees are planted with agriculture crops and allowed to grow and eventually convert the fields to forest (King 1968), was used successfully to establish big-leaf mahogany on Forest Service lands 50 to 70 years ago. Although the economic conditions necessary for this system are unlikely to occur again, the experience showed that big-leaf mahogany could be established by the taungya method and that long-term growth was roughly comparable with plantations established by other techniques (Weaver 1989). One of the most popular methods of mahogany establishment in recent years has been by line planting. This system, in which mahogany trees are planted in 2- to 3-m-wide strips cut at wide spacing through secondary forests of low economic value, has been used successfully since 1950 to establish mahogany in several hundred hectares of secondary forest in Puerto Rico (Bauer 1987).

The various planting trials have indicated rather strongly that small-leaf mahogany should not be planted in wet forest (rainfall 2000–3000 mm yr^{-1}) and only on excessively drained areas in the moist forest (rainfall 1000–2000 mm yr^{-1}). All the small-leaf mahogany plantations attempted on wet forests have failed. Plantations of small-leaf mahogany in moist forests have only been able to survive the competition of secondary forest trees on rocky hills or where given frequent cleanings. Likewise, big-leaf mahogany has not done well in the dry forest (rainfall >1000 mm yr^{-1}) (Nobles and Briscoe 1966). The Mexican and Central American provenances have done better in moist than in wet forests, particularly in limestone areas. The mahoganies have proved fairly tolerant of different topography, drainage, and soil type. Seedlings planted on severely eroded old fields, however, have almost always grown slowly and survived poorly (Marrero 1947).

The hybrid mahogany has grown faster than the small-leaf mahogany on all but the driest microsites in the region, as well as or better than the big-leaf mahogany in the moist forest, and even surprisingly well in the wet forest. Part of the explanation for this growth response is that (at least theoretically) the hybrid propagates in a classic Mendelian ratio of $1:2:1$ of big-leaf, hybrid, and small-leaf seedlings (Marquetti et al. 1975). The environment and competition tend to eliminate individuals in plantations not adapted to the site, leaving a stand attuned to local conditions.

Growth Rates

Several publications report heights, diameters, and basal areas at various ages and growth rates over time for small-leaf, big-leaf, and hybrid mahoganies (Holdridge and Marrero 1941; Wadsworth 1947; Marrero 1950b; Tropical Forest Experiment Station 1951, 1953, 1955; Wadsworth 1960; Briscoe 1962; Institute of Tropical Forestry 1963; Bauer 1987; Weaver and Francis 1988; Francis 1991). With few exceptions, big-leaf mahogany dominants and codominants average more than 1 cm of diameter growth per year through small sawlog sizes. Small-leaf mahogany dominant and codominant trees may average as little as 0.25 cm to as much as 1.5 cm of diameter growth per year over long periods, depending on site conditions and stocking. Height and diameter growth over a long period, diameter distributions, merchantable bole parameters, thicknesses of bark and sapwood, and a heartwood volume table have been produced for a 58-year-old big-leaf mahogany plantation in a moist forest area in central Puerto Rico (Wadsworth et al., Chapter 17, this volume). By use of plantation-grown hybrid mahogany trees from the Luquillo Experimental Forest in Puerto Rico, volume tables have also been developed for stem volume, branch volume, total tree volume, and local volume (Bauer and Gillespie 1990). Of numerous factors of soil and site tested, only the depth of the A_1 horizon correlated significantly with big-leaf mahogany tree height, and it was only able to account for 10% of the variation in growth rates observed (Ewel 1963).

Genetics

The hybrid mahogany has always generated a great deal of interest both in Puerto Rico and abroad and, as a result, has stimulated several scientific investigations. The hybrid is particularly attractive because it is more drought resistant and resists shoot borer better than big-leaf mahogany, has better form than small-leaf mahogany, and grows faster on some sites than either parent (Whitmore and Hinojosa 1977). Briscoe and Lamb (1962) compared leaves of seedling progeny of big-leaf, small-leaf, and putative hybrid mother trees. They concluded that big-leaf and small-leaf mahogany seedlings could be distinguished by gross measurements of leaflets that mother trees of both species and putative hybrid mother trees yielded mixed progeny and that big-leaf, small-leaf, and Pacific coast mahoganies can be distinguished on the basis of venation. Seed weight of hybrid mahogany is also intermediate between big-leaf and small-leaf mahogany (Whitmore and Hinojosa 1977). Seedlings of the hybrid and both pure species were planted in several locations in St. Croix. After 7 years, average heights were as follows: hybrids, 4.8 m; big-leaf, 4.6 m; and small-leaf, 3.0 m. In this same test, percentage mortality averaged 34% for hybrids, 63% for big-leaf, and 44% for small-leaf (Nobles and Briscoe 1966).

A recently completed molecular genetics study—partially supported by the Institute—that used random amplified polymorphic DNA markers clearly showed small-leaf, Pacific coast, and big-leaf mahoganies as distinct species, with small-leaf and Pacific coast mahoganies sharing the lowest polymorphic bands (Abrams 1995). The situation with hybrid mahogany remained enigmatic in that it shared relatively few polymorphic bands with the parent species and produced several bands not present in either parent.

In 1964 and 1965, seed collections throughout México, Central America, and the Caribbean were used to establish a provenance test from 11 collections of big-leaf mahogany, 7 of Pacific coast mahogany, and 1 of small-leaf mahogany. The provenance trial was originally established with 11 planting sites in Puerto Rico and 2 in St. Croix. Measurements and comparisons of provenances after planting and at 6 years old have been published (Geary et al. 1973). At 20 years old, three of the locations were evaluated (Glogiewicz 1986); by then, small-leaf and Pacific coast mahoganies had begun to suffer high mortality on the high rainfall sites evaluated. Sources from southern México and Nicaragua showed slight advantages over other sources. No evaluation for all currently viable sites has been made recently, but inspections showed that mortality for all species continued to be high in the wetter sites; big-leaf mahogany from the Northern Hemisphere appeared to be growing much better on the drier sites where it had been tested.

Shoot Borer

The damage caused by the mahogany shoot borer, one of the most controversial issues associated with establishing mahogany plantations in the mainland Neotropics, has not been a serious deterrent to mahogany management in Puerto Rico. Although no one denies damage by the insect, I have not encountered any reference nor do I know of any mahogany plantation that has failed from this cause alone. In fact, Whitmore (1976) claimed not to know of any trees in Puerto Rico killed by the shoot borer. He further maintained that the shoot borer causes forked trees, will attack seedlings in shade as well as in full sun, is less successful in finding its new hosts in dispersed plantings in stands, and attacks big-leaf mahogany much more frequently than small-leaf mahogany. He also recommended controlling the insect chemically in the nursery. A total of 58%, 11%, and 18% of line-planted mahoganies established in three different years in the Caribbean National Forest demonstrated some degree of shoot borer damage. No significant differences in shoot borer damage were noted by crown class or basal area class (Weaver and Bauer 1986). Total incidence of attack was not significantly different between big-leaf and hybrid mahoganies, but incidence of multiple attacks was almost three times higher in big-leaf than hybrid mahoganies (Bauer 1987). Although several other insects may feed on mahogany trees (Martorell 1975), possibly the most damaging to young trees (Bauer 1987) is the ambrosia beetle, *Xylosandus compactus* (Eichoff).

In the Urban Environment

Mahoganies (all species lumped and ranging in diameters from 10 to 100 cm), along with 11 other species growing at various distances from streets in San Juan, Puerto Rico, and Mérida, México, were evaluated for the probability of their roots damaging sidewalks and curbs. Mahogany was the third of 12 species most likely to damage sidewalks, but its probability of damaging curbs lay near the median for the 12 species evaluated. Logistic equations, with the two dependent variables tree diameter and distance to sidewalk or curb, for predicting the probability of damage to the structures are presented in Francis et al. (1996).

Effects of High Winds

In areas of high rainfall, young big-leaf mahogany trees are reported to suffer blowdown loss (Holdridge and Marrero 1941). Damage during the severe hurricanes in 1956 and 1989 indicates that large big-leaf mahogany trees are susceptible to windthrow in Puerto Rico (Basnet et al. 1992). Small-leaf mahogany appears to be much more wind resistant (Wadsworth and Englerth 1959). Francis and Alemañy (Chapter 4, this volume) found differences in resistance to storm damage between mahogany species and differences between provenances of big-leaf mahogany.

Ecosystem Studies

Big-leaf mahogany plantations 17 and 50 years old were compared with paired secondary forest stands (Lugo 1992). By 17 years of age, native tree species were invading the understory of the plantation; at 50 years old, the plantation had almost as many species as the adjacent secondary forest (Lugo 1992). The plantations had greater aboveground but lower total root biomass than did paired secondary stands. Big-leaf mahogany nutrient cycling was compared with that of nine other species of trees in small planted blocks (Lugo et al. 1990). Mahogany had a relatively small leaf mass, produced relatively little litter, and consequently accumulated low totals of nutrients in the leaves. The concentration of nutrients in leaves and litter tended to be average or higher than in the other species, however. The soil under the big-leaf mahogany was higher in total N and P and exchangeable K than was typical for the group, but not unusual in regards to pH, organic matter, total K, Ca, Mg, Mn, Fe, and Al, exchangeable Ca, Mg, Al, and Na, effective cation-exchange capacity (CEC); percentage of Al saturation, and available Mn, P, and Fe.

Wood and Wood Products

Research in Puerto Rico has yielded useful observations about mahogany wood and wood products. The heartwood of big-leaf mahoganies was rated susceptible to attack by the West Indian dry-wood termite, *Cryptotermes brevis* (Walker), but small-leaf mahogany rated highly resistant to attack in a five-tier rating system (Wolcott 1946). Untreated, small, round posts of small-leaf mahogany remained serviceable for 2.1 years before being destroyed, mostly by rot. Preservative-treated posts gave excellent service (13.8 years) (Chudnoff and Goytía 1972). A test of incising on big-leaf mahogany posts showed that it decreased drying time slightly, greatly increased penetration and retention of preservative, and resulted in moderate decreases in modulus of rupture (Chudnoff and Goytía 1967).

Institute scientists studied the variation in wood density by bole position in plantation-grown big-leaf mahoganies (Briscoe et al. 1963). Densities ranged from 0.36 to $0.65\,\mathrm{g\,cm^{-3}}$ (moisture content not given), and differences were found by height, distance from the pith, and cardinal direction on the trunk. To determine the variation in mahogany wood density, Institute personnel collected wood samples of big-leaf mahogany and Pacific coast mahogany trees from 24 locations from central México to Panamá (Boone and Chudnoff 1970). Big-leaf mahogany sapwood and heartwood averaged 0.62 and $0.61\,\mathrm{g\,cm^{-3}}$, and Pacific coast mahogany sapwood and heartwood averaged 0.76 and $0.78\,\mathrm{g\,cm^{-3}}$, at 12% moisture content. Wood densities of the big-leaf mahogany trees from which seeds were collected previously were correlated with wood densities of the progeny then growing in Puerto Rico (Chudnoff and Geary 1973). The investigators found no evidence that wood density was a strongly inherited trait in mahogany. I compared oven-dry heartwood densities of plantation-grown, small-leaf, hybrid, and big-leaf mahoganies and found mean densities of 0.58, 0.55, and $0.47\,\mathrm{g\,cm^{-3}}$. The small-leaf and hybrid densities were not significantly different from each other, but both were significantly greater than mean big-leaf heartwood density (Francis, Chapter 18, this volume).

Acknowledgments. This study was performed in cooperation with the University of Puerto Rico.

Literature Cited

Abrams, J.S. 1995. *Taxonomic Study of* Swietenia *spp. Using Restriction Fragment Length Polymorphism, Random Amplified Polymorphic DNA, and Morphological Traits.* Master's thesis, Alabama Agricultural and Mechanical University, Normal, AL.

Basnet, K., Likens, G.E., Scatena, F.N., and Lugo, A.E. 1992. Hurricane Hugo: damage to a tropical rain forest on Puerto Rico. *Journal of Tropical Forestry* **8**:47–55.

Bauer, G.P. 1987. Swietenia macrophylla *and* Swietenia macrophylla × S. mahagoni *Development and Growth: The Nursery Phase and the Establishment Phase in Line Planting, in the Caribbean National Forest, Puerto Rico.* Master's thesis, State College of New York, College of Environmental Science and Forestry. Syracuse, NY.

Bauer, G.P., and Gillespie, A.J.R. 1990. *Volume Tables for Young Plantation-Grown Hybrid Mahogany* (Swietenia macrophylla × S. mahagoni) *in the Luquillo Experimental Forest of Puerto Rico.* Research Paper SO-257. U.S. Department of Agriculture, Forest Service, Southern Forest Experiment Station, New Orleans, LA.

Boone, R.S., and Chudnoff, M. 1970. Variations in wood density of the mahoganies of México and Central America. *Turrialba* **20**(3):369–371.

Briscoe, C.B. 1962. *Tree Diameter Growth in the Dry Limestone Hills.* Tropical Forestry Note 12. U.S. Department of Agriculture, Forest Service, Tropical Forestry Research Center, Río Piedras, PR.

Briscoe, C.B., and Lamb, B. 1962. Leaf size in *Swietenia. Caribbean Forester* **23**(2): 112–115.

Briscoe, C.B., Harris, J.B., and Wyckoff, D. 1963. Variation of specific gravity in plantation-grown trees of big-leaf mahogany. *Caribbean Forester* **24**(2):67–74.

Brush, V.D. 1925a. Condición de las Plantaciones Forestales en los Bosques Insulares de Maricao y Guánica. *Revista de Agricultura de Puerto Rico* **14**(1):88–96.

Brush, V.D. 1925b. Progreso y desenvolvimiento del plan del plantel del Servicio Forestal Insular. *Revista de Agricultura de Puerto Rico* **14**(1):113–131.

Chudnoff, M., and Geary, T.F. 1973. On the heritability of wood density in *Swietenia macrophylla. Turrialba* **23**(3):359–361.

Chudnoff, M., and Goytía, E. 1967. *The Effect of Incising on Drying, Treatability, and Bending Strength of Posts.* Research Paper ITF-5. U.S. Department of Agriculture, Forest Service, Institute of Tropical Forestry, Río Piedras, PR.

Chudnoff, M., and Goytía, E. 1972. *Preservative Treatments and Service Life of Fence Posts in Puerto Rico (1972 Progress Report).* Research Paper ITF-12. U.S. Department of Agriculture, Forest Service, Institute of Tropical Forestry, Río Piedras, PR.

Ewel, J.J. 1963. Height growth of big-leaf mahogany. *Caribbean Forester* **24**(1):34–35.

Fernández Ledón, J. 1869. Reseña forestal de la Isla de Puerto Rico. Fomento de Puerto Rico, Legajo 350, Expediente 10. Archivo Histórico Nacional, Madrid, Spain.

Francis, J.K. 1991. Swietenia mahagoni *Jacq. West Indies Mahogany.* Research Note SO-ITF-SM-46. U.S. Department of Agriculture, Forest Service, Southern Forest Experiment Station, New Orleans, LA.

Francis, J.K. 1993. *Leucaena leucocephala* established by direct seeding in prepared seed spots under difficult conditions. *Nitrogen Fixing Tree Research Reports* **11**:91–93.

Francis, J.K. 1995. Forest plantations in Puerto Rico. In *Tropical Forests: Management and Ecology*, eds. A.E. Lugo and C. Lowe, pp. 210–223. Springer-Verlag, New York.

Francis, J.K., and Liogier, H.A. 1991. *Naturalized Exotic Tree Species in Puerto Rico.* General Technical Report SO-82. U.S. Department of Agriculture, Forest Service, Southern Forest Experiment Station, New Orleans, LA.

Francis, J.K., and Rodríguez, A. 1993. *Seeds of Puerto Rican Trees and Shrubs: Second Installment.* Research Note SO-374. U.S. Department of Agriculture, Forest Service, Southern Forest Experiment Station, New Orleans, LA.

Francis, J.K., Parresol, B.R., and Marín de Patiño, J. 1996. Probability of damage to sidewalks and curbs by street trees in the tropics. *Journal of Arboriculture* **22**(4):193–197.

Geary, T.F., Barres, H., and Ybarra-Coronado, R. 1973. *Seed Source Variation in Puerto Rico and Virgin Islands Grown Mahoganies.* Research Paper ITF-17. U.S. Department of Agriculture, Forest Service, Institute of Tropical Forestry, Río Piedras, PR.

Glogiewicz, J.S. 1986. *Performance of Mexican, Central American, and West Indian provenances of* Swietenia *grown in Puerto Rico.* Master's thesis, State University of New York, Syracuse, NY.

Holdridge, L.R., and Marrero, J. 1941. Preliminary notes on the silviculture of the big-leaf mahogany. *Caribbean Forester* **2**(1):20–23.

Institute of Tropical Forestry. 1963. Annual report for 1962. *Caribbean Forester* **24**(1):1–17.

King, K.F.S. 1968. Agri-silviculture (the taungya system). Department of Forestry, University of Ibadan, Ibadan, Nigeria.

Lugo, A.E. 1992. Comparison of tropical tree plantations with secondary forests of similar age. *Ecological Monographs* **62**(1):1–41.

Lugo, A.E., Cuevas, E., and Sánchez, M.J. 1990. Nutrients and mass in litter and top soil of ten tropical tree plantations. *Plant and Soil* **125**:263–280.

Marquetti, J.R., Gains, M.A., León Acosta, J.L., and Monteagudo, R. 1975. Algunos aspectos del comportamiento genético de las Swietenias. *Baracoa* **5**(1/2):1–27.

Marrero, J. 1943. A seed storage study of some tropical hardwoods. *Caribbean Forester* **4**(3):99–106.

Marrero, J. 1947. *A Survey of the Forest Plantations in the Caribbean National Forest.* Master's thesis, University of Michigan, School of Forestry and Conservation, Ann Arbor, MI.

Marrero, J. 1949. Tree seed data from Puerto Rico. *Caribbean Forester* **10**(1):11–30.

Marrero, J. 1950a. Reforestation of degraded lands in Puerto Rico. *Caribbean Forester* **11**(1):3–15.

Marrero, J. 1950b. Results of forest planting in the insular forests of Puerto Rico. *Caribbean Forester* **11**(3):107–147.

Martorell, L.F. 1975. *Annotated Food Plant Catalog of the Insects of Puerto Rico.* University of Puerto Rico, Agriculture Experiment Station, Department of Entomology, Río Piedras, PR.

May, D.W. 1914. Report of the Porto Rico Agricultural Experiment Station, 1914. U.S. Department of Agriculture, Puerto Rico Agricultural Experiment Station, Mayagüez, PR.

Nobles, R.W., and Briscoe, C.B. 1966. *Height Growth of Mahogany Seedlings, St. Croix, Virgin Islands.* Research Forest Notes ITF-10. U.S. Department of Agriculture, Forest Service, Tropical Forest Research Center, Río Piedras, PR.

Renato de Grosourdy, D. 1864. *El médico Botánico Criollo*, Vol. 2. Libreria de Francisco Brachet, París.

Stehle, H. 1958. Les mahoganys des Antilles Francaises et le *S. aubrevilleana* Stehle et Cusin. Nov. Spec. Bul. de la Soc. Bot. de France. *Memoirs* **1956–57**:41–51.

Tropical Forest Experiment Station. 1951. Eleventh annual report. *Caribbean Forester* **12**(1):1–17.

Tropical Forest Experiment Station. 1953. Thirteenth annual report. *Caribbean Forester* **14**(1/2):1–33.

Tropical Forest Research Center. 1955. Fifteenth annual report. *Caribbean Forester* **16**(1/2):1–11.

Wadsworth, F.H. 1947. The development of *Swietenia mahagoni* Jacq. on St. Croix. *Caribbean Forester* **8**(2):161–162.

Wadsworth, F.H. 1960. Records of forest plantation growth in México, the West Indies, and Central and South America. *Caribbean Forester* **21**(supplement).

Wadsworth, F.H., and Englerth, G.H. 1959. Effects of the 1956 hurricane on forests in Puerto Rico. *Caribbean Forester* **20**(1/2):38–51.

Weaver, P.L. 1989. Taungya plantings in Puerto Rico. *Journal of Forestry* **87**(3):37–41.

Weaver, P.L., and Bauer, G.P. 1986. Growth, survival and shoot borer damage in mahogany plantings in the Luquillo Forest in Puerto Rico. *Turrialba* **36**(4): 509–522.

Weaver, P.L., and Francis, J.K. 1988. Growth of teak, mahogany, and Spanish cedar on St. Croix, U.S. Virgin Islands. *Turrialba* **38**(4):309–317.

Whitmore, J.L. 1976. Myths regarding *Hypsipyla* and its host plants. In *Studies on the Shootborer* Hypsipyla grandella *(Zeller) Lep. Pyralidae*, ed. J.L. Whitmore, pp. 54–55. CATIE Miscellaneous Publication 1, Vol. 3. Centro Agronómico Tropical de Investigación y Enseñanza, Turrialba, Costa Rica.

Whitmore, J.L., and Hinojosa, G. 1977. *Mahogany* (Swietenia) *Hybrids*. Research Paper ITF-23. U.S. Department of Agriculture, Forest Service, Institute of Tropical Forestry, Río Piedras, PR.

Wolcott, G.N. 1946. A list of woods arranged according to their resistance to the attack of the West Indian dry-wood termite, *Cryptotermes brevis* (Walker). *Caribbean Forester* **7**(4):329–334.

17. Fifty-Nine-Year Performance of Planted Big-Leaf Mahogany (*Swietenia macrophylla* King) in Puerto Rico

Frank H. Wadsworth, Edgardo González González, Julio C. Figueroa Colón, and Javier Lugo Pérez

Abstract. This chapter adds to the sparse literature on the performance of planted big-leaf mahogany (*Swietenia macrophylla* King) at advanced age. The site is in latitude 18° N on a terrace over limestone receiving an average of 185 cm of rainfall annually. The growth period was from 1937 to 1996. Mixed, planted *Senna siamea* Irwin & Barnaby and *Calophyllum calaba* L. were mostly girdled between 1977 and 1984. The chapter documents mahogany dbh growth and "efficiency" (annual basal area increment in percent) relative to tree canopy position, the more rapidly growing trees, and abundant naturally regenerated mahogany and associated timber species. A stem analysis provides provisional regressions for taper and heartwood volume outside and inside bark and heartwood alone. The canopy was at 30 m. The mean dbh of the mahoganies at age 59 was 76 cm, with a maximum of 123 cm. The plantation basal area at age 59 was $23\,m^2\,ha^{-1}$, of which $22\,m^2\,ha^{-1}$ were mahogany. Inside-bark merchantable volume was about $114\,m^3\,h^{-1}$, of which about $98\,m^3\,ha^{-1}$ was heartwood. The deduced rotation to 60 cm dbh for dominant and codominant trees is between 29 and 48 years.

Keywords: *Swietenia*, Mahogany, Growth, Puerto Rico, Heartwood

Introduction

The continuing decline in the area of tropical forests (Anonymous 1993) precludes a predictable future area of natural forests for production of mahogany (*Swietenia macrophylla* King). Concentration of production in plantations, a practice long established in both hemispheres of the tropics (Nelson-Smith 1941; Griffith 1946; Voorg 1948; Barnard 1949; Marie 1949; Liu and Yang 1952; Anonymous 1956a; Busby 1967), can be expected to supply an increasing proportion of the mahogany timber available in the future. The greater yield potential of plantations as compared to mixed forests economizes land and also may relieve pressure on natural forests.

Reports on growth of young plantations of big-leaf mahogany are legion (Anonymous 1956b; González 1970; Kawahara et al. 1981; Soesilotomo 1992; Higuchi et al. 1994; Tillier 1995). There are few reports on growth to maturity and none was found on the productivity of heartwood alone, the economically attractive product.

This chapter reports on tree growth in a 59-year-old, 4-ha plantation. Using measurements of more than 250 trees, we report on tree form and relations between topography, tree crown position in the canopy, age, growth performance, and within-site natural regeneration.

Methods

Site Selection

The plantation studied is in 18° N latitude, at 200 m above sea level in north-western Puerto Rico (Fig. 17.1). It is within the Puerto Rican municipality of San Sebastian near the south shore of Guajataca Reservoir (see Fig. 17.1). The land, of the government of Puerto Rico, is held in usufruct by the Puerto Rico Council of the Boy Scouts of America (Harold Hernández, 1996, personal communication) as the Guajataka Scout Reservation.

The Holdridge (1978) life zone is subtropical moist forest (Ewel and Whitmore 1973). The mean annual rainfall at Guajataca Dam, at the same elevation 3 km northwest of the site, is 185 cm, with extremes of 132 cm and 259 cm during the 1977 to 1996 period. At San Sebastian, 8 km southwest of the site at an elevation of 170 m the mean temperature is 25°C, with extremes of 13°C and 37°C. Monthly means (Anonymous 1976–1995) range as shown in Figure 17.2.

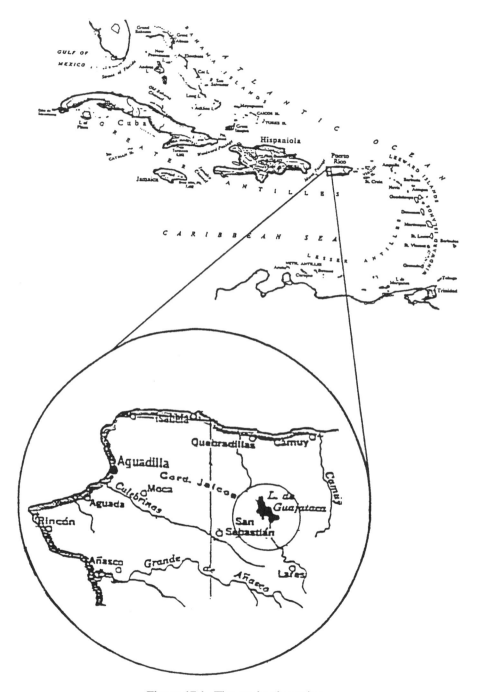

Figure 17.1. The study plantation.

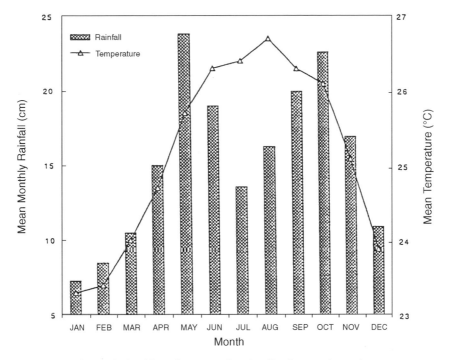

Figure 17.2. Climatic means for the Guajataca plantation.

Geologically, the region is karst, underlain by Tertiary calcareous parent material at a depth averaging 60 cm (Roberts 1942). The soil is Santa Clara clay, an Inceptosol, a fine mixed isohyperthermic, typic eutropept (Lugo-López et al. 1995). The top 15-cm layer is grayish-brown, neutral, granular plastic clay. The yellowish-brown alkaline subsoil is plastic heavy clay, which at a depth of 30 to 50 cm phases to grayish-yellow, more friable clay or clay loam (Roberts 1942).

Some 200 hectares (ha) of publicly expropriated lands around the Guajataca Reservoir, formerly mostly in sugar cane, were planted to trees by the Forest Service from 1935 to 1938 (Anonymous 1936–1942; Marrero 1950). Different tree species were planted in succession following poor survival, leading to mixed composition. Spanish cedar (*Cedrela odorata* L.), a native species, generally failed, as it had elsewhere in the Caribbean (Beard 1942). *Petitia domingensis* Jacq., also native, survived poorly and grew slowly. Other species exotic to the site were more successful: big-leaf mahogany from Venezuelan seed, *Calophyllum calaba* L., and *Senna siamea* Irwin & Barnaby. Because of the presence of surviving trees of other species, the spacing of mahogany was irregular.

By the eighth year, the largest mahogany was 30.7 cm in diameter at breast height (dbh) and 15 m tall. At age 18, the largest mahogany had a

Table 17.1. Plantation Diameter at Breast Height Distributions for the Plantation at 39 and 59 Years in Guajataca, Puerto Rico

Diameter Class (cm)	Age 39 Years	Age 59 Years		
Number of trees ha^{-1}				
10–19	1	0	41	41
21–39	11	2	43	45
41–59	14	8	4	12
61–79	16	13	0	13
81–99	4	10	0	13
>100	0	6	0	6
Totals	46[a]	39[a]	88[b]	127
Basal area ha^{-1} (m^2)	12[a]	18[a]	5[b]	23

[a] Big-leaf mahogany.
[b] Other species.

dbh of 58.2 cm and a height of 18 m, and natural seedlings had appeared (Anonymous 1956b). By that time, *S. siamea* had grown to 51 cm dbh, with a few trees competing for canopy space with the mahogany. A tabulation of the number of trees by diameter classes at ages 39 (1977) and 59 (1996) appears as Table 17.1.

Data Collection

By 1984, when the plantation was 47 years old, canopy trees of *C. calaba* and *S. senna* were girdled to leave a canopy primarily of big-leaf mahogany. In the same year, 73 mahoganies were selected in a tract of 1.61 ha (Fig. 17.3) for repeated measurement to monitor relative stem growth. These mahoganies were spot-painted at breast height and tagged, and taped for diameter. They have been remeasured during March or April of each subsequent year.

In 1981, we undertook a more detailed study of 104 mahogany trees throughout the 4.1 ha (Fig. 17.3). The trees were classified by crown position in the canopy: dominant, codominant, intermediate, and suppressed (Smith 1986). Crown radii were determined by measuring from the stump out to a suspended plumb bob beneath the periphery at four perpendicular bearings. Merchantable heights to 20 cm outside-stem diameter or to major forks were estimated with an optical range finder. Upper-stem diameters were estimated with a Wheeler pentaprism. Duplicate measurements of bark and sapwood thickness, using a bark gauge and increment borer, were made at heights attainable from the ground above the butt swell and related to corresponding taped stem diameter measurements.

In 1990, 145 naturally regenerated trees of 10 cm dbh or more beneath the plantation in the 1.61-ha area were tagged and measured for dbh. Of these, 94 were mahoganies. Their stem diameters have been remeasured in

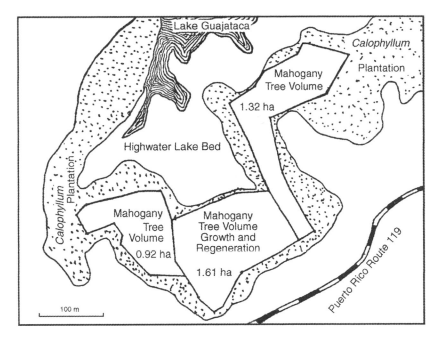

Figure 17.3. The study site layout.

every subsequent year. The result is an accounting of the original trees from age 39 to 59 years and of the associated natural regeneration for the last 6 years. The 1996 summary in Table 17.1 includes these trees.

In 1995, natural regeneration of less than 10 cm dbh was counted in five 10- by 10-m plots distributed at random in the 1.61-ha area. Two size-classes were recognized: seedlings to breast height (1.3 m) and saplings from breast height to 9.9-cm dbh. The results are summarized in Table 17.2.

In summary, for tree growth comparisons, we used 62 of the 73 planted trees that survived from 1977 to 1996 within the 1.61-ha area. Tree crown and volume measurements were based on 104 trees from the 4.1 ha. Second-generation natural regeneration of 10 cm dbh or more in 1990 included 94 mahogany trees of more than 10 cm dbh and a sample of 6840 mahogany saplings and seedlings, located in the 1.61-ha area.

Analysis

Where crown position in the canopy was recorded, growth rates were compared. The growth significance of crown diameter relative to stem diameter was explored.

Tree increment was documented in terms of dbh, stem basal-area, and efficiency, or annual basal-area increment as a percentage of mean basal area (Dawkins 1963). Heartwood diameter was deduced by subtracting

Table 17.2. Natural Regeneration in a Big-Leaf Mahogany Plantation in Puerto Rico

Species[a]	Family	Trees per Hectare		
		Seedlings[b]	Saplings[c]	Poles[d]
Timber species				
Andira inermis	Fabaceae	6,875	0	7
Calophyllum calaba	Guttiferae	30,000	320	3
Cupania americana	Sapindaceae	625	0	0
Guarea guidonia	Meliaceae	832	1,300	7
Homalium mamosum	Flacourtaceae	0	0	1
Ocotea lencoxylon	Lauraceae	208	0	0
Ocotea sintenisii	Lauraceae	625	0	2
Petitia domingensis	Verbenaceae	0	0	1
Big-leaf mahogany	Meliaceae	136,875	2,000	57
Tabebuia heterophylla	Bignoniaceae	0	0	4
Zanthoxylum martinicensis	Rutaceae	0	0	1
Subtotal		176,040	3,620	83
Nontimber species		74,583	2,000	5
Total		250,623	5,620	88
Big-leaf mahogany (%)		55	36	65

[a] Nomenclature follows Lioger and Martorell (1982).
[b] Seedlings (height ≤ 1.3 m); sampled in five 10- × 10-m plots.
[c] Saplings (dbh ≤ 9.9 cm); sample area = 1.61 ha.
[d] Poles (dbh ≥ 10.0 cm); sample area = 1.61 ha.

from the outside-bark diameter double the corresponding bark and sapwood thicknesses (Table 17.3). Upper heartwood diameters were derived in the same manner from the pentaprism estimates of upper outside-bark diameter (Table 17.4). These were then related to dbh by a regression reflecting diameter reductions due to merchantable height (Table 17.5). Merchantable stem volumes outside bark, inside bark, and of heartwood alone were computed by Huber's formula.

Because climbing or felling the trees was impractical, the relations found between stem diameter, bark thickness, and sapwood thickness in the lower part of the merchantable stem was also applied at the upper level of merchantability. The derived volume regressions, although indicative, are therefore considered only provisional. Volume expressions are from multiple regressions based on the product of dbh squared and merchantable height.

Results and Discussion

Tree Form

Tables 17.3 and 17.4 summarize findings concerning tree form. For persons familiar with this species elsewhere, the merchantable heights may appear low. Major forks rather than taper dictate merchantable heights in many of

Table 17.3. Relationships Between Diameter at Breast Height (dbh) and Merchantable Height, with Bark and Sapwood Thickness, in Plantation Grown Big-Leaf Mahogany (*Swietenia macrophylla*) in Puerto Rico

dbh (cm)	Merchantable Height[a] (m)	Bark Thickness[b] (cm)	Sapwood Thickness[c] (cm)
20	—	1.1–1.3[d]	1.7–2.3[d]
40	11.6	1.3–1.5	2.2–2.9
60	11.3	1.5–1.9	2.5–3.6
80	8.3	1.7–2.2	2.8–4.3

[a] Height to lowest major stem fork.
[b] Bark thickness (cm) = 0.0121 dbh (cm) + 0.985; $r^2 = 0.94$.
[c] Sapwood thickness (cm) = 0.026 dbh (cm) + 1.491; $r^2 = 0.79$.
[d] 95% probability limits.

the trees of this plantation. Of 104 trees, 53 have major forks in the first 10 m of bole. The mean merchantable height of all the planted trees, as seen in Table 17.3, is only about 11 m.

The forking, although commonly in the first 10 m, is nevertheless generally higher than the level of attacks by the shoot borer (*Hypsipyla grandella*). It is instead laid to the fact that the mahoganies outgrew the other tree species with which they had been interplanted, were widely separated, and thereafter fully exposed from an early age. In support of this supposition, of 88 pole-sized mahoganies in the natural regeneration beneath the planted tree overstory, only 5% have a significant fork in the first 10 m of bole. This sharp difference appears to result from the shaded environment through which the second generation has grown, discouraging side branching. Fewer shoot borer attacks under shade are also a common observation. In addition, any of these shaded trees that may have been attacked might have again formed single leaders because of the

Table 17.4. Relations Between Diameter at Breast Height (dbh) and Merchantable Height, with Heartwood Diameter in Plantation-Grown Big-Leaf Mahogany in Puerto Rico[a]

dbh (cm)	Merchantable Height (m)[b]					
	6	8	10	12	16	20
Upper heartwood diameter (cm)[c]						
40	32	26	22	20	—	—
60	45	35	30	26	21	—
80	57	45	37	32	26	22

[a] Sample size was 104 trees.
[b] Merchantable height = height to lowest major stem fork.
[c] Upper heartwood diameter (cm) = 3.79 dbh (cm)/m-ht**(m) + 6.9; $r^2 = 0.71$.

Table 17.5. Relations Between Diameter at Breast Height (dbh) (outside bark) and Merchantable Height, with Volume for Plantation-Grown Big-Leaf Mahogany in Puerto Rico[a]

dbh (cm)	Merchantable Height (m)					
	6	8	10	12	16	20
40	0.57[b]	0.77	0.95			
	0.55[c]	0.73	0.90			
	0.37[d]	0.51	0.66			
60	1.21	1.63	2.04	2.45	3.27	
	1.21	1.60	2.00	2.39	3.19	
	0.91	1.23	1.55	1.87		
80	3.10	4.13	5.16	6.19	8.26	10.32
	2.13	2.83	3.54	4.24	5.64	7.06
	1.66	2.23	2.81	3.38	4.53	5.77

[a] Sample size was 95 trees. Data are volume (m^3).
[b] Total volume = $0.567 \ [dbh \ (m)^2 \times ht \ (m)] + 0.0467$; $r^2 = 0.94$.
[c] Volume without bark = $0.549 \ [dbh \ (m)^2 \times ht \ (m)] + 0.023$; $r^2 = 0.95$.
[d] Heartwood volume = $0.448 \ [dbh \ (m)^2 \times ht \ (m)] - 0.0608$; $r^2 = 0.93$.

stimulus for vertical regrowth resulting from illumination only from directly above.

Crown spread of 51 mahoganies without major forks in the first 10 m (and therefore more like forest-grown trees) averaged 17.4 times dbh. Bark and sapwood thickness (Table 17.3) increased with the diameter of the stem. Sapwood thickness is roughly double that of bark. Supporting regressions accompany Table 17.3.

Table 17.5 shows volumes derived for the trees without major forks. The heartwood volume ranges from 67% to 82% of the inside bark volume, depending on log diameter and length. Because it is only the heartwood that is widely prized, this indication of the percentage of heartwood, particularly the increasing proportion with log size, should be of broad interest.

Stem Growth

Table 17.6 contrasts the growth of trees on different topography for the period before and after the first measurement at age 39. It is seen that, although the 59-year averages of the stem diameters are similar on the three topographic sites, the trees grew more rapidly on the slopes and in the valley during the first 39 years. In the subsequent 20 years, this relation was reversed without an obvious explanation. Increment was greater in the second period in all three topographies (ridge, +133%; slope, +49%; valley,

Table 17.6. Increment in Two Growth Periods for Plantation-Grown Big-Leaf Mahogany on Differing Topography

	Ridge	Increment (dbh)[a] (*Basal Area*)[b] Slope	Valley
0–40 years	*1.34*	*1.53*	*1.58*
	0.53	0.70	0.86
Mean final dbh (cm)	52.2	59.8	61.6
41–59 years	*1.30*	*1.03*	*0.82*
	1.24	1.04	0.97
Mean final dbh (cm)	75.6	78.3	76.3

[a] dbh increment = $cm\,yr^{-1}$.
[b] Basal area increment = $cm^2\,yr^{-1}$.

+13%). This change allowed the trees on the ridges nearly to catch up with those on the other two sites.

Table 17.7 compares mean tree growth relative to canopy position. A sharp contrast appears between the growth rate of dominants and intermediate trees, the crowns of both of which receive some direct sunlight. This is evident in their relative stem diameters attained after 43 and 59 years and even more so in their stem growth in the most recent 20 years. The dominants grew in stem diameter at more than four times the rate of the oppressed trees and even in basal area percent at triple their rate. The derived rotations presented in Table 17.7 also contrast sharply, suggesting the potential for shortening rotations by liberating immature trees to improve the canopy position of their crowns. The correlation between the 1982 crown diameter:dbh ratio and the 1977 to 1996 dbh growth proved weak, the coefficient being only 0.11.

Table 17.7. Relations Between Tree Crown Position in the Canopy and Diameter Growth at 43 and 59 Years in Plantation-Grown Big-Leaf Mahogany in Puerto Rico

Crown Canopy Position[a]	n	Mean dbh (cm) 43 years	Mean dbh (cm) 59 years	Growth[b] dbh $cm\,yr^{-1}$	Growth[b] Basal area[c] (% yr^{-1})	Rotation[d] (years)
Dominant	19	73	100	1.8	4.1	29–38
Codominant	28	58	76	1.2	3.5	36–48
Intermediate	14	43	55	0.8	3.2	46–74
Suppressed	7	28	34	0.4	1.3	66–122

[a] Crown position at age 43 years.
[b] Growth period = 15 years, between ages 43 and 59 years.
[c] Mean annual basal-area increment as a percentage of basal area at the beginning of growth period.
[d] Years needed to attain 60 cm dbh ($p = 0.67$).

Table 17.8. Diameter Increment in Different Growth Periods for Selected Fast-Growing Trees of Plantation-Grown Big-Leaf Mahogany in Puerto Rico

Age, Growth Period (years)	Tree 1	Tree 7	Tree 45	Mean
dbh (cm)				
39	98.9	65.3	77.0	58.5
57	121.9	105.3	118.3	76.8
Mean annual dbh increment (cm yr^{-1})				
0–39	2.5	1.7	2.0	1.5
42–45	1.5	1.7	2.2	1.0
45–48	0.9	2.3	2.4	1.1
48–51	1.2	3.0	2.2	1.1
51–54	1.3	2.0	2.3	0.9
54–57	1.5	3.1	3.1	1.1
39–57	1.3	2.2	2.3	1.0
Mean annual basal-area increment efficiency (%)[a]				
42–45		4.8		3.1
45–48		5.9		3.3
48–51		7.1		3.1
51–54		4.2		2.5
54–57		6.1		3.0
39–57		4.9		2.8

[a] Data are mean annual basal area increment as percent of basal area at beginning of growth period.

Table 17.8 shows the exceptional growth of three mahoganies. Each of these trees, at 59 years, is a dominant with a dbh of more than 1 m. Outstanding during the first 39 years was tree number 1, with a mean annual dbh growth of 2.5 cm. In the next 20 years, trees 7 and 45 both accelerated stem growth to exceed 2 cm per year. The mean growth efficiency (basal area percent per year) of these three trees for the 20-year period was 4.9%. Because of the premium value of large trees and the increasing percent of heartwood found with log size, this increment is even more impressive than is indicated solely by stem growth.

Natural Regeneration

Table 17.2 summarizes diversity in the natural regeneration in the 1.61-ha plantation based on a count of all trees of more than 10 cm dbh and a 2.5% sample in five 10- × 10-m plots for those smaller. About a dozen timber tree species are in the regeneration. Of the 88 poles of more than 10 cm dbh per hectare, 57 poles, almost two-thirds, are mahoganies. This density of mahogany pole-sized trees is more than double the number of the overstory mahoganies. The sampling of seedlings was taken in August, a season when the current year's mahogany seedlings form a carpet (27-cm average spacing) under the trees, which largely disappears during the dry season the

Tabla 17.9. Growth of Selected Big-Leaf Mahogany Regeneration Under Plantation-Grown Big-Leaf Mahogany in Puerto Rico

dbh (cm)		Growth		
		dbh	Basal area[a]	
1990	1996	(cm yr^{-1})	(% yr^{-1})	Tree Code
	17.9	1.0	13.2	45 fd
12.0	19.5	1.2	15.0	62 g
12.7	21.9	1.5	20.0	62 r
13.7	21.0	1.2	13.4	3 c
16.3	22.6	1.0	10.5	46 u
17.0	23.5	1.1	1.0	62 l
17.3	24.8	1.2	11.4	45 e
18.6	25.5	1.2	10.2	62 t
20.4	29.2	1.5	11.4	62 h
20.9	30.9	1.7	12.4	62 p
25.8	37.6	1.9	12.0	45 fc
32.2	43.6	1.9	9.8	45 ga

[a] Mean annual basal-area increment as a percentage of basal area at the beginning of the growth period.

next spring. Nevertheless, the saplings above breast height, 2000 ha^{-1}, are less prone to disappear and constitute additional abundant potential replacement of mahogany.

By 1996, natural regeneration within the plantation now produced a good stand of straight mahogany poles up to 30 cm dbh and 20 m in height. Although mahogany is at home with complete exposure, these regenerating mahoganies beneath a mahogany overstory have been growing rapidly. Table 17.9 summarizes 6-year growth records for 12 selected trees initially between 12 and 32 cm dbh. Their average growth in dbh is more than 1 cm yr^{-1} and in basal area more than 10% yr^{-1}.

Of the saplings, about two-thirds are of species other than mahogany, foretelling a potential for biodiversity enrichment. The regeneration of native timber tree species was 39,165 seedlings, 1620 saplings, and 26 poles per hectare.

Plantation Performance

Tables 17.1 and 17.2 summarize the performance of the trees as a plantation. Having started mixed, only recently has the canopy become nearly continuous with crowns of mahoganies. Table 17.2 shows the degree of remaining mixture. A regular progression of trees through the diameter classes during the 20-year period of measurement is evident.

Table 17.1 further shows the plantation's development by 1977 and by 1996. The distance between all trees in 1996 averaged about 9 m and, between planted mahoganies alone, about 16 m. The basal area has been within the range of 15 to $25 \, \mathrm{m^2 \, ha^{-1}}$, a productive level for broadleaf tropical forests (Silva 1989). Applying the average ratio of crown diameter to stem diameter at breast height (17.4:1), by 1996 the mahogany crowns covered about 70% of the forest area.

Table 17.10 shows volume increment in the plantation during three periods of measurement. During the first 39 years, half of which came before the mahoganies reached 30 cm dbh, mean annual increment per unit of forest area was low. During the next 20 years, increment almost doubled, raising the mean for the 59 years. Of the total volume of mahogany, 76% was heartwood; of the inside-bark volume, it was 79%. If, instead of a basal area of $23 \, \mathrm{m^2 \, ha^{-1}}$, the plantation had attained a possible $30 \, \mathrm{m^2 \, ha^{-1}}$, the mean annual yield of heartwood might have reached $3 \, \mathrm{m^3 \, ha^{-1} \, yr^{-2}}$.

Table 17.11 presents a compendium of mahogany increment from several sources. From these plantations, mean annual basal-area increment of dominant and codominant trees could be compared. The growth of the trees in the Guajataca plantation compares well with the others listed.

An indication of the site productivity potential is seen in a prediction published for this species in the Philippines (Revilla et al. 1976). Tree height at Guajataca at age 59, about 30 m, corresponds to at least their site 25 (25 m at age 40). Fully stocked, this site in the Philippines was capable of producing annually $10^3 \, \mathrm{ha^{-1}}$ inside bark from the 25th to the 40th year. Inside-bark annual yields on this site, predicted there by the 40th year, average $6.6 \, \mathrm{m^3 \, ha^{-1}}$. Were this to be 75% heartwood, the annual heartwood production potential would be about $5 \, \mathrm{m^3 \, ha^{-1}}$.

Table 17.10. Volume Increments for Plantation-Grown Big-Leaf Mahogany in Puerto Rico

Growth Period (years)	Big-Leaf Mahogany	Other Species	Total	Annual Yield
Outside bark				
39	77		77	2.0
39–58	42	40	82	4.3
59	119	40	159	2.7
Inside bark				
39	73		73	1.9
39–58	41	34	75	3.9
59	114	34	148	2.6
Heartwood				
39	56		56	1.4
39–58	34	13	47	2.5
59	90	13	103	1.8

Data are volume $(\mathrm{m^3 \, ha^{-1}})$ for individuals >30 cm dbh.

Table 17.11. Basal-Area Increments for Dominant and Codominant Plantation-Grown Big-Leaf Mahogany in Puerto Rico[a]

Age (years)	Elevation (m)	Annual Rainfall (cm)	Basal-Area Increment $(cm^2\,yr^{-1})$	Natural Regeneration	Location
20	180	230	51	None	Puerto Rico
20	110	250	63	Scarce	Perú
24	200	200	75	Abundant	Puerto Rico
24	100	300	36	Abundant	Perú
26	200	190	33	Abundant	Puerto Rico[b]
28	80	170	41	—	Philippines[c]
30	10	220	34	Scarce	Belize
32	480	400	61	Abundant	Martinique
36	480	400	59	Scarce	Martinique
43	200	190	75	Abundant	Puerto Rico[b]
44	150	300	50	Abundant	Perú
59	200	190	100	Abundant	Puerto Rico[b]

[a] Adapted from Wadsworth (1960).
[b] This study.
[c] Valera (1962).

Conclusions

- Forking in the open-grown mahogany in the first 10 m of bole characteristic of the open-grown planted trees is not found in natural mahogany regeneration beneath the old trees.
- Heartwood in mahogany trees averaging 76 cm dbh was about 75% of the inside-bark volume.
- The site, despite shallow soil, produces trees to 30 m in height, possibly capable of producing $5\,m^3\,ha^{-1}\,yr^{-1}$ of mahogany heartwood.
- Canopy position of tree crowns is sharply related to tree growth. Early liberation is suggested.
- Growth data support 60 cm dbh mahogany rotations between 29 to 48 years for trees that attain dominant or codominant canopy positions.
- Individual trees of consistent, exceptionally rapid growth, 2 cm dbh per year, were found. The cause merits more research for a full explanation.
- Regeneration of mahogany beneath a thinned plantation has been profuse and has produced trees 30 cm dbh, some of which have been growing in basal area at an annual rate of nearly 5%.

Acknowledgments. This study was conducted in cooperation with the University of Puerto Rico. The authors are indebted to many supporters; first, to the Puerto Rico Council of the Boy Scouts of America for preserving the plantation and making it available for the study, then to some 300 16-to 20-year-old members of the Boy Scout Nature Team who, during the past 18 years, counted, measured, and liberated the mahoganies of the plantation. Jeffrey Glogiewicz, then graduate student at the University of Michigan, mapped the plantation and participated in the study of crown sizes and canopy position classification. Carlos Rivera of the International Institute of Tropical Forestry (since retired) assisted with identifying the regeneration samples. The 1995 and 1996 rainfall records were supplied by Leandro A. Faura of the Puerto Rico Autoridad de Energía Eléctrica. Computation support was provided by María Rivera and Brynne Bryan of the International Institute of Tropical Forestry.

Literature Cited

Anonymous. 1936–1942. Public forest planting records. Caribbean National Forest/Puerto Rico Reconstruction Administration/Puerto Rico Department of Agriculture. Stored in the library of the International Institute of Tropical Forestry, Río Piedras, PR.

Anonymous. 1956a. Comparative growth of *Swietenia mahagoni* and *S. macrophylla* in the Seychelles. Victoria, Seychelles, Seychelles Department of Agriculture Annual Report, 1955.

Anonymous. 1956b. The broadleaf mahogany (*Swietenia macrophylla* King). *Caribbean Forester* **17**(1/2):4–5.

Anonymous. 1976–1995. Climatological data, Puerto Rico and the Virgin Islands, Vols. 22–41. National Oceanic and Atmospheric Administration (NOAA), National Climatic Center, Asheville, NC.

Anonymous. 1993. Forest resources assessment, 1990. Tropical countries. FAO Forestry Paper 112. United Nations Food and Agricultural Organization, Rome, Italy.

Barnard, R.C. 1949. Promising exotics. *Malayan Forester* **12**(4):212.

Beard, J.S. 1942. Summary of silvicultural experience with cedar *Cedrela odorata* Roem in Trinidad. *Caribbean Forester* **3**(3):91–102.

Busby, R.J.N. 1967. Reforestation in Fiji with large-leaf mahogany. In *Papers from the Ninth British Commonwealth Forestry Conference*, 1968, New Delhi, India. Suva, Fiji, Department of Forestry, Research Division, Suva Fiji.

Dawkins, H.C. 1963. *The Productivity of Tropical High Forests and Their Reaction to Controllable Environments*. Doctoral dissertation, University of Oxford, Department of Forestry, Commonwealth Forestry Institute, Oxford, UK.

Ewel, J.J., and Whitmore, J.L. 1973. *The Ecological Life Zones of Puerto Rico and the U.S. Virgin Islands*. Forestry Research Paper ITF-18. U.S. Department of Agriculture, Forest Service, Institute of Tropical Forestry, Río Piedras, PR.

González, R.M. 1970. The yield of forest plantations in the tropics. Universidad Nacional Agraria La Molina (Perú). Departamento de Publicaciones **3**(1/2):109–121.

Griffith, A.L. 1946. The stomates and early growth of some timber trees of the Malabar Coast. Indian Forest Records (n.s.). *Silviculture* **6**(2):62–92.

Higuchi, N., dos Santos, J.M., Imanaga, M., and Yoshida, S. 1994. The aboveground biomass estimate for Amazonian dense tropical moist forests. Kagoshima University (Japan). *Memoirs of the Faculty of Agriculture* **30**:43–54.

Holdridge, L.R. 1978. Ecología basada en zonas de vida. International Institute of Agricultural Sciences, San Jose, Costa Rica.

Kawahara, T., Kanazawa, Y., and Sakurai, S. 1981. Biomass and net production of man-made forests in the Philippines. *Journal of the Japanese Forestry Society* **63**(9):320–327

Lioger, H.A., and Martorell, L.F. 1982. Flora of Puerto Rico and adjacent islands: a systematic synopsis. Editorial de la Universidad de Puerto Rico, Río Piedras, PR.

Liu, S.H., and Yang, P.L. 1952. Studies on the increment of four introduced species in Heng-Chun. *Taiwan Forest Research Institute Bulletin 30*.

Lugo-López, M.A., Beinroth, F.H., Vick, R.L., Acevedo, G., and Vázquez, M.A. 1995. Updated taxonomic classification of the soils of Puerto Rico. Mayagüez, Puerto Rico, University of Puerto Rico, Mayagüez Campus, College of Agricultural Sciences, Agricultural Experiment Station, Mayagüez, PR.

Marie, E. 1949. Afforestation with *S. macrophylla*. *Caribbean Forester* **10**(3): 205–222.

Marrero, J. 1950. Results of forest planting in the insular forests of Puerto Rico. *Caribbean Forester* **11**(3):107–147.

Nelson-Smith, J.H. 1941. The formation and management of mahogany plantations at Silk Grass Forest Reserve. *Caribbean Forester* **3**:75–78.

Revilla, A.V., Bonita, M.L., and Dimapilis, L.L. 1976. A yield prediction model for *Swietenia macrophylla* King plantations. *Pterocarpus* **2**(2):172–179.

Roberts, R.C. 1942. *Soil Survey of Puerto Rico*. U.S. Department of Agriculture and the University of Puerto Rico Agricultural Experiment Station, Río Piedras, PR.

Silva, J.N.M. 1989. The behaviour of the tropical forest of the Brazilian Amazon after logging. PhD thesis, Green College, Oxford, UK.

Smith, D.M. 1986. The practice of silviculture, eighth ed. John Wiley & Sons, New York.

Soesilotomo, P.S. 1992. The potential of mahogany in and outside forest areas in East Java. *Duta-Rimba* **18**:141–142.

Tillier, S. 1995. Bigleaf mahogany in Martinique. *Bois et Forêts des Tropiques* **244**:55–66.

Valera, M.Z. 1962. The study of the diameter growth of thinned large-leaf mahogany (*Swietenia macrophylla* King) stand in the Makiling National Park. *Philippine Journal of Forestry* **18**(1/4):29–39.

Voorg, C.N.A. de. 1948. The Janlappa forest plantations. *Tectona* **38**(2):63–76.

Wadsworth, F.H. 1960. Records of forest plantation growth in Mexico, the West Indies, and Central and South America. *Caribbean Forester* **21** (Supplement).

18. Wood Densities of Mahoganies in Puerto Rican Plantations

John K. Francis

Abstract. The wood density of mahogany is a critical determinant of its working properties and strength. The wood densities of the big-leaf and small-leaf mahogany are known principally from old-growth trees rather than from the plantation-grown trees now beginning to enter the market. The wood densities of their hybrid have not been reported. Wood density samples of heartwood of three trees each from six plantations of the two species and their hybrid (54 sample trees) were collected by auger across Puerto Rico. These species tended to be planted in somewhat differing rainfall regimes. Mean annual precipitation of the plantation areas ranged from 810 to 2800 mm yr^{-1}, and the plantation ages ranged from 16 to 59 years. The heartwood densities were analyzed and compared, using analysis of covariance and Tukey's studentized range test. The class variable, species, and the continuous variable, mean annual precipitation, were significant, but age was not. The means and standard deviations for oven-dry heartwood densities of small-leaf mahogany (0.577 ± 0.059 g cm^{-3}) and the hybrid (0.546 ± 0.060 g cm^{-3}) were not significantly different from each other, but were significantly greater than those for big-leaf mahogany (0.470 ± 0.052 g cm^{-3}).

Keywords: Mahogany, Heartwood density, Puerto Rico, Plantations, *Swietenia*

Introduction

Wood densities from old-growth trees of big-leaf and small-leaf mahogany in various countries are available (Heck 1937; Kynoch and Norton 1938; Swabey 1941; Chudnoff and Geary 1973) and, to a limited extent, for plantation-grown trees in Puerto Rico (Briscoe et al. 1963). Little attention had been paid to the wood density of hybrid mahogany and how it compares to those of the parent species. Because wood density is such an important determinant of wood machinability and strength, the heartwood density of the three types of mahogany grown in Puerto Rico is of great interest to managers.

Methods

Six plantations, each of the three mahogany taxa, were sampled (Fig. 18.1). Heartwood was collected from three trees in each plantation, for a total of 54 trees sampled. The samples were collected and their densities determined by the auger method (Francis 1994) because it is nondestructive, cheap, and simple. Briefly described, sample trees were bored with an auger bit until a color change in the chips indicated that the heartwood had been reached. The depth of the hole was measured, boring was continued, and the chips collected in a plastic bag until a depth of one-half the tree diameter was reached, and then the depth of the hole was measured again. At each step, the hole was cleaned of residual chips with a small spatula. The chips were oven-dried and the wood density calculated from the dry weight of the chips extracted and the volume of the hole from which they were taken.

Ages of the plantations ranged from 16 to 59 years; a range of sizes and crown classes were included in the sample. Mean annual precipitation in the plantation areas varied from 810 to 2800 mm yr^{-1}. All analyses used SAS Institute (1988) statistical software. Analysis of covariance was used to establish significance of the treatment (mahogany type), using type as a class variable and age and mean annual precipitation as continuous covariants. Minimum acceptable probability of error was set at 5%. I used Tukey's studentized range test to distinguish differences in means of heartwood density among treatments.

Results and Discussion

Mahogany type proved to be a highly significant determinant of heartwood density. Because small-leaf mahogany tended to be planted (and to have survived) in drier areas and big-leaf mahogany in wetter areas, mean annual precipitation was a highly significant covariant (Table 18.1). Precipitation seemed to be less important to hybrid mahogany, which can grow fairly well across the range of rainfall tolerated by the genus. Age did not significantly affect mahogany heartwood density. Bartlett's test for homogeneity of variance (Snedecor and Cochran 1967) indicated that differences do exist among treatments (mahogany types).

A mean standard error of 0.002664, and a critical value for the studentized range of 3.418, gave a minimum significant difference of 0.0416. The

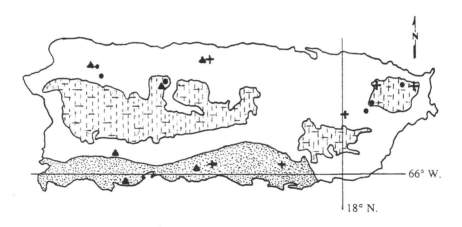

Figure 18.1. Plantations sampled during the study in the rainfall regimes of Puerto Rico.

Table 18.1. Analysis of Covariance of Heartwood Densities of Samples from Small-Leaf, Hybrid, and Big-Leaf Mahogany Grown in Plantations in Puerto Rico

Sources	df	Sum of Squares	Mean square	F value	Probability
Model	(4)	(0.156065)	(0.039916)	14.64	0.0001
Type	2	0.109580	0.054790	20.56	0.0001
Age	1	0.001856	1.001856	0.70	0.4080
Rainfall	1	0.144629	0.044629	16.75	0.0002
Error	49	0.130550	0.002664		
Total	53	0.286615			

mean heartwood densities ($g\,cm^{-3}$) and their Tukey groupings are as follows (means followed by the same letter are not significantly different):

Small-leaf	Hybrid	Big-leaf
$0.577 \pm 0.059a$	$0.546 \pm 0.060a$	$0.470 \pm 0.052b$

Heartwood densities of the hybrid mahogany fall between the parent species, but the hybrid densities were more similar to small-leaf than big-leaf mahogany. The hybrid appears to be somewhat more variable than at least the small-leaf parent. Confirming what is already generally accepted by managers in the West Indies, nothing about the heartwood densities of plantation-grown hybrid mahoganies would discourage their continued use in planting programs.

Acknowledgments. This study was performed in cooperation with the University of Puerto Rico.

Literature Cited

Briscoe, C.B., Harris, J.B., and Wyckoff, D. 1963. Variation of specific gravity in plantation-grown trees of bigleaf mahogany. *Caribbean Forester* **24**(2):67–74.

Chudnoff, M., and Geary, T.F. 1973. On the heritability of wood density in *Swietenia macrophylla*. *Turrialba* **23**(3):359–361.

Francis, J.K. 1994. Simple and inexpensive method of extracting wood density samples from tropical hardwoods. *Tree Planters' Notes* **45**(1):10–12.

Heck, G.E. 1937. *Average Strength and Related Properties of Five Foreign Woods Tested at the Forest Products Laboratory.* Laboratory Report R1139. U.S. Department of Agriculture, Forest Service, Forest Products Laboratory, Madison, WI.

Kynoch, W., and Norton, N.A. 1938. *Mechanical Properties of Certain Tropical Woods, Chiefly from South America.* Bulletin 7. School of Forestry and Conservation, University of Michigan, Ann Arbor, MI.

SAS Institute. 1988. *SAS/STAT User's Guide.* Release 6.03. SAS Institute, Inc., Cary, NC.

Snedecor, G.W., and Cochran, W.G. 1967. *Statistical Methods*, sixth edition. Iowa State University Press, Ames, IA.

Swabey, C. 1941. *The Principal Timbers of Jamaica.* Bulletin 29 (New Series). Department of Science and Agriculture, Kingston, Jamaica.

19. Evaluation of Four Taungya Permanent Big-Leaf Mahogany Plots, Aged 36 to 38 Years, in Belize

Richard C. Ennion

Abstract. From 1955 to 1964, 591 ha of big-leaf mahogany plantations were established by the taungya agroforestry system in Columbia River Forest Reserve, Belize, Central America. Eleven permanent plots were originally established and 4 have survived: 2 in the 1956 plantation and 2 in the 1958 plantation. The diameter and height of individual trees in these plots were periodically remeasured until 1994, when the trees were 36 to 38 years old. The permanent plots are all unthinned, so rates of stocking have remained fairly constant throughout the period. Results showed that diameter mean annual increment (MAI) per plot at 36 and 38 years was 0.7 to $0.8 \, \text{cm} \, \text{yr}^{-1}$ and that diameter periodic annual increment (PAI) has been decreasing in all plots since age 20 to 25 years. Annual volume MAI per plot at 36 and 38 years ranged from 4.3 to $5.3 \, \text{m}^3 \, \text{ha}^{-1} \, \text{yr}^{-1}$ in the more densely stocked stands (252–288 stems ha^{-1}) to 2.0 to $2.6 \, \text{m}^3 \, \text{ha}^{-1}$ in the less densely stocked stands (138–119 stems ha^{-1}). Plot mean diameter growth was unaffected by stocking density, which suggested that between-tree competition in the most densely stocked stand (288 stems ha^{-1}) was not sufficiently intense to depress mean diameter growth. On the better site, rotation length (maximum yield rotation) was about 38 years.

Keywords: *Swietenia macrophylla*, Mahogany, Rotation, Stocking, Taungya, Competition, Diameter, Permanent, Volume, Belize

Introduction

From 1955 to 1964, 591 ha of big-leaf mahogany plantations were established by the taungya agroforestry system in Columbia River Forest Reserve, Belize, Central America. Eleven permanent plots were originally established in the plantations. This chapter describes the results from 4 surviving permanent plots that have been remeasured 7 to 11 times, last in 1994. The objective of the permanent plots, at the time of establishment, was to collect data on big-leaf mahogany growth by measuring individual tree height and diameter (Research Division File Note 1963).

Taungya (a Burmese word meaning a cultivated hill plot) is a tropical agroforestry system with a long history (Weaver 1989). During the 1940s and 1950s, taungya systems were used to establish big-leaf mahogany plantations on a large scale in Belize and many Caribbean countries (Mayhew and Newton 1998). Mahogany is typically planted with crops of maize and bananas, which are cultivated for a few years before tree growth precludes agricultural production (Mayhew and Newton 1998). The system gives good results because the plantations are well managed during the first 2 years or while crop production is maintained (Soubieux 1983), a critical period for establishing mahogany seedlings (Mayhew and Newton 1998).

Continuous inventories of big-leaf mahogany plantations (with two or more sets of stand measurements from the same area) have been reported from several countries around the world (Mayhew and Newton 1998); these authors identified several common deficiencies that have prevented reliable use of a significant portion of the data to describe relations between key variables, however. The deficiencies included unorthodox silvicultural treatments, management that led to long-term reduction of tree productivity, and the use of nonstandard parameters. The data from the permanent plots, although variable in several respects, have none of the deficiencies identified by Mayhew and Newton. Unfortunately, despite the reliability of the data, the small number of permanent plots that survived prevents using the information to prepare reliable yield models. The data could be used, however, to indicate volume production and mean diameter for a range of stocking densities and site types over a 38-year period. The permanent plots were also used to gauge the success of taungya as an agroforestry system suitable for establishing big-leaf mahogany plantations in Belize.

The Study Site

The Columbia River Forest Reserve is in southern Belize in the Toledo District. Guatemala borders to the south and west, and the Caribbean Sea is to the east. The four surviving big-leaf mahogany permanent plots are close to the southern border of the Reserve. Two permanent plots are in the 1956 plantation (P-1 and P-4), and two are in the 1958 plantation (P-2 and P-3). Altitude varies from 20 to 400 m (King et al. 1986).

Permanent plots P-1 and P-4 (1956 plantation) are in the Toledo Uplands Land System (Uplands). The main rock type consists of mudstone, sandstone, limestone, and occasional conglomerates, although sandstone dominates because of heavy weathering of the mudstone. The pattern of soils in the Upland land system is complex. Soils are mainly dark or reddish brown, of intermediate depth, well drained, moderately fertile, fine to medium texture, leached, and acidic (but not excessively so, with topsoil pH 6 and subsoil rarely less than pH 5.2). Base saturation rarely exceeds 50% (King et al. 1986).

Permanent plots P-2 and P-3 (1958 plantation) are in the Xpicilha Hills Land System (the Hills). The main rock type is Cretaceous limestone. Karst landscape dominates, with densely dissected and steeply sloping hills. The 1958 plantation is in a very limited area of rolling topography, which does not contain karst development. Generally, soils in the Hills are shallow and tend to have high expandable clay content. They are typically very stony within 30 cm of the surface, and the underlying material is limestone rubble. Pockets of fine earth may be washed down to considerable depth and can be exploited by tree roots. Soils on the rolling topography are usually deeper, although limestone is still likely to be encountered at a depth of less than 1.5 m. The pH is nearly neutral, and base saturation is high (King et al. 1986).

At 16°N latitude, the study area is in a region where the trade winds predominate and the climate is tropical moist-wet (King et al. 1986). Mean annual rainfall for the Reserve is about 3000 mm yr^{-1}, with a wet season in which 90% of the annual rainfall extends from May to January, July being the wettest month. Annual deviations in total rainfall are small but monthly deviations are more marked. The 21-year mean monthly maximum temperature for May is 31°C and 27°C for December. The mean monthly minimum temperature for February is 31°C, and for June, 22°C. Maximum temperatures rarely exceed 38°C or fall below 7°C. Humidity, at 80%, remains high all year (King et al. 1986). Belize is periodically subject to hurricanes and tropical storms (Friesner 1993). The 1956 plantations are the oldest under study; since they were established, eight hurricanes and one tropical storm have struck Belize. Of these, three hurricanes have affected the study area (Abby in 1960, Hattie in 1961, and Francelina in 1969). Information on these hurricanes and their effects on the plantations is incomplete. The 1961 Forest Department annual report describes damage by Hattie in the Reserve as being slight and concentrated in pockets. Unfortunately, the effects of the other hurricanes on the Reserve were not recorded. The plantations may be assumed to have suffered little permanent damage, however.

Six ecological zones have been identified in Belize (Hartshorn et al. 1984). The Reserve falls close to the boundary between two of them: tropical wet-transition to subtropical, and subtropical wet. Wright's vegetation classification (which is a function of climate, soil type, and drainage) was used to describe the natural vegetation in the study area (Wright et al. 1959). Two vegetation types affect the plantations. Natural vegetation in the 1958 plantation is broadleaf evergreen forest with occasional lime-loving species. The 1956 plantation falls on the boundary between two types, that found in the 1958 plantation and an area of evergreen seasonal broadleaf forest with few lime-loving species. The forest vegetation before the plantations were established included big-leaf mahogany, *Terminalia amazonica*, and *T. sapotilla* (Research Division File Note 1963).

The Big-Leaf Mahogany Plantations

From 1955 to 1964, 591 ha of big-leaf mahogany were planted by the Forest Department with the taungya system in an area adjacent to the southern boundary of the Reserve. The Department annually licensed an area of land to the local Indians, who cut and burned the existing vegetation and planted maize in the prepared site. Mahogany seeds were sown into the maize crop, but the precise method varied from year to year. Between 1955 and 1957, the mahogany seeds were sown at a rate of two seeds per planting hole (peg), with 750 to 900 pegs ha^{-1}. In 1958, the seeds were sown in groups of six or eight seeds with seed spaced about 45 cm apart; group density varied considerably, from 120 to 544 groups ha^{-1}. This system was thought likely to give a better distribution of elite stems compared with line planting because growth rates in big-leaf mahogany were so variable. Group planting proved to be very successful, so it was used for all subsequent planting (Belize Forestry Department Annual Reports 1958).

The plantations were weeded by hand from 1955 to 1964. The 1955 to 1957 plantations were cleaned four times during this period, but younger plantations were cleaned only twice. The Department's annual report does not say whether this difference was due to less weed competition or to financial constraints. After 1964, the plantations received no further maintenance. None of the plantations have received any silvicultural treatments since they were established; they were then unthinned.

The Big-Leaf Mahogany Permanent Plots

Plot History

Records describing the history and maintenance of the 11 permanent plots are unfortunately incomplete; specifically, details on the sampling design used to locate plots in the plantations were not described.

Eleven permanent plots were originally established; 6 were abandoned in 1973 because of poor growth and survival. After 1973, data collection continued in 5 of the permanent plots because they were well stocked and growing well, and the data collected were believed to be reliable. These 5 permanent plots were periodically remeasured until 1994. Permanent plot CR26B is 1.25 ha, but only a sample of trees were measured since its establishment. Unfortunately, the sampling rationale was not recorded. The data from this plot usefully chart the growth of individual trees but cannot be used to predict yield per unit area. Consequently, I have not included the data from this plot.

All plots were cleaned by cutting all nonwoody vegetation before measurement. Several other tree species that became established between plot measurements were left uncut; thus, by 1994, most plots contained several large hardwood species. To facilitate the 1994 remeasurement, all vegetation less than 10 cm in diameter was cut. Additional plot maintenance consisted of repainting the tree identification numbers and diameter bands on the bole of each tree. Bands, painted at 1.3 m from the ground, were used to locate the precise point of diameter remeasurement on each tree. New identification signs and plot markers were also erected in each plot.

Like the mahogany plantations, none of the permanent plots have received any silvicultural treatments, except for the periodic removal of small vegetation. No plots have been thinned.

Description of the Permanent Plots

Plots P-1, P-2, and P-3 are 0.2 ha (40.2 m by 50.3 m) each; plot P-4 is 0.12 ha (40.2 m by 30.2 m). Before the permanent plots were established, they were cleaned to release the mahogany from competing vegetation. Elevation of plots is about 180 m.

Plot P-1, established in May 1963 in the 1956 plantation, is directly adjacent to P-4. Big-leaf mahogany seed were sown at the rate of two seeds per peg, with pegs spaced at 3.05 m by 3.66 m, or 896 pegs ha^{-1}. The Forest

Department's 1956 annual report gives survival at 36%; underripe seeds were thought to be partly responsible for this low figure, but stocking at 322 stems ha^{-1} was considered adequate. In 1994, the plot contained two mature Spanish cedars, which may have been planted during the establishment phase to replace lost mahogany. Before plots were established, three mahoganies were cut at the ground because of poor form, and five 3-m-tall trees were not measured. By 1994 (13 years after the previous measurement), an understory of a palm (*Orbignya cohune*) and *Spondias mombin* had developed, and one *Ficus* sp. was dominant.

Plot P-2, established in May 1963 in the 1958 plantation, appears to have been line-planted, but the original plot description does not confirm this pattern. Big-leaf mahogany seeds were sown at a rate of two seeds per station, and field measurements suggest that spacing was about 3.1 by 6.0 m, or 537 stations ha^{-1} (spacing between rows was consistent but between-tree spacing ranged from 2.7 to 3.6 m; the mean figure has been used). When established, the plot contained 24 trees or 119 trees ha^{-1} (22% survival). Before plots were established, 9 trees were cut and 15 trees more than 3 m tall were not measured. By 1994, ingrowth of other species was significant. An overstory of *Schizolobium parahybum*, *Zanthoxylum mayanum*, *Ficus* sp., and *Spondias mombin* had developed, together with an understory of *O. cohune*, *Guazuma ulmifolia*, and *Pouteria campechiana*.

Plot P-3, established in May 1963 in the 1958 plantation, was sown with big-leaf mahogany seed in groups of five, 45 cm apart (Forest Department annual report for 1958 refers to eight seeds per group), with groups spaced 9.1 m by 9.1 m, or 120 groups ha^{-1}. Within each group, the best tree was marked and measured. Any other mahogany trees in a group were not measured. By 1973, 9 of the original 24 selected trees were dead (only 1 tree per group was originally selected). In five of these groups, another stem was selected and marked as a crop tree; four groups were found to contain no surviving trees. Only 1 tree per group was selected; no other mahogany trees in the group were marked or felled. In 1994, several of these "other" trees were larger than those originally selected, and 1 tree previously not measured was 45 cm in diameter at 1.3 m (dbh). In 1994, the diameter of all mahogany in the plot was measured. By 1994, an overstory of *Schizolobium parahybum*, and *Ficus* sp. had developed, along with an understory of *O. cohune* and *Spondias mombin*.

Plot P-4, established in February 1968 in the 1956 plantation, is directly adjacent to P-1. In 1968, 12 stems were selected and measured (the plot contained about 40 stems in total). The decision to select only 12 trees was not explained, but the trees were probably expected to form the final crop. From 1973 on, all plot trees were measured. By 1994, an overstory of one *Schizolobium parahybum* and one *Spondias mombin*, along with an understory of *O. cohune*, had developed. The large *Ficus* sp. growing in P-1 also influenced this plot.

Methods

Data Collection

Diameter at 1.3 m was recorded in all measurement years (before 1981, girth, rather than diameter, was measured). Total height was measured before 1981. In 1981 and 1994, top height (mean total height of the 100 largest diameter trees ha^{-1}) was measured. Additionally, in 1994, height was measured to the first branch that forms the crown.

Data Analysis

From 1963 to 1981, data were collected in imperial units, and girth (rather than diameter) was measured until 1976. All data were converted to metric units, and girth was converted to diameter. Volume was calculated by using a single tree volume equation taken from work on plantation-grown mahogany in Sri Lanka (Mayhew and Newton 1998):

$$V = 0.056 - 0.01421D + 0.001036D^2$$

where V is volume, overbark, for main stem to 10-cm-top diameter, or to the last 2-m log, regardless of branches, and D is dbh, diameter (in cm) at 1.3 m.

The volume equation is based on a sample size of 319 trees with a dbh range of 10 to 90 cm. A relascope was used to measure every 2 m up the main stem. Mayhew and Newton (1998) give a high reliability to this equation compared to other single tree volume equations.

For each permanent plot, diameter and volume mean annual increment (MAI) and periodic annual increment (PAI) were calculated. The diameter of each tree was measured during each remeasurement. Individual tree data were used to calculate plot mean diameter, stand basal area, and volume. Mean annual increment is the mean rate of growth in diameter or volume, plotted against age (for example, the total volume produced to date, divided by age, plotted against age). Periodic annual increment is the current growth rate against age (for example, the difference in total volume between two measurement years, divided by the period between measurements, plotted against age).

Data Reliability

The data from these permanent plots were sufficiently consistent to suggest that the growth of individual trees, over a span of 26 to 31 years, can be reliably traced. The number of trees per plot does fluctuate from year to year, however; unfortunately, the precise reason for this fluctuation was not explained in the plot histories. Likely, some measured trees died and the small number of mahogany not previously measured were incorporated as plot trees. Some trees might have been missed in some years and "found" during subsequent measurements.

The quality of tree height measurements was erratic. In some years, height was not measured; in others, a "mean" height was measured; and, in still others, the data appear to be erroneous. Beginning in 1981, top height only was measured. Because of problems with the data, we decided to exclude results for height from the analysis.

Results

For all plots, mean diameter at 36 and 38 years was 26.2 to 30.1 cm, with standard error ranging from 1.4 to 3.3. In all plots, variability in mean diameter increased with plot age. For all plots, diameter MAI at age 36 and 38 was 0.7 to 0.8 cm yr^{-1} and PAI was 0.3 to 0.6 cm yr^{-1}. Diameter PAI had been decreasing in all plots since age 20 to 25 years. A maximum rate of diameter increment was 2.2 cm yr^{-1} in P-4 at age 13 years (Table 19.1).

Total stand basal area is strongly associated with stocking density throughout the measurement period. At 38 years, basal area for plots P-1 and P-4 (252 and 288 stems ha^{-1}) was 20.3 and 23.3 m^3 ha^{-1}; at age 36 years, however, basal area for plots P-2 and P-3 (119 and 138 stems ha^{-1}) was 10.4 and 8.7 m^3 ha^{-1}. Total standing volume at 36 to 38 years was 93.9 to 200.0 m^3 ha^{-1} (Table 19.1).

Volume MAI at age 36 and 38 years differed between plots (Fig. 19.1). In the two plots with the highest stocking, P-1 and P-4 (252 and 288 stems ha^{-1}), MAI was 4.3 and 5.3 m^3 ha^{-1}. In P-2 and P-3 (119 and 138 stems ha^{-1}), MAI was 2.6 and 2.0 m^3 ha^{-1}. The PAI has been decreasing in P-1 since age 18 (Fig. 19.1a), in P-3 since age 16 (Fig. 19.1c), and in P-4 since age 25 (Fig. 19.1d). It continues to rise in P-2 (Fig. 19.1c), however, although the pattern is variable. The peak of the MAI curve measures the maximum sustained yield of a stand averaged per year over a whole rotation. In P-1 and P-4, maximum mean annual increment (MMAI) is at about 38 years. In P-2 and P-3, MMAI is yet to occur (Fig. 19.1).

Diameter distribution in all plots tends to be skewed to the left (Fig. 19.2). The proportion of stems greater than 30 cm in diameter was 40% to 51% in P-1, P-2, and P-4, and 25% in P-3. Plot P-3 was group-planted; where more than one tree survived per group, competition between trees may be higher than the low stocking of 100 stems ha^{-1} suggested. In such groups, competition between trees will be high, which may explain the more positively left-diameter distribution in this plot. Where only one tree per group survived, such trees would have less competition, and the diameter increment should be greater.

The incidence of the mahogany shoot borer was not recorded in these plots. To gauge possible effects of early damage on current tree form, crown height was measured in 1994. Shoot borers attack the apical shoot, causing a loss of form and increment; although damage may be severe, it rarely kills the tree (Newton 1993). In all plots, mean crown height ranged from 42%

Table 19.1. Growth and Yield for Plots P-1, P-2, P-3, and P-4

Age yr	No. Trees ha^{-1}	Mean dbh (cm)	Mean dbh Standard Error	dbh MAL (cm)	dbh PAL (cm)	Basal area m^{3}ha^{-1}	Volume m^{3}ha^{-1}	Volume MAI m^{3}ha^{-1}yr^{-1}	Volume PAI m^{3}ha^{-1}yr^{-1}
P-1									
7	321	5.7	0.2	0.8	0.8	0.9	0.1	0.0	0.0
8	326	6.6	0.3	0.8	0.8	1.5	0.5	0.1	0.4
9	326	8.3	0.3	0.9	1.7	2.0	2.6	0.3	2.2
10	326	9.3	0.4	0.9	1.0	2.6	5.4	0.5	2.8
12	316	10.8	0.4	0.9	0.8	3.3	10.1	0.9	2.4
14	302	13.2	0.6	0.9	1.2	4.7	19.8	1.4	4.8
17	302	16.9	0.6	1.0	1.2	8.0	45.0	2.7	8.4
18	316	17.8	0.8	1.0	1.0	9.3	55.5	3.1	10.5
20	316	19.6	0.9	1.0	0.9	11.3	73.2	3.7	8.9
25	287	24.2	1.1	1.0	0.9	15.2	110.7	4.4	7.5
38	252	29.9	1.4	0.8	0.4	20.3	163.9	4.3	4.1
P-2									
5	119	4.1	0.3	0.8	0.8	0.2	0.0	0.0	0.0
6	124	4.9	0.4	0.8	0.9	0.3	0.0	0.0	0.0
7	124	6.2	0.5	0.9	1.3	0.4	0.0	0.0	0.0
10	124	8.5	0.7	0.9	0.8	0.8	2.0	0.2	0.7
12	109	10.3	0.9	0.9	0.9	1.1	3.7	0.3	0.8
14	109	12.5	1.2	0.9	1.1	1.6	7.2	0.5	1.8
16	114	15.0	1.5	0.9	1.3	2.4	13.8	0.9	3.3
18	119	17.6	1.6	1.0	1.3	3.4	22.2	1.2	4.2
23	119	21.4	2.2	0.9	0.8	5.3	40.3	1.8	3.6
36	119	29.4	3.3	0.8	0.6	10.4	93.9	2.6	4.1
P-3									
5	119	4.9	0.3	1.0	1.0	0.3	0.0	0.0	0.0
6	119	5.5	0.4	0.9	0.6	0.3	0.0	0.0	0.0
8	114	7.2	0.5	0.9	0.8	0.5	0.5	0.1	0.2
10	109	8.1	0.6	0.8	0.7	0.6	1.0	0.1	0.3
12	79	10.6	0.9	0.9	1.2	(Not calculated)			
14	74	12.8	1.2	0.9	1.1	(Not calculated)			
15	99	14.2	1.2	1.0	1.4	1.8	8.6	0.6	1.5
16	99	15.7	1.3	1.0	1.5	2.1	11.8	0.7	3.2
23	99	22.5	2.0	1.0	1.0	4.5	33.0	1.4	3.1
36	138	26.2	2.0	0.7	0.3	8.7	71.0	2.0	2.9
P-4									
17	305	15.9	1.0	0.9	0.9	6.8	37.7	2.2	2.2
18	346	16.2	1.0	0.9	0.3	8.3	48.4	2.7	10.7
20	338	18.5	1.2	0.9	1.2	10.5	67.8	3.4	9.7
25	329	23.5	1.4	0.9	1.0	16.3	123.0	4.9	11.0
38	288	30.1	1.9	0.8	0.5	23.3	200.0	5.3	5.9

dbh, diameter at breast height; MAI, mean annual increment; PAI, periodic annual increment.
P-1 and P-4 are in the 1956 plantation.
P-2 and P-3 are in the 1958 plantation.

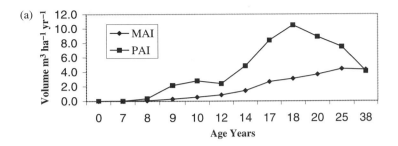

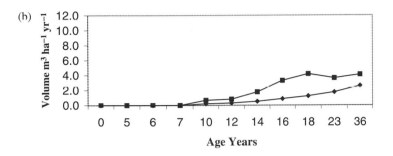

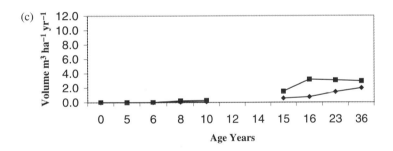

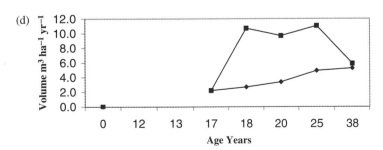

Figure 19.1a–d. Volume mean annual increment (MAI) and periodic annual increment (PAI) for all *Swietenia macrophylla* stems in (**a**) P-1, (**b**) P-2, (**c**) P-3, and (**d**) P-4.

(a)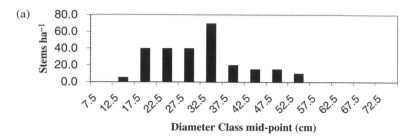

(b)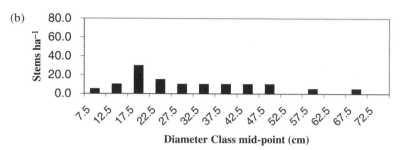

(c)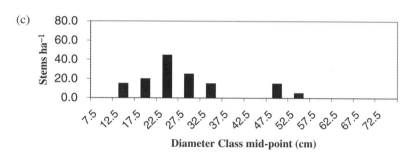

(d)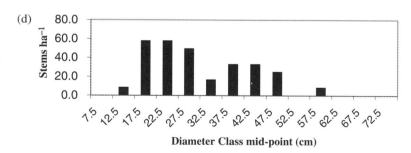

Figure 19.2a–d. Diameter distribution of all *Swietenia macrophylla* stems in (**a**) P-1 at 38 years, (**b**) P-2 at 36 years, (**c**) P-3 at 36 years, and (**d**) P-4 at 38 years.

to 51% of total tree height, and tree form was generally good. Despite the likely incidence of shoot borers, little evidence suggested that these plantations have suffered long-term damage.

Rates of stocking differed among the plots, ranging from 100 to 320 stems ha^{-1} (minimum and maximum figures since establishment). Despite differences in stocking, mean diameter among plots was about the same (mean of 28.9 cm). Results suggested that between-tree competition was not sufficiently intense at higher stocking to depress mean diameter growth. Similar results were found in Sri Lanka (J. Mayhew, Edinburgh University, UK, personal communication), where stocking of 300 stems ha^{-1} were found to be too low for stocking density to be a limiting factor for growth.

Discussion

The surviving permanent plots have demonstrated that taungya plantations of big-leaf mahogany can be successfully established with minimal maintenance. Of the original 11 permanent plots, however, 6 were abandoned because of poor growth and survival. The success measured in 4 permanent plots should be qualified by the failures in others. Where plots and plantations survive, however, the stocking and growth of the big-leaf mahogany is impressive.

Although the Forestry Department described group planting as a very successful means of establishing big-leaf mahogany, the single permanent plot established by this method (P-3) did not show superior growth compared to plots established by line planting. Where several trees per group survived, group planting may prove to be undesirable, if the stand is unlikely to be thinned.

If the objective of the plantation is to maximize yield, the rotation length is the time taken to reach MMAI in volume. Volume data for P-1 and P-4 suggested a rotation length of about 38 years. Rotation lengths based on maximum yield are determined by growth rate and thus site quality (Fattah 1992). My results suggested that site quality is higher in P-1 and P-4 (Upland System), where the soil tends to be deeper and fertility higher than P-2 and P-3 (Hill System).

Acknowledgments. I thank the Research Division team for assistance in the field and O. Sabido and N. Bird for valuable discussion.

Literature Cited

Belize Forest Department Annual Reports. 1958. National Archives, Belmopan, Belize, Central America.

Fattah, H.A. 1992. Mahogany forestry in Indonesia. In *Mahogany Workshop: Review and Implications of CITES*, 3–4 February 1992. Tropical Forest Foundation, Washington, DC.

Friesner, J. 1993. *Hurricanes and the Forests of Belize*. FPMP Occasional Series 1. Forest Department, Belmopan, Belize, Central America.

Hartshorn, G., Nicolait, L., Hartshorn, L., Bevier, G., Brightman, R., Cal, J., Cawich, A., Davidson, W., Dubois, R., Dyer, C., Gibson, J., Hawley, W., Leonard, J., Nicolait, R., Weyer, D., White, H., and Wright, C. 1984. Belize: Country environmental profile: a field study. Report by Robert Nicolait & Associates, Ltd., for USAID. Forest Department, Belmopan, Belize, Central America.

King, R.B., Baille, I.C., Bissett, P.G., Grimble, R.J., and Johnson, G.L. 1986. Land resources survey of Toledo District, Belize. Overseas Development Natural Resources Institute. National Archives, Belmopan, Belize.

Mayhew, J.E., and Newton, A.C. 1998. *The Silviculture of Mahogany*. CABI Publishing, Wallingford, UK.

Newton, A. 1992. Prospects of growing mahogany in plantations. In *The Proceedings of the Second Pan American Furniture Manufacturers Symposium on Tropical Hardwoods*, November 3–5, 1992. Center for Environmental Study, Grand Rapids, Michigan.

Research Division File Note. 1963. The mahogany increment plots, Columbia River Forest Reserve. Melinda Forest Station, Melinda, Stann Creek District, Belize, Central America.

Soubieux, J.M. 1983. Croissance et production du mahogany (*Swietenia macrophylla* King) en peuplements artificiels en Guadeloupe. Ecole Nationale des Ingenieurs des Travaux des Eaux et Forêts, Direction Régionale pour la Guadeloupe, Office National des Forêts, Guadelope.

Weaver, P.L. 1989. Taungya plantings in Puerto Rico: assessing the growth of mahogany and María stands. *Journal of Forestry* **87**(3):37–39.

Wright, A.C.S., Romney, D.H., Arbuckle, R.H., and Vial, V.E. 1959. Land in British Honduras. National Archives, Belmopan, Belize, Central America.

20. Growth of Small-Leaf Mahogany Crop Trees in St. Croix, U.S. Virgin Islands

John K. Francis

Abstract. Small-leaf mahogany, introduced to St. Croix in the U.S. Virgin Islands more than 200 years ago, has naturalized widely and now dominates Estate Thomas, a small USDA Forest Service property characterized by shallow, very stony soil. In 1962, 109 mahogany trees were liberated from competition and measured periodically until 1996. The study trees averaged 20 cm in diameter at the start of the study and 29 cm at the final measurement, when the total height averaged 11 m and the merchantable height averaged 4 m. Although the study trees increased in diameter very slowly ($0.25 \, \text{cm} \, \text{yr}^{-1}$) during the 34-year period, they grew about 50% faster than the unthinned control trees. Mortality during the period was 17%, most of it the result of a hurricane that struck in 1989. The trees are minimally merchantable now at about 82 years old and would be about 170 years old at a more favorable diameter of 0.5 m.

Keywords: *Swietenia mahagoni*, Growth

Introduction

Small-leaf mahogany, native to the Bahamas through Hispaniola, was introduced to St. Croix, U.S. Virgin Islands, about 1790 (Weaver and Francis 1988), presumably from a Jamaican seed source. The species can live more than 200 years and grows steadily throughout its life. Although it is normally a medium-sized tree, small-leaf mahogany occasionally reaches more than 2 m in diameter and 30 m in height in St. Croix (Francis 1991). One of the most notable features of the species is its ability to compete on difficult sites, particularly dry and rocky limestone hills. This chapter details a study of the growth of mahogany trees on such a site.

Estate Thomas, a 60-ha tract near Christiansted, St. Croix, consists of low limestone hills with small valleys between them. The hills with natural stands have shallow, very stony or skeletal clay loam soil over porous limestone. The area receives about 1100 mm of mean annual precipitation (Wadsworth 1947). The reforestation history of the property can only be guessed. Apparently, after cultivation ceased, the estate was grazed and natural reforestation of the hills began. Small-leaf mahogany, which seeded in from a few trees outside the property, dominated the resulting stand. Wadsworth (1947) states that the area had received little or no disturbance in 30 years. He found an average of 6.1 rings cm^{-1} in eight trees of various crown classes.

Methods

In 1962, a small-leaf mahogany crop tree study was established in which 109 trees were released from competition by clearing all vegetation in a circle with diameter equal to 30 times the diameter at breast height (dbh) of each tree. Woody vegetation that sprouted afterward was poisoned. To gauge competition, 1 to 8 mahogany trees from the untreated stand near each study (crop) tree were selected as controls. The dbh of both study trees and control trees was measured annually in 1962 through 1969 and 1971 through 1973. A final dbh measurement was made in 1996; merchantable height (to major branching) and total height were also measured. The data were tabulated with the Microsoft Excel program. Lacking a specifically developed volume model for small-leaf mahogany, I structured a simple generic formula and used it to approximate merchantable outside-bark volume: merchantable bole volume = basal area × (merchantable height − 20 cm) × 0.9.

Results and Discussion

None of the study trees died between 1962 and 1973. By 1996, 19 trees (17.4%) had died, most of them apparently as a result of Hurricane Hugo, which struck the area in 1989. Nearly all the survivors had evidence of major limb breakage and probably most have lost height as a result of the storm. The average height of the survivors at the final measurement was 10.6 ± 0.1 m (mean $\pm$ SE). Diameters of the 109 study trees at the start of the study averaged 20.3 ± 0.3 cm. The survivors at the end of the study averaged 28.9 ± 0.6 cm, an average annual diameter increment during the 34-year study period of 0.25 ± 0.00 cm yr^{-1}. The higher average diameter growth during the first 11 years of 0.28 ± 0.02 cm yr^{-1} indicated a slight slowing of growth in the last 23 years of the study. Because the most rapid growth rates were during the first decade or two of a small-leaf mahogany tree's life (Francis 1991), growth was probably faster (0.42 cm yr^{-1} on average) in the 48 years or so of the stand's life before the study began or that the trees are actually older than estimated.

The average commercial height of the survivors in 1996 was 4.2 ± 0.02 m, which indicates an approximate average outside-bark merchantable volume of 0.25 ± 0.00 m^3 tree^{-1}. This figure is somewhat lower than the figure given by a local volume table (based on dbh only) constructed for hybrid mahogany in Puerto Rico (Bauer and Gillespie 1990), but probably would be accounted for by the shorter stature of the St. Croix trees. In reality, accurately predicting merchantable volume would be even more complicated; only heartwood volume of small-leaf mahogany is valuable, and some usable volume can be harvested from limbs for the local carving, craft, and furniture market. The design of this particular study did not permit calculating basal areas or wood volumes per hectare. Measurements in 1986 (prehurricane) of naturally regenerated stands on the property, in which small-leaf mahogany dominated, revealed a mean basal area of 27 m^3 ha^{-1} and bole volumes of about 109 m^3 ha^{-1} (Weaver and Francis 1988).

The control trees averaged somewhat smaller than the crop trees, 16.3 ± 0.1 cm at the beginning of the experiment and 20.2 ± 0.2 cm at its conclusion, for a mean diameter increment of 0.16 ± 0.00 cm yr^{-1}. This increment indicates that diameter growth of small-leaf mahogany in similar habitats can be increased by 50% by crop tree thinning. This result must be taken with a degree of caution, however. The control trees averaged a somewhat smaller diameter than the crop trees and may have had, on average, an inferior crown class and therefore less vigor than the crop trees at the start of the experiment. At the same time, the control trees, which circled the crop trees, probably benefited in a small way by the thinning because a portion of the crowns of many of them bordered on the space created by the thinning.

Mean annual diameter increments of 19 small-leaf mahogany plantations around the world ranged from 0.24 to 1.46 cm yr^{-1} (Francis 1991). The control trees in this study are considerably slower growing then the slowest small-leaf mahogany plantation (Guánica, Puerto Rico) reported in the literature (Francis 1991). Small-leaf mahogany trees in a nearby 26-year-old plantation on the valley bottom had an average annual growth rate of 0.86 cm yr^{-1} in diameter, 0.44 m yr^{-1} in height, and 6 m^3 ha^{-1} annually in stem volume increment (Weaver and Francis 1988). If the mahogany stand of this study started about when the United States purchased the Virgin Islands in 1914, it would have been 82 years old in 1996 and had an average diameter of 29 cm. If the target merchantable dbh is arbitrarily set at 0.5 m and an average future growth rate of 0.24 cm can be assumed, a rotation of 170 years is a reasonable expectation. The current diameters are large enough for most uses by the local wood-carving and custom furniture industry. This growth is too slow to consider managing small-leaf mahogany on similar terrain for profit even if the wood is highly valuable. One hundred or 200 ha, well stocked with small-leaf mahogany whose existence can be justified for other benefits, is probably enough to supply the small local industry.

Literature Cited

Bauer, G.P., and Gillespie, A.J.R. 1990. *Volume Tables for Young Plantation-Grown Hybrid Mahogany* (Swietenia macrophylla × S. mahagoni) *in the Luquillo Experimental Forest of Puerto Rico*. Research Paper SO-257. U.S. Department of Agriculture, Forest Service, Southern Forest Experiment Station, New Orleans, LA.

Francis, J.K. 1991. Swietenia mahagoni *Jacq. West Indies mahogany*. Research Note SO-ITF-SM-46. U.S. Department of Agriculture, Forest Service, Southern Forest Experiment Station, New Orleans, LA.

Wadsworth, F.H. 1947. The development of *Swietenia mahagoni* Jacq. on St. Croix. *Caribbean Forester* **8**(2):161–162.

Weaver, P.L., and Francis, J.K. 1998. Growth of teak, mahogany, and Spanish cedar on St. Croix, U.S. Virgin Islands. *Turrialba* **38**(4):308–317.

4.   Shoot Borer

21. Taxonomy, Ecology, and Control of *Hypsipyla* Shoot Borers of Meliaceae

Robert B. Floyd, Caroline Hauxwell, Manon Griffiths, Marianne Horak, Don P.A. Sands, Martin R. Speight, Allan D. Watt, and F. Ross Wylie

Abstract. *Hypsipyla grandella* and *Hypsipyla robusta* are serious pests of species of the subfamily Swietenioideae of the family Meliaceae in virtually every moist tropical region of the world. An international workshop reviewed the ecology and control of *Hypsipyla* shoot borers of Meliaceae, identified promising control methods, and set priorities for future research. The conclusions of the workshop are presented with specific recommendations for research in aspects of the taxonomy, biology, and ecology of *Hypsipyla*, and pest management options that use host plant resistance and chemical, biological, and silvicultural control.

Keywords: Integrated pest management, *Hypsipyla robusta*, *Hypsipyla grandella*, Host plant resistance, Chemical control, Biological control

Introduction

Plantations of tree species in the subfamily Swietenioideae of the family Meliaceae, hereafter referred to as "Swietenioideae," have been very difficult to grow economically in areas inhabited by the shoot borers, *Hypsipyla robusta* Moore in tropical Asia and Africa or *Hypsipyla grandella* (Zeller) in tropical America. Concerted research efforts in the 1970s resulted in two major reviews (Grijpma 1973; Whitmore 1976a,b), focusing principally on the shoot borer in Central and South America. Newton et al. (1993) reviewed the prospects for shoot borer control with a particular emphasis on that part of the world. Because of advances in techniques and approaches to insect pest management and the continuing high priority for various countries to grow high-grade timber of Swietenioideae species, a workshop in Kandy, Sri Lanka, in 1996, reviewed the ecology and control of the shoot borers of Meliaceae.

The aims of the workshop were these:

• Review the biology and management of *Hypsipyla* spp.
• Identify successful or promising control methods
• Prioritize areas for future research
• Facilitate discussion and international collaboration on shoot borer research

Reviews of research themes, scientific papers, country reports and summaries of the workshop outcomes are given in the workshop proceedings (Floyd and Hauxwell 2001). Our chapter provides summaries of the country reports, mainly from Asia and the Pacific; reviews of each research theme; and outcomes of the workshop sessions on each theme in which promising research findings and future research priorities were identified. The research themes were taxonomy, biology, and ecology of the genus; host plant resistance; chemical control and pheromones; biological control; silvicultural control; and integrated pest management.

Hypsipyla robusta Damage on Swietenioideae in Asia and the Pacific

The known geographic distribution of *H. robusta*, as recorded by Entwistle (1967), has now been extended (see country reports in Floyd and Hauxwell 2001). The major changes in range since Entwistle's 1967 paper are confirmed records from Bangladesh, China, Vietnam, Laos, Thailand, the Philippines, Cote d'Ivoire, and Tonga. The report that *H. robusta* is widespread in Tonga (Waterhouse 1997) is of particular interest because the species is absent from Fiji, the main island group between Tonga and the closest occurrence of the species (Vanuatu). Furthermore, Fiji has large areas of big-leaf mahogany plantations. Numerous species of Swietenioideae in several countries are attacked by shoot borers, and the exotic big-leaf and small-leaf mahogany and various native species of *Toona* are almost invariably heavily attacked (Table 21.1). Species of *Khaya* and *Cedrela* (both exotics) are generally less damaged. The generalization that *Hypsipyla* app. attacks native Swietenioideae to a greater extent than exotic species (Grijpma 1974) is not supported by these observations.

Taxonomy, Biology, and Ecology of *Hypsipyla* spp.

Despite the pantropical economic importance of *H. robusta* and *H. grandella*, many aspects of the taxonomy, biology, and ecology of these species remain unknown although a more complete understanding is fundamental to developing and implementing an integrated pest management strategy.

The current taxonomic knowledge of the genus *Hypsipyla* Ragonot is adequate for tropical America yet quite poor for tropical Asia and Africa (Horak 2001). Eleven species of *Hypsipyla* are currently recognized, 4 from tropical America and 7 from tropical Asia and Africa. The taxonomy of the tropical American species is well documented, although intraspecific variations in ecology, host–plant interactions, and particularly pest status may be considerable. Some biological information is available on the main pest species, *H. grandella*, as well as on *H. ferrealis* (Hampson) and two species of the related genus *Sematoneura* Ragonot, all of which also feed on species of Swietenioideae. By contrast, the taxonomy of tropical Asian and African species is poorly understood, with several species known only from their type specimens. The taxonomic status of the different populations of *H. robusta* remains unresolved. Two closely related genera, *Catopyla* Bradley and *Eugyroptera* Bradley, whose larvae also feed on species of Swietenioideae, have been described from Nigeria.

A thorough taxonomic investigation of *Hypsipyla* and related genera, particularly of tropical Asian and African species, is required. This review should clarify recognized differences in the morphology, biology, behavior, and pheromone composition between populations currently treated as *H. robusta* and species in related genera with similar biology in Central America and Africa. This review would require collecting specimens throughout the geographic range of what is presently considered to be *H. robusta*, from all host species and all plant parts it feeds on, particularly the fruits that seem to be the exclusive hosts for some of the genera related to *Hypsipyla*. Our knowledge of the host range of *Hypsipyla* is also far from complete. In addition to morphological examination, molecular and biochemical techniques may be required to assist in determining species.

The biology and ecology of *Hypsipyla* have been the subject of only a few studies, with most investigations strongly directed toward pest management (Griffiths 2001). In particular, the biology and population

Table 21.1. Damage by *Hypsipyla robusta* Moore on species of Meliaceae (subfamily Swietenioideae) grown in various African, Asian, and Pacific countries[a,b]

Tree Species	Ghana	Bang	Sri L	India	Phil	Viet	Laos	Thai	Mala	Indon	PNG	Sol I
Spanish cedar	0							+	+			0
Cedrela lilloi C.DC.			+				+	+				
Chukrasia tabularis A. Juss.		+				+	+	++	+			
Entandrophragma angolense (Welw.) C.DC	+											
Entandrophragma cylindricum (Sprague)	+											
Entandrophragma utile (Dawe & Sprague)	+											
Khaya anthotheca (Welw.) C.DC.	+									+		
Khaya grandiflora C.DC.										+		
Khaya ivorensis A. Chev.X									+			
Khaya nyasica Stapf. ex Baker												
Khaya senegalensis (Desr.) A. Juss.			+						0	+		
Pacific coast mahogany	++					+						
Big-leaf mahogany	++	++	+	+	+	+	+	++	++	++	+	++
Small-leaf mahogany		++	+	+				++	+++	+		
Toona ciliata Roem.		+	+	+			+	++	+++	+	+	
Toona calantas Merr. & Rolfe					+				+			
Toona sinensis (A. Juss.) M. Roem						+	+					
Toona sureni (Blume) Merr						+		+	++	+	+	
Xylocarpus moluccensis (Lam.) M. Roem.				+				++				

Bang, Bangladesh; Sri L, Sri Lanka; Phil, Philippines; Viet, Vietnam; Laos, Lao PDR; Thai, Thailand; Mala, Malaysia; Indon, Indonesia; PNG, Papua New Guinea; Sol I, Solomon Islands; Aust, Australia.

[a] Countries with a single plus sign against all species recorded for that country indicate damage has been observed but with no indication of relative severity of damage.

[b] Empty cells in the table indicate the absence of records of a tree species in a country. The number of plus signs indicates severity of damage, and a zero (0) indicates no damage observed.

dynamics of *Hypsipyla* in natural stands and its feeding behavior on non-shoot plant parts have received little attention. The movement of both adults and larvae is not well understood, but this behavior is central to life history studies and the design of control strategies because it influences mating, dispersal, and locating hosts. Movement by larvae has a strong influence on larval survival and the distribution and severity of damage on the host plant. Sites of pupation have been identified, but the relative importance of the different sites and their effect on mortality rates has not been determined. The incidence, nature, and triggers of the quiescent phase are not known, despite the importance of these factors in influencing the seasonality and population dynamics of the different species.

Many difficulties need to be overcome in studying the biology and ecology of *Hypsipyla*. For instance, adults are nocturnal and do not readily come to light, population densities are generally low, and larvae feed in concealed places. Often, only one larva per tree is required to render the timber unmarketable. Rearing moths in captivity, especially obtaining mating and oviposition behavior under laboratory conditions, has proved to be difficult. Successful rearing methods, including developing an artificial diet and maximizing mating and oviposition success, need to be established. Such studies may also provide insight into the effect of environmental influences on mating and oviposition under natural conditions.

In summary, research priorities in taxonomy, biology, and ecology are the following:

- Taxonomic revision of the genus *Hypsipyla* and closely allied genera, particularly from tropical Asia and Africa
- Aspects of poorly understood *Hypsipyla* biology with direct relevance to different pest management options (such as mating behavior, dispersal, host location, oviposition, larval movement, pupation, and dormancy)
- Host–plant relations, population regulation, and behavior in natural forests, particularly as related to host tree phenology
- Practical and effective rearing techniques that include mating in captivity and feeding on an artificial diet

Host Plant Resistance

Watt et al. (2001) reviewed evidence for different forms of resistance in Meliaceae to shoot borers, in particular the resistance of Spanish cedar and big-leaf mahogany to *H. grandella*. Significant genetic variation in resistance to attack does appear in these two species, mainly taking the form of tolerance, but Spanish cedar may also vary in nonpreference and antibiosis. Deploying resistant planting stock in silvicultural or agroforestry systems that also encourage natural biological control or otherwise minimize the abundance and effects of shoot borers was recommended by Watt et al. (2001) as the most promising option for overcoming shoot borer damage.

Besides noting differences in resistance to shoot borers, the workshop also identified several other promising findings:

- The possibility that even low resistance can be valuable when integrated with biological and other methods of control
- The differences in plant chemistry in different provenances of Spanish cedar, the correlation between proanthocyanidin concentration and resistance, and, thus, the potential for selecting resistant plant material on the basis of plant chemistry
- The relation between the age of trees and both the apparent susceptibility to attack and its expression in terms of damage
- The possibility that the environment (for example, shade) can influence resistance
- The way volatile chemicals in host plants affect adult insect behavior

Discussions on priorities for future research centered on establishing tree species and provenance trials on a range of species and in a range of countries and regions. Much discussion also focused on methods, particularly the need for standardized protocols for damage assessment. Developing methods for capturing genetic gains (breeding and genetic manipulation) and multiplying resistant stock using clonal propagation or seed orchards will be important, although these were not seen as immediate priorities because of the limited amount of *Hypsipyla*-resistant plant material. Nevertheless, we hope that the multiplication and deployment of resistant Spanish cedar will proceed in the near future, as resistance in this species currently shows the most promise.

Other priority areas for research identified were studies of the biochemical basis for resistance and entomological aspects of resistance such as the phases of host selection.

Chemical Control and Pheromones

Wylie (2001) reviewed research and operational experience with chemical control of shoot borers, research that now spans about eight decades and has involved at least 23 countries throughout the tropics. Despite this effort, no chemical or application technology has yet been developed that will provide reliable, cost-effective, and environmentally sound protection for any of the high-value meliaceous tree species for the period necessary to produce a marketable stem. Reasons for this relate mainly to the biology of the insects, the nature of the damage they cause, constraints imposed by climate, and the period of protection required, which may be up to 5 years from planting. Wylie (2001) concluded that the future role of chemical pesticides in shoot borer control would continue to be protecting nursery stock or as part of a program of integrated pest management. The use of chemical pesticides alone was generally believed unlikely to solve the shoot borer problem.

Several new-generation compounds are available commercially that would merit testing against shoot borers, particularly in controlled-release formulations. Two such compounds suggested were imidocloprid (Bayer) and fipronil (Rhône-Poulenc), which have contact and systemic action. Further research was recommended on the use of antifeedants and of natural plant compounds.

Apart from the vagaries of climate, insect behavior, and equipment performance that commonly plague insecticide trials, a factor that characterizes much of the work on chemical control of shoot borer around the world—and hindered interpretation of the results—has been the lack of uniformity in screening procedures. The wide range of formulations and application techniques used, the different methods of assessing attack severity and chemical efficacy, and the lack of controls has made comparisons between trials very difficult. In future chemical trials, standardizing test methods, assessing, and reporting is essential. Determining the economic threshold for shoot borer attacks in stands will aid in decisions on the use of chemical pesticides.

Little is known about the volatile components of the shoot borer sex pheromone glands (Bellas 2001). Three components have been identified in secretions from *H. robusta* from a culture in France believed to have originated from West Africa. Because this species has such a wide and discontinuous geographic distribution, determining the pheromone composition from a range of locations is important. Preliminary studies on Australian populations have shown the presence of the same compounds, among others, but in different ratios. Three different compounds have been identified from the *H. grandella* ovipositor tip.

Among the most promising findings (Bellas 2001) were the apparent differences in composition of pheromones in various populations of *H. robusta* and between this species and *H. grandella*. This difference could have important implications for the taxonomy of *H. robusta*, given its disjunct distribution. Pheromones were seen as potentially useful tools for monitoring shoot borer abundance, although further work is needed to determine their composition and develop suitable lures. The remarkable ability of shoot borers to find isolated and distant host trees suggests that chemoreception is probably very well developed and important in the insect's behavior. Using the apparently well-developed chemoreceptive ability of these insects to attract and trap adults with volatiles from the shoot tips of host trees as attractants might be possible.

The main research priorities relating to chemical control and pheromones identified by the workshop were the following:

- Determine economic thresholds for shoot borer damage in stands
- Screen new biologically active compounds and formulations against shoot borers, including antifeedants and natural plant compounds, particularly in controlled-release formulations in nurseries and plantations
- Identify ovipositional stimuli and investigate the role of chemoreception in shoot borers
- Investigate pheromone differences (components and ratios) between and within the genus as taxonomic tools or for population monitoring
- Develop standard procedures for screening pesticides against shoot borers

Biological Control

Pathogens

Entomopathogens are a diverse group of organisms that includes fungi, viruses, bacteria, protozoa, and nematodes. Some have been successfully used for controlling various insect pests, particularly Lepidoptera, in forestry and agriculture. These pathogens have been introduced against other forest pest species as "classical" biological control agents, by augmentive release, or as inundative applications as biopesticides. Shoot borers remain a difficult target for microbial control, however. Hauxwell et al. (2001b) reviewed the published information on incidence of pathogens in the forest and shoot borer infection by entomopathogens. In addition, the results of collections from Ghana and Costa Rica and cross-infection with a baculovirus were presented.

The shoot borers are cryptic, occur at low densities, and have low damage thresholds. They are both temporally and spatially patchy, and the susceptible part of the host plant grows rapidly. Consequently, as with chemical insecticides, applications are difficult to deliver to the target. Unlike chemical insecticides, however, entomopathogens (with the probable exception of *Bacillus thuringiensis*) can replicate in their host and can be transmitted between hosts by several mechanisms that might be enhanced by targeted introductions or silvicultural management. Potential strategies for use of entomopathogens against shoot borers were discussed.

Little practical work has been done with shoot borer pathogens. Some, notably fungi originating from other species, can infect and kill the larvae, but few pathogens have been isolated from the genus. Collecting and identifying pathogens that might be manipulated as bioinsecticides is an important first step. Persistent, vertically transmitted pathogens (that is, transmitted between host life stages) that reduce the population of the pest might be introduced and spread from one population to another as in classical biological control. Alternatively, silvicultural treatments might be manipulated to enhance pathogen abundance. In particular, mixed-species plantings may create a favorable microclimate because pathogens are more effective under conditions of increased humidity, reduced ultraviolet light, or both. The effects of silvicultural treatments on pathogen incidence should be monitored.

These research priorities were identified by the workshop:

- Establish a base collection of shoot borer pathogens
- Evaluate pathogens to determine how they infect and affect shoot borers, especially in relation to the host's biology
- Identify vertically transmitted, population-persistent pathogens

Parasitoids and Predators

Agents for classical biological control of exotic arthropod pests are usually selected from natural enemies developing on the pest or closely related species in their native range. When native pests such as the shoot borers are not regulated by natural enemies, predators, pathogens, or parasitoids from related exotic species may be effective when introduced free of their own enemies. Some shoot borer parasitoids may be sufficiently host specific to be acceptable for introducing, from one country to another, as biological control agents for related species (Sands and Murphy 2001).

Previous attempts at biological control of the shoot borers have not been successful (Sands and Murphy 2001). The generalist egg parasitoid *Trichogramma chilonis* Ishii failed to establish when released to control *H. robusta* in Madras, India. Most parasitoid species released against *H. grandella* in the Caribbean have also failed to establish. Many natural enemies (such as *Apanteles, Cotesia, and Dolichogenidea*), however, are related to species known to be effective biological control agents for other pests. Several other parasitoid groups contain potentially valuable agents for introduction if freed of their natural enemies. Inundative releases of native parasitoids, although the method may lead to control, are unlikely to be economically viable for the shoot borers; however, the natural insect enemies and their potential as biological control agents are discussed in Sands and Murphy (2001).

The workshop highlighted the importance of further taxonomic studies for the major groups of parasitoids because many of them could not be identified. Natural enemies of shoot borers should be further surveyed, emphasizing countries or localities where the moths are less serious pests and where natural enemies may be reducing abundance (such as the Solomon Islands).

Extensive studies on the natural enemies of *H. robusta* in India and *H. grandella* in Latin America have provided a basis for selecting the most promising agents for biological control programs. Many attempts to control *H. grandella* by introducing parasitoids of *H. robusta* into Central America, the Caribbean, and Brazil, however, have been unsuccessful (Melo 1990;

Newton et al. 1993), possibly because of narrow host specificity or environmental differences.

More detailed studies are needed to quantify the effects of each parasitoid species on shoot borer populations at selected sites. Although the importance of life table studies was recognized, the logistics and time required to gather data were considered to be serious constraints. Very little is known about egg parasitoids, a priority area requiring much more research. Many of the parasitoid genera known to attack the eggs of other Lepidoptera are not recorded for shoot borers.

None of the shoot borer predators recorded is likely to be sufficiently host specific to be suitable as exotic biological control agents. Native predators, however, might be encouraged or introduced locally to minimize the density of the immature stages of the shoot borers. Further work is needed to identify ants that have been used to control other pests and may be amenable to establishing in forest plantations (Khoo 2001).

These research priorities were identified by the workshop:

- Pursue taxonomic studies on the major groups of shoot borer parasitoids
- Collect additional parasitoids (including egg parasitoids), especially in countries not previously surveyed and where shoot borer abundance is low
- Evaluate manipulating native predators, particularly ants
- Quantify the effects of arthropod natural enemies of the shoot borer to help identify promising species

Silvicultural Control

Several silvicultural treatments have been used to reduce shoot borer damage. Much of the information available is anecdotal, however; trials are often unreplicated and results have been inconsistent, creating what Whitmore (1976a,b) described as "myths." Consequently, guidelines that give effective, consistent results are not available, and an experimental analysis of the different silvicultural treatments is needed. Hauxwell et al. (2001a) reviewed the range of silvicultural treatments and discussed the relative importance of mechanisms by which they may contribute to shoot borer control. For example, mixed plantings containing non-Swietenioideae species may hinder the host-finding ability of adult moths, or cover crops or planting density may affect the persistence and effectiveness of shoot borer regulation by natural enemies.

Silvicultural techniques were recognised as having considerable potential for reducing the intensity of shoot borer damage. The workshop groups highlighted two areas as particularly promising:

- Promoting recovery after attack in open-site plantations by selecting favorable sites (fertile soil and adequate drainage), producing vigorous nursery stock, and form pruning

- Reducing the intensity of attack in mixed-species plantings by using timber nurse crops and shade

The workshop recognized that, for all the silvicultural options, understanding of the mechanisms by which they may affect both the plant and the insect is lacking, as are sound experimental data on which reliable recommendations could be based. Analysis of the economic costs and benefits of silvicultural treatments in different geographic regions is also needed.

Research priorities on silvicultural control included the following:

- Regionally evaluating the effects of silvicultural practices on shoot borer control by using replicated, controlled, and standardized experiments
- Experimentally evaluating mixed-species plantations (nurse crops, agroforestry, and enrichment or line planting) and, in particular, quantifying effects of shade on physical and chemical plant responses and on insect behavior and survival
- Quantifying site effects, particularly drainage and nutrients such as calcium, on plant susceptibility to and recovery from attack
- Examining the effects of silvicultural treatments on natural enemies
- Analyzing the economics of silvicultural options, including considering market values of the timber produced and returns on investment to industry

Integrated Pest Management

Integrated pest management includes making decisions at the planning stage followed by careful monitoring, and eventually the complementary use of several pest control tactics, if and when required, that enable a crop to be grown economically. No single tactic is likely to be successful in controlling shoot borers. A great deal of research has been carried out on many types of pest management, and still some species of Swietenioideae cannot be grown successfully when shoot borers are present. Shoot borers are classic low-density pests in that even one or two attacks on young trees may render their future timber production uneconomic. We are now at a position where knowledge about and experience with individual strategies should be combined, incorporating the best points of each.

Speight and Cory (2001) considered examples of integrated pest management in tropical forestry to illustrate how this philosophy may be put into practice and refer the lessons learned to the particular problems of shoot borers. For successful integration to be developed, fundamental problems must be overcome. These problems include the major knowledge gaps already described here, limitations of chemical and biological control, and—above all—the lack of a mechanism for international collaboration with the central coordination needed for a multidisciplinary approach. The fundamental key is the acquisition of substantial funding for research and development on an international scale.

Workshop delegates considered what would be the most appropriate components of an integrated pest management system for their geographic region (Oceania, Asia, Latin America, and Africa). The outcome was a strong focus on minimizing pest damage rather than a cure. The major issues identified were these:

- Good genetic stock to be procured and deployed
- Optimal nursery methods to rear healthy transplants
- Favorable site selection in terms of nutrition, drainage, and slope
- Promotion of fast early growth of trees by fertilizing soil, weeding, using growth tubes, and pruning
- Appropriate silvicultural practices that may include more resistant genotypes and mixed species planting
- Curative pest management as an emergency measure, with options that may include local chemical control and inundative release of natural enemies, if available, and monitoring the crops to detect the sudden appearance of insect damage
- Advisory or extension services to support growers using an integrated strategy

Note that not one of these issues can so far be supported by sound scientific research.

Conclusion

The top five research objectives that the workshop participants collectively identified, in order of importance, were the following:

- Establish species, provenance, and clonal resistance trials in different countries and regions within these countries
- Evaluate the use and efficacy of mixed-species plantations (nurse crops, agroforestry) and assess the effects of shade
- Investigate the taxonomy of shoot borer species by collecting them from throughout their geographic range, from all their host species, and from all the plant parts they feed on, using morphological, molecular, and biochemical techniques to identify species
- Investigate shoot borer biology and ecology in native forests, including behavior (dispersal and host location), mating, oviposition, larval movement, pupation, and dormancy
- Screen new biologically active compounds and formulations against shoot borer, including antifeedants and natural plant compounds, particularly in controlled-release formulations, in nurseries and plantations

The complete list of research priorities, presented in Floyd and Hauxwell (2001), consists of research on specific control methods as well as research of a more general nature, or research areas that underpin specific control methods.

Considering control methods alone, the perceived priorities from the full list, in order, were the following:

- Host resistance
- Mixed-species plantations and agroforestry
- Biologically active compounds (kairomones and novel insecticides)
- Site manipulation (particularly soil nutrients and drainage)
- Other silvicultural methods (to be developed)
- Manipulation of the action of predators
- Pathogens of insects
- Introduced parasitoids
- Pheromones (to be developed)
- Traditional insecticides

In conclusion, the workshop supported developing control by using selected silvicultural and agroforestry practices as being more promising than approaches using traditional insecticides or biological control through predators, pathogens, and parasitoids. The prospect of novel methods of control, through kairomones and new insecticides, however, was seen as a high priority. The need for an integrated approach to reducing damage caused by shoot borers to species of Swietenioideae was strongly endorsed by all workshop participants.

Acknowledgments. The workshop in Kandy, Sri Lanka, on August 20–23, 1996, was organized by CSIRO Entomology, Queensland Forestry Research Institute, and the Oxford University Department of Zoology, with major support from the Australian Centre for International Agricultural Research (ACIAR), as well as the Department for International Development (DFID), AusAID, British Council, and the Australian Rural Industries Research and Development Corporation. The workshop was attended by 40 delegates from about 20 countries encompassing Oceania, Southeast Asia, South Asia, Africa, and Central and South America.

Literature Cited

Bellas, T. 2001. Semiochemicals of *Hypsipyla* shoot borers. In Hypsipyla *Shoot Borers in Meliaceae. ACIAR Proceedings No. 97*, eds. R.B. Floyd and C. Hauxwell, pp. 116–117. Australian Centre for International Agricultural Research (ACIAR), Canberra, Australia.

Entwistle, P.F. 1967. The current situation on shoot, fruit and collar borers of the Meliaceae. *Proceedings of the Ninth British Commonwealth Forestry Conference*, Commonwealth Forestry Institute, Oxford, UK.

Floyd, R.B., and Hauxwell, C. 2001. Hypsipyla *Shoot Borers in Meliaceae. ACIAR Proceedings Series 97.* ACIAR, Canberra, Australia.

Griffiths, M. 2001. The biology and ecology of *Hypsipyla* shoot borers. In Hypsipyla *Shoot Borers in Meliaceae. ACIAR Proceedings No. 97*, eds. R.B. Floyd and C. Hauxwell, pp. 74–80. ACIAR, Canberra, Australia.

Grijpma, P. 1973. Proceedings of the first symposium on integrated control of *Hypsipyla*. In *Studies on the Shoot-Borer*, Hypsipyla grandella *(Zeller) (Lepidoptera: Pyralidae), Vol. I.* IICA Miscellaneous Publications 101. CATIE, Turrialba, Costa Rica.

Grijpma, P. 1974. *Contributions to an Integrated Control Programme of* Hypsipyla grandella *(Zeller) in Costa Rica*, p. 147. Doctoral thesis, Agricultural University, Wageningen, The Netherlands.

Hauxwell, C., Mayhew, J., and Newton, A. 2001a. Silvicultural management of *Hypsipyla*. In Hypsipyla *Shoot Borers in Meliaceae. ACIAR Proceedings No. 97*, eds. R.B. Floyd and C. Hauxwell, pp. 151–163. ACIAR, Canberra, Australia.

Hauxwell, C., Vargas, C., and Opuni-Frimpong, E. 2001b. Entomopathogens for control of *Hypsipyla*. In Hypsipyla *Shoot Borers in Meliaceae. ACIAR Proceedings No. 97*, eds. R.B. Floyd and C. Hauxwell, pp. 131–139. ACIAR, Canberra, Australia.

Horak, M. 2001. Current status of the taxonomy of *Hypsipyla* Ragonot (Pyralidae: Phycitinae). In Hypsipyla *Shoot Borers in Meliaceae. ACIAR Proceedings No. 97*, eds. R.B. Floyd and C. Hauxwell, pp. 69–73. ACIAR, Canberra, Australia.

Khoo, S.G. 2001. *Hypsipyla* shoot borers of the Meliaceae in Malaysia. In Hypsipyla *Shoot Borers in Meliaceae. ACIAR Proceedings No. 97*, eds. R.B. Floyd and C. Hauxwell, pp. 24–30. ACIAR, Canberra, Australia.

Melo, L.A.S. 1990. Controle biologico de pragos agricolas no Brasil. Investiga fo agrfria (INIA). *Cabo Verde* **3**:35–38.

Newton, A.C., Baker, P., Ramnarine, S., Mesen, J.F., and Leakey, R.R.B. 1993. The mahogany shoot borer—prospects for control. *Forest Ecology and Management* **57**:301–328.

Sands, D.P.A., and Murphy, S.T. 2001. Prospects for biological control of *Hypsipyla* spp. with insect agents. In Hypsipyla *Shoot Borers in Meliaceae. ACIAR Proceedings No. 97*, eds. R.B. Floyd and C. Hauxwell, pp. 121–130. ACIAR, Canberra, Australia.

Speight, M.R., and Cory, J.S. 2001. The integrated pest management of *Hypsipyla* shoot borers. In Hypsipyla *Shoot Borers in Meliaceae. ACIAR Proceedings No. 97*, eds. R.B. Floyd and C. Hauxwell, pp. 169–174. ACIAR, Canberra, Australia.

Waterhouse, D.F. 1997. *The Major Invertebrate Pests and Weeds of Agriculture and Plantation Forestry in the Southern and Western Pacific*. Australian Centre for International Agricultural Research, Canberra, Australia.

Watt, A.D., Newton, A.C., and Cornelius, J.P. 2001. Resistance in mahoganies to *Hypsipyla* species—a basis for integrated pest management. In Hypsipyla *Shoot Borers in Meliaceae. ACIAR Proceedings No. 97*, eds. R.B. Floyd and C. Hauxwell, pp. 89–95. ACIAR, Canberra, Australia.

Whitmore, J.L. 1976a. *Studies on the Shoot Borer* Hypsipyla grandella *(Zeller) Lepidoptera: Pyralidae, Vol. II.* IICA Miscellaneous Publications 101. Centro Agronómico Tropical de Investigación y Enseñanza (CATIE), Turrialba, Costa Rica.

Whitmore, J.L. 1976b. *Studies on the Shoot Borer* Hypsipyla grandella *(Zeller) Lepidoptera: Pyralidae, Vol. III.* IICA Miscellaneous Publications 101. CATIE, Turrialba, Costa Rica.

Wylie, F.R. 2001. Control of *Hypsipyla* spp. shoot borers with chemical pesticides: a review. In Hypsipyla *Shoot Borers in Meliaceae. ACIAR Proceedings No. 97*, eds. R.B. Floyd and C. Hauxwell, pp. 109–117. ACIAR, Canberra, Australia.

22. Resistance to the Shoot Borer in Mahoganies

Allan D. Watt, Adrian C. Newton,
and Jonathan P. Cornelius

Abstract. Considerable research effort on *Hypsipyla* shoot borers has failed to produce effective methods of control. Deploying pest-resistant planting stock as a basis for managing these pests has not been considered until recently, however. We review evidence for the existence of resistance to shoot borers in Meliaceae, with particular emphasis on research in Costa Rica on resistance to *Hypsipyla grandella* in big-leaf mahogany and Spanish cedar. This research has shown appreciable genetic variation in resistance to attack by shoot borers in these tree species. Strategies for future research are discussed, and we conclude that the best option for successful shoot borer management lies in using resistant planting stock in silvicultural or agroforestry systems to encourage natural biological control or otherwise minimize the abundance and adverse effects of shoot borers.

Keywords: *Hypsipyla*, Shoot borers, *Swietenia macrophylla*, *Cedrela odorata*, Genetic resistance, Pest management

Introduction

Mahogany shoot borer species, *Hypsipyla grandella* and *Hypsipyla robusta*, have severely restricted reforestation programs with Spanish cedar, big-leaf mahogany, *Toona ciliata*, *Khaya ivorensis*, and other Meliaceae species in many parts of the tropics (Entwistle 1967; Wagner et al. 1991; Newton et al. 1993a). The larvae of these pyralid moths destroy the terminal shoots of the host plant by boring into the pith, which can result in a highly branched tree of little economic value (Newton et al. 1993a).

Considerable research effort on these pests (Grijpma 1974; Whitmore 1976a,b) has failed to produce effective methods of control (Newton et al. 1993a). Thus, the development of mahogany plantations has been restricted, and the demand for timber from mahogany and other Meliaceae has led to increasing pressure on naturally growing Meliaceae.

Recent research, discussed in this chapter, on resistance to *H. grandella* in Spanish cedar and big-leaf mahogany has shown that deploying pest-resistant planting stock could form an effective basis for managing this shoot borer (Newton et al. 1993b, 1995, 1998, 1999). Not only might this strategy provide a means for these tree species to be grown in forest plantations or agroforestry systems, but it could also alleviate the pressure on naturally growing Meliaceae.

We describe research on resistance to shoot borers in Meliaceae species, with particular reference to recent work on resistance to *H. grandella* in Spanish cedar and big-leaf mahogany, and discuss future prospects for developing and deploying resistance to mahogany shoot borers. Three forms of resistance may be acting (Van Emden 1987): nonpreference, in which a plant is not preferred for colonization, oviposition, or feeding by insect pests; antibiosis, where insects on a plant species take longer to develop than they would on other plants, suffer greater rates of mortality, grow more slowly, or produce fewer offspring; and tolerance, in which a plant shows an increased capacity to recover from insect attack.

Resistance to Shoot Borers Among Meliaceae Species

Although *Hypsipyla* spp. only attack Meliaceae, most members of this family, including *Guarea* spp., *Melia* spp., and other desirable timber tree species, are not attacked by them. That these apparently completely resistant species are suitable plantation species has frequently been suggested (Newton et al. 1993a).

In the Swietenioideae, nonnative species have often been reported resistant to the native species of *Hypsipyla*. Thus, the Neotropical big-leaf mahogany, susceptible to *H. grandella* but apparently resistant to *H. robusta*, has been successfully established in plantations in Southeast Asia and the south Pacific (Evans 1982). Similarly, *Toona ciliata*, native to Asia and Africa and susceptible to *H. robusta*, has been reported resistant to *H. grandella* (Whitmore 1976a). Recent reports of damage by *Hypsipyla* spp., however, suggest that nonnative Swietenioideae are susceptible to both species (Floyd and Hauxwell 1997).

Attempts to establish plantations of nonnative Meliaceae have met with mixed results (Sánchez et al. 1976; Evans 1982), the failures being more the result of incompatibility of the tree species with local growing conditions than shoot borer attack (Newton et al. 1993b). The use of nonnative Meliaceae is not, therefore, the simple universal answer to the shoot borer problem; sustainable approaches to managing native Meliaceae must be found.

Resistance to Shoot Borers in Spanish Cedar and Big-Leaf Mahogany

Despite repeated suggestion that variation in resistance to shoot borers may be present in Meliaceae species (Roberts 1966; Grijpma 1976), screening for genetic variation in resistance has seldom been attempted. Some information on resistance in Spanish cedar has been obtained from international provenance trials, however (Whitmore 1978; Chaplin 1980; McCarter 1986, 1988). Although survival in these trials has usually been poor and the trees have been heavily attacked by shoot borers, a few provenances in these trials have shown apparent resistance to shoot borer attack. In each case, tolerance was demonstrated by pronounced vigor and by producing a new, single, strong leading shoot after attack.

Thus, sufficient information is published to warrant a closer examination of resistance to shoot borers in Meliaceae species. An investigation was therefore designed to assess genetic variation in characteristics conferring pest resistance in Spanish cedar and big-leaf mahogany by using seedling screening trials started in 1990 at the Tropical Agronomic Centre for Research and Higher Education, Costa Rica (CATIE), as a joint initiative with ITE (Institute of Terrestrial Ecology, Edinburgh) (Newton et al. 1995, 1998, 1999).

Seed of Spanish cedar was collected from trees in four localities in Costa Rica: Carmona (latitude 9°60′ N, longitude 85°15′ W), Hojancha (10°4′ N, 85°25′ W), and Cañas (10°25′ N, 85°6′ W), all dry-zone localities; San Carlos (10°22′ N, 84°28′ W), a wet-zone locality; and from one locality in Trinidad, St. Andrew, 10°28′ N, 61°5′ W. The trees were selected based on stem straightness and lack of forking. Big-leaf mahogany seed was obtained from bulked collections from five provenances: Haiti (19°42′ N, 72°24′ W), Trinidad (10°28′ N, 61°5′ W), Honduras (15°20′ N, 84°24′ W), and two from Puerto Rico, Guajataca (18°22′ N, 67°0′ W) and Juana Diaz (18°0′ N, 66°31′ W). Details of seed origins and seedling establishment are given in Newton et al. (1995).

Two field trials screening Spanish cedar and big-leaf mahogany were established at CATIE during February 1991 (see Newton et al. 1995). Seedlings of Spanish cedar were arranged by family (25 families divided equally among the five provenances) in fully randomized 5-tree-row plots, in nine replicate blocks, and seedlings of big-leaf mahogany were arranged in fully randomized square plots of 25 trees in five replicate blocks.

Each tree in both experiments was measured for the following:

• Height, measured after 26, 56, 88, and 141 weeks for Spanish cedar, and after 26, 56, 88, and 177 weeks for big-leaf mahogany
• Leaf phenology
• Incidence of shoot borer attack at 14-day intervals, for 84 weeks from April 1991
• Height to first branching (at 141 or 177 weeks)
• Number of damage loci, indicated by forking (at 141 or 177 weeks, a measure of serious attack, rather than just the number of attacks)

The results of these trials are fully described by Newton et al. (1995, 1999). Genetic variation in height growth was recorded for both tree species, differences between both provenances and families tending to become more pronounced with time. At the final assessments, Spanish cedar mean height ranged from 183 cm in Hojancha to 501 cm in San Carlos (Fig. 22.1a). The mean height of big-leaf mahogany at the end of the trial varied from 211 cm in Dirici, Haiti, to 267 cm in Guajataca, Puerto Rico (Fig. 22.1b).

Genetic variation in Spanish cedar phenology was also observed, particularly in leaf abscission during the dry season. Trees from San Carlos and Trinidad were more densely foliated than were those of the other three provenances. The majority of big-leaf mahogany trees had foliage throughout the experiment.

In Spanish cedar, two pronounced peaks in attack were observed, one each year (Newton et al. 1998). At the first peak in Spanish cedar, the San Carlos provenance was least attacked, but these trees and those from the Trinidad provenance had the most attacks during the second peak (10–14 attacks per tree; Fig. 22.2a). We observed a single peak of attack in big-leaf mahogany in the second year of the trial. In the two big-leaf mahogany provenances in Puerto Rico, the number of attacks per tree at the peak ranged from 0.6 in Juana Diaz to 1.3 in Guajataca (Fig. 22.2b). Thus, Spanish cedar had more attacks than did big-leaf mahogany.

The number of damage loci in Spanish cedar, assessed after 141 weeks, was significantly affected by provenance. The mean number of damage loci per tree ranged from 1.55 in Hojancha to 2.64 in Trinidad. The number of damage loci in big-leaf mahogany, assessed after 177 weeks, ranged from 1.9 per tree in Juana Diaz, Puerto Rico, to 2.9 per tree in Trinidad (Fig. 22.3a). The other Puerto Rican provenance, Guajataca, which was most

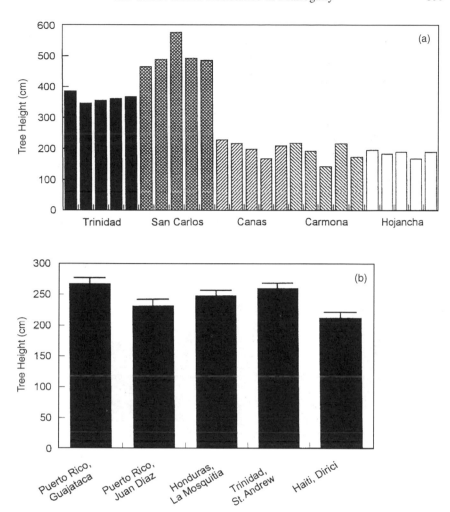

Figure 22.1a,b. The height of (**a**) Spanish cedar families (each denoted by a family number) within different provenances (see key on figure) and of (**b**) big-leaf mahogany provenances in the third year of the trials.

heavily attacked at the peak in year 2, had about the same number of damaging attacks as the Juana Diaz provenance did (Fig. 22.3b).

The height at the first damaging attack in Spanish cedar (after 141 weeks) was also significantly affected by provenance. The height ranged from 70 cm in Hojancha to 240 cm in San Carlos (Fig. 22.4a). In big-leaf mahogany, the height at first serious attack (after 177 weeks) ranged from 110 cm in Dirici, Haiti, to 140 cm in Trinidad (Fig. 22.4b). The Guajataca provenance had a mean height at first attack of 130 cm.

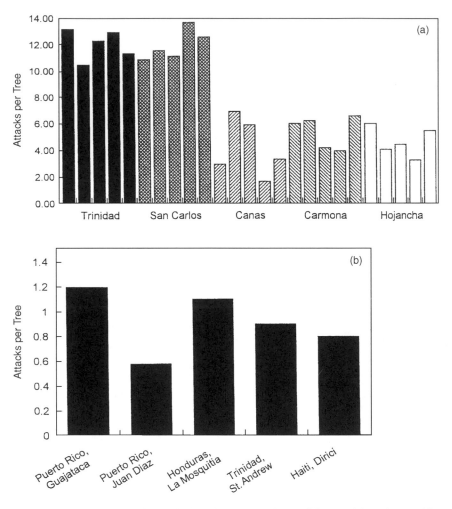

Figure 22.2a,b. The numbers of shoot borer attacks on (**a**) Spanish cedar families within different provenances and (**b**) big-leaf mahogany provenances in the second year of the trials.

The trials in Costa Rica generally showed that the rate of attack to big-leaf mahogany and Spanish cedar was similar, particularly in the number of damaged loci and the height of the first serious attack. In terms of variation within species, the results from the Spanish cedar trial were particularly promising. The San Carlos provenance trees showed nonpreference through a lower incidence of shoot borer attack in the first year of growth than was experienced by trees from other provenances. In addition, although trees from San Carlos were heavily attacked during their second year of growth, they showed a greater degree of tolerance than other provenances in having fewer damaging attacks than expected for their height and

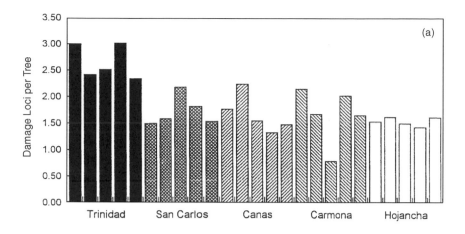

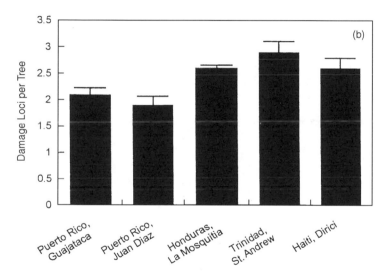

Figure 22.3a,b. The total number of damage loci resulting from attacks 1991–1993 to (**a**) Spanish cedar families and (**b**) big-leaf mahogany provenances.

reaching a greater height before their first damaging attack. Trees from the Trinidad provenance also showed tolerance to *H. grandella* attack, but they did so by producing several vigorously growing stems; San Carlos trees tended to respond to attack by producing a single main stem that showed stronger apical dominance. Apical dominance is an important characteristic in the response of trees to pests such as the mahogany shoot borers; it is also a characteristic that can be selected for in seedling decapitation tests (Newton et al. 1995).

Although the finding that many shoot borer attacks led to few damaging attacks may have been largely a result of tolerance, antibiosis may also have

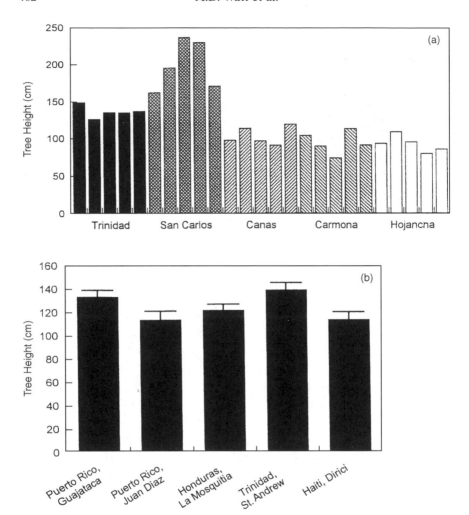

Figure 22.4a,b. The height of the first damaging shoot borer attack for (**a**) Spanish cedar families and (**b**) big-leaf mahogany provenances.

been responsible; that is, some of the attacks may have been reduced in severity because of greater larval mortality in the San Carlos trees. Support for this possibility comes from the greater concentrations of proantho-cyanidins (condensed tannins) in the foliage of San Carlos trees relative to trees from other provenances (Newton et al. 1999).

None of the big-leaf mahogany provenances tested showed the same degree of resistance to shoot borers as did the San Carlos Spanish cedar provenance. Further screening of big-leaf mahogany, with a much larger number of provenances, is clearly needed to evaluate the potential of resistance in this species.

Conclusions

Recent research in Costa Rica found genetic variation in resistance to attack by *H. grandella* in Spanish cedar and big-leaf mahogany. The results with Spanish cedar were particularly promising. As with previous provenance trials, the basis for resistance appeared to be tolerance, but this study also demonstrated variation in nonpreference and antibiosis in Spanish cedar.

These results should serve to encourage further studies on resistance to shoot borers in both species and in other Meliaceae, such as *Khaya* spp. in West Africa. Two points should be emphasized, however. First, the recent research in Costa Rica has been unique in combining regular assessments of attack (similar to research by Yamazaki et al. 1990, 1992) with assessing growth, form, and damage. Future research should include monitoring the insects, preferably both adult abundance and larval attack, and assessing the damage they cause.

Second, resistance to insect pests is a characteristic to be cherished and used to maximum effect. Thus, research on resistance should go hand in hand with research on other potential control methods, so that resistant mahoganies, once identified, can be deployed in an integrated management strategy. The greatest potential for successfully managing shoot borers lies in incorporating resistant planting stock in appropriate silvicultural or agroforestry systems (Newton et al. 1993a). These systems include those that encourage natural biological control by predators and parasitic insects or otherwise reduce the abundance and effect of shoot borers. We do not support the view that effective control will be achieved by any single method of controlling shoot borers, be it chemical, silvicultural, resistance, or any other method. We strongly recommend an integrated approach.

Acknowledgments. The CATIE/ITE link project was financed by the UK Overseas Development Administration. The advice and encouragement of Roger Leakey and Francisco Mesén are gratefully acknowledged.

Literature Cited

Chaplin, G.E. 1980. Progress with provenance exploration and seed collection of *Cedrela* spp. Proceedings, Eleventh British Commonwealth Forestry Conference, Commonwealth Forestry Institute, Oxford, September 1980, UK, pp. 1–17.

Entwistle, P.F. 1967. The current situation on shoot, fruit and collar borers of the Meliaceae. Proceedings, Ninth British Commonwealth Forestry Conference. Commonwealth Forestry Institute, January 1967, Oxford, UK.

Evans, J. 1982. *Plantation Forestry in the Tropics*. Clarendon Press, Oxford.

Floyd, R., and Hauxwell, C., eds. 1997. Hypsipyla *Shoot Borers in Meliaceae*. ACIAR, Canberra, Australia.

Grijpma, P. 1974. *Contributions to an Integrated Control Programme of* Hypsipyla grandella *(Zeller) in Costa Rica*. Doctoral thesis, Agricultural University Wageningen, The Netherlands.

Grijpma, P. 1976. Resistance of Meliaceae against the shoot borer *Hypsipyla* with particular reference to *Toona ciliata* M. J. Roem var. *australis* (F. v. Muell.) CDC. In *Tropical Trees: Variatioon, Breeding and Conservation*, eds. J. Burley and B.T. Styles, pp. 69–78. Linnaean Society, London.

McCarter, P.S. 1986. The evaluation of the international provenance trials of *Cordia alliodora* and *Cedrela* spp. Annual Report to the UK Overseas Development Administration, Oxford Forestry Institute, Oxford, UK.

McCarter, P.S. 1988. Report on a visit to Colombia and Ecuador. Report to the UK Overseas Development Administration, Oxford Forestry Institute, Oxford, UK.

Newton, A.C., Baker, P., Ramnarine, S., Mesen, J.F., and Leakey, R.R.B. 1993a. The mahogany shoot borer—prospects for control. *Forest Ecology and Management* **57**:301–328.

Newton, A.C., Leakey, R.R.B., and Mesen, J.F. 1993b. Genetic variation in mahoganies—its importance, capture and utilization. *Biodiversity and Conservation* **2**:114–126.

Newton, A.C., Cornelius, J.P., Mesen, J.F., and Leakey, R.R.B. 1995. Genetic variation in apical dominance of *Cedrela odorata* seedlings in response to decapitation. *Silvae Genetica* **44**:146–150.

Newton, A.C., Cornelius, J.P., Mesen, J.F., Corea, E.A., and Watt, A.D. 1998. Variation in attack by the mahogany shoot borer, *Hypsipyla grandella* (Lepidoptera: Pyralidae) in relation to host growth and phenology. *Bulletin of Entomological Research* **88**:319–326.

Newton, A.C., Watt., A.D., Cornelius, J.P., Mesen, J.F., and Corea, E.A. 1999. Genetic variation in host susceptibility to attack by the mahogany shoot borer, *Hypsipyla grandella* (Zeller). *Agricultural and Forest Entomology* **1**:11–18.

Roberts, H. 1966. A survey of the important shoot, stem, wood, flower and fruit boring insects of the Meliaceae in Nigeria. *Nigerian Forestry Information Bulletin (New Series)* **15**:38.

Sánchez, J.C., Holsten, E.H., and Whitmore, J.L. 1976. Compartamiento de 5 especies de Meliaceae en Florencia Sur, Turrialba. In *Studies on the Shoot Borer* Hypsipyla grandella *(Zeller) Lepidoptera: Pyralidae, Vol. III*, ed. J.L. Whitmore, pp. 97–103. IICA Miscellaneous Publications 101. CATIE, Turrialba, Costa Rica.

Van Emden, H.F. 1987. Cultural methods: the plant. In *Integrated Pest Management*, eds. A.J. Burn, T.H. Coaker, and P.C. Jepson, pp. 27–68. Academic Press, London.

Wagner, M.R., Atuahene, S.K.N., and Cobbinah, J.R. 1991. *Forest Entomology in West Tropical Africa: Forest Insects of Ghana*. Kluwer, Dordrecht.

Whitmore, J.L. 1976a. Myths regarding *Hypsipyla* and its host plants. In *Studies on the Shoot Borer* Hypsipyla grandella *(Zeller) Lepidoptera Pyralidae, Vol. III*. IICA Miscellaneous Publications 101. CATIE, Turrialba, Costa Rica.

Whitmore, J.L. 1976b. *Studies on the Shoot Borer* Hypsipyla grandella *(Zeller) Lepidoptera: Pyralidae. Volume II. IICA Miscellaneous Publications 101*. CATIE, Turrialba, Costa Rica.

Whitmore, J.L. 1978. Cedrela *provenance trial in Puerto Rico and St Croix: Establishment phase*. Research Note ITF 16. U.S. Department of Agriculture, Forest Service, Río Piedras, PR.

Yamazaki, S., Taketani, A., Fujita, K., Vasques, C., and Ikeda, T. 1990. Ecology of *Hypsipyla grandella* and its seasonal changes in population density in Peruvian Amazon forest. *Japan Agricultural Research Quarterly* **24**:149–155.

Yamazaki, S., Ikeda, T., Taketani, A., Pacheco, C.V., and Sato, T. 1992. Attack by the mahogany shoot borer, *Hypsipyla grandella* Zeller (Lepidoptera, Pyralidae), on the meliaceous trees in the Peruvian Amazon. *Applied Entomology and Zoology* **27**:31–38.

5. Epilogue

23. Recent Literature on Big-Leaf Mahogany

Ariel E. Lugo

Introduction

Since this book was assembled, several publications relevant to the conservation of big-leaf mahogany have appeared. I thought it would be helpful to highlight some of these publications, hoping this volume will be as useful as possible for those interested in the conservation of this magnificent species. For practical purposes, I classified publications by subject matter into several topics. For each topic, I list the relevant reference in alphabetical order by lead author and provide a short description of its content. I also include reports on small-leaf mahogany.

Mahogany Habitat

Moran, E.F., Brondizio, E.S., Tucker, J.M., da Silva-Forsberg, M.C., McCracken, S., and Falesi, I. 2000. *Forest Ecology and Management* **139**:93–108.

This article focuses on the effects of soil fertility and land use on forest succession in Amazônia. The relevance to conservation of big-leaf mahogany rests on the temporal analysis of satellite images to show forest recovery and land uses after agricultural use was abandoned in Amazônia. That Amazonian forests recover after deforested lands are abandoned—as they do in the Caribbean and elsewhere in the world—is good news to the regrowth of big-leaf mahogany, which grows well in abandoned farmlands and pastures.

Pérez García, E.A., Malave, J., and Gallardo, C. 2001. Vegetación y flora de la región de Nizanda, Istmo de Tehuantepec, Oaxaca, México. *Acta Botánica Mexicana* **56**:19–88.
This article is a description of the vegetation and flora of the Nizanda region in the isthmus of Tehuantepec, Juchitán District, Oaxaca, México. They report *Swietenia humilis* in sabana and selva mediana vegetation.

Community Structure and Composition

Schulze, M.D., and Whitacre, D.F. 1999. A classification and ordination of the tree community of Tikal National Park, Petén, Guatemala. *Bulletin of the Florida Museum of Natural History* **41**:169–297.
This monograph is on classification and ordination of the tree communities of Tikal National Park, Petén, Guatemala. The authors sampled 294 tree plots of 0.041 ha each placed systematically along topographic gradients. Their measurements included trees, saplings, and environmental factors. This data-rich monograph has valuable insights into the regeneration and survival of big-leaf mahogany in this part of Central America. It contains information for species associated with mahogany as well as their distribution along topographic gradients.

Physiology

Barker, M.G., and Pérez-Salicrup, D. 2000. Comparative water relations of mature mahogany (*Swietenia macrophylla*) trees with and without lianas in a subhumid, seasonally dry forest in Bolivia. *Tree Physiology* **20**:1167–1174.
The high loads of lianas on mature big-leaf mahogany trees do not affect the mahogany's water relations. Although lianas were more effective than big-leaf mahogany in acquiring water, both plants were able to secure water for their functioning, even at the end of the dry season. The study was conducted in a subhumid, seasonally dry forest in Bolivia.

Genetics

Gillies, A.C.M., Navarro, C., Lowe, A.J., Newton, A.C., Hernández, M., Wilson, J., and Cornelius, J.P. 1999. Genetic diversity in Mesoamerican populations of mahogany (*Swietenia macrophylla*), assessed using RAPD's. *Heredity* **83**:722–732.

This article is a wide-ranging survey of molecular variation in big-leaf mahogany covering populations from Mexico to Panama and Belize. The authors found distinct big-leaf mahogany populations in Mexico and high genetic diversity in populations in Panama. Selective logging explained 24% of the total variation in genetic diversity ($P = 0.033$) when recent logging was assessed according to a scale of 0.00 to 0.60 (virgin forest = 0.00, lightly logged = 0.30, moderately logged = 0.48, and heavily logged = 0.60; the basis for this scale and the criteria for classifying plots are not given). The genetic diversity contained in big-leaf mahogany was distributed across the whole of its range, however. The within-population component of genetic diversity was 80%, and 20% was maintained between populations.

Lepsch-Cunha, N., and Mori, S.A. 1999. Reproductive phenology and mating potential in a low density tree population of *Couratari multiflora* (Lecythidaceae) in central Amazonia. *Journal of Tropical Ecology* **15**:97–121.
This article focuses on the reproductive phenology and mating potential of *Couratari multiflora*, a tropical tree in Central Amazônia. The research is relevant to big-leaf mahogany because this rare canopy tree has mature tree density similar to big-leaf mahogany. The study contributes to understanding the ecology and genetics of rare trees that exert considerable influence in their communities. Trees of this species, as those of big-leaf mahogany, were 100% outcrossed, in spite of the 1-km distance between trees.

Silviculture

Francis, J.K. 2000. Comparison of hurricane damage to several species of urban trees in San Juan, Puerto Rico. *Journal of Arboriculture* **26**:189–197.
This article is an assessment of the effects of Hurricane Georges on urban trees in San Juan, Puerto Rico. Data include small-leaf mahogany, the most resistant species among those studied.

Francis, J.K., and Lowe, C.A., eds. 2000. *Bioecología de árboles nativos y exóticos de Puerto Rico y las Indias Occidentales*. U.S. Department of Agriculture Forest Service International Institute of Tropical Forestry, Río Piedras, PR. General Technical Report IITF-15.
This compendium of silviculture information for 101 tropical tree species is in Spanish, but write-ups for individual species are also available in English. Descriptions for big-leaf and small-leaf mahogany are included that summarize information available for each species.

Kalil Filho, A.N., Hoffmann, H.A., Cortezzi Graça, M.E., and Rodrigues Tavares, F. 2000. Enxertia de mongo em toona para a indução de resistência à *Hypsipyla grandella* (Zeller, 1948) no mogno sul-americano

(*Swietenia macrophylla*). *Boletin Pesquizas Flona, Colombo* **41**:74–78 (jul./dez.).
This article reports the results of grafting experiments that sought to increase the resistance of big-leaf mahogany to *Hypsipyla grandella*. They grafted big-leaf mahogany with *Toona ciliata*, *T. ciliata* with big-leaf mahogany, and big-leaf mahogany with itself as a control. After 4 to 5 months, the combination of big-leaf mahogany on *T. ciliata* had the best survival, but eventually all the experimental plants died and only the controls survived (98% after 10 months).

Kammesheidt, L., Torres Lezama, A., Franco, W., and Plonczak, M. 2001. History of logging and silvicultural treatments in the western Venezuelan plain forests and the prospect for sustainable management. *Forest Ecology and Management* **148**:1–20.
This article reviews logging activities since the 1920s in Venezuela. Big-leaf mahogany was a preferred timber species, and the article contains production data in the State of Portuguesa and Venezuela for the period 1930 to 1969. Logging rates for big-leaf mahogany during the period of 1972 and 1997 in Ticoporo ranged from 863 to 2047 m³/yr and a total extraction of 28,692 m³ or 5% of all the extraction in that concession area. Big-leaf mahogany mature forests in Caparo have a density of 1 tree/ha with 3.27 m³/ha of bole volume.

Lemos Filho, J.P., and Duarte, R.J. 2001. Germination and longevity of *Swietenia macrophylla* King seeds—mahogany (Meliaceae). *R. Árvore, Viçosa*-MG **25**:125–130.
This article reports the results of experiments with seed germination and their longevity under different forms of storage. Germination of recently collected seed was not different under dark or continuous white light in spite of a higher germination velocity in darkness. This approach allows big-leaf mahogany to coexist with late-succession species in tropical forests. Seeds stored under low temperature presented a larger germination percentage. The presence of the tegument reduced germination after a short storage time. Seeds stored 1 year in a refrigerator without the integument exhibited 90% germination.

Mayhew, J.E., and Newton, A.C. 1998. The silviculture of mahogany. CABI Publishing, Wallingford, Oxon, UK.
This book is an excellent summary of the state of knowledge on the silviculture of big-leaf mahogany. It has a strong focus on plantation forestry as a strategy to lower harvesting pressure on natural stands of the species.

Morris, M.H., Negreros-Castillo, P., and Mize, C. 2000. Sowing date, shade, and irrigation affect big-leaf mahogany (*Swietenia macrophylla* King). *Forest Ecology and Management* **132**:173–181.

This article reports the results of experiments with seed germination in a nursery at Quintana Roo, Mexico. Germination increases linearly with increasing shade. Irrigated seed have significantly higher germination than nonirrigated seed. Most seed sown in August did not germinate.

Mostacedo, C.B., and Fredericksen, T.S. 1999. Regeneration status of important tropical forest tree species in Bolivia: assessment and recommendations. *Forest Ecology and Management* **124**:263–273.
This article assesses the regeneration status of 68 tree species in Bolivian forests. The regeneration capacity of big-leaf mahogany is poor. Solutions to remedy the situation are available but they are expensive.

Conservation

Blundell, A.G., and Rodan, B.D. 2001. United States imports of bigleaf mahogany (*S. macrophylla* King) under CITES Appendix III. Information document 1. CITES Mahogany Working Group, Santa Cruz, Bolivia, October 3–5, 2001.
This report analyzes data on big-leaf mahogany imports to the United States of America (USA) and the implementation of the Convention on International Trade of Endangered Species (CITES) for 1997–1999. In 1999, nearly 80% of the mahogany sawnwood imports recorded in the USA was accompanied by the necessary CITES permits/certificates of origin. The report contains suggestions for further improving the implementation of protocols at the borders.

Greenpeace. 2001. Partners in mahogany crime. Amazon at the mercy of "gentlemen's agreements." Available at: Greenpeace International, 1016 DW Amsterdam, The Netherlands. www.greenpeace.org.
This report, obtainable through the Internet, contains the results of a Greenpeace investigation on the illegal logging of big-leaf mahogany in Brazil. It quotes the value of big-leaf mahogany wood ("green gold") as $1600.00/m^3. The investigation focused on Pará, and the report includes the names of those alleged to be involved in the illegal trade of mahogany. The report contains recommendations for the conservation of big-leaf mahogany.

ISTF News. 2001. Mahogany trade report issues, page 9 (March). International Society of Tropical Foresters, 5400 Grosvenor Lane, Bethesda, MD 20814, USA.
This is a news summary of a trade report on big-leaf mahogany issued by TRAFFIC.

Kvist, L.P., and Nebel, G. 2001. A review of Peruvian flood plain forests: ecosystems, inhabitants and resource use. *Forest Ecology and Management* **150**:3–26.

This article reports big-leaf mahogany as "heavily depleted" in floodplain forests of the lower Río Ucayali and lower Río Marañon in Peru. In 1996, the extraction of big-leaf mahogany roundwood and sawnwood, respectively, was 31,868 and 22,861 m^3 in the Loreto Department and 14,654 m^3 of sawnwood in the Ucayali Department of Peru. These rates of extraction represent, respectively, 18%, 34%, and 12% of the total extraction from floodplains in those departments.

Kvist, L.P., Andersen, M.K., Stagegaard, J., Hesselsøe, M., and Llapapasca, C. 2001. Extraction from woody forest plants in flood plain communities in Amazonian Peru: use, choice, evaluation and conservation status of resources. *Forest Ecology and Management* **150**:147–174.
This article reports that mahogany provides between 10% and 50% of the sawnwood demand on Peruvian floodplain forests. The species is listed as "locally depleted." National and international urban consumers demand big-leaf mahogany wood by name rather than other woods with similar characteristics.

Lugo, A.E. 1999. Point-counterpoint on the conservation of big-leaf mahogany. General Technical Report WO-64. U.S. Department of Agriculture Forest Service, Washington, DC.
This literature review focuses on 14 controversial issues about conserving big-leaf mahogany. The manuscript takes a point-counterpoint strategy to highlight the contrasting views on each topic. It also examines the issue of genetic erosion and conservation of genetic variation in big-leaf mahogany and provides science-based conservation strategies for the species.

Robbins, C.S. 1999. Mahogany matters: the U.S. market for mahogany and its implications for the conservation of the species. TRAFFIC, North America. Washington, DC.
This comprehensive report describes the commercial trade of big-leaf mahogany in Latin America and the concerns it raises for conserving the species. The focus is on the U.S. market. The report contains useful recommendations for improving controls over harvesting and trade of mahogany so that the species might be conserved.

Rosado, O. 1999. *Diagnóstico de la caoba* (*Swietenia macrophylla* King) Belize. PROARCA/CAPAS. Tropical Science Center, San José, Costa Rica.
This comprehensive assessment covers the current and historic situation for big-leaf mahogany in Belize. The document was used to lead a discussion during a national workshop held in Belmopan, Belize, in August 1999.

Weaver, P.L., and Bauer, G.P. 2000. Major Meliaceae in Nicaragua. General Technical Report IITF GTR 10. U.S. Department of Agriculture Forest Service, International Institute of Tropical Forestry, Río Piedras, PR.

This comprehensive assessment describes the situation of the Meliaceae in Nicaragua. The authors reviewed and summarized a large volume of published and unpublished material and review the history of exploitation and conservation in Nicaragua.

Zimmerman, B., Peres, C.A., Malcom, J.R., and Turner, T. 2001. Conservation and development alliances with the Kayapó of south-eastern Amazonia, a tropical forest indigenous people. *Environmental Conservation* **28**(1): 10–22.
This article describes conservation efforts in the Kayapó Indian territories of Pará and Mato Grosso states of Brazil's Amazon basin. It focuses on the preservation of 13,000,000 ha of forest and cerrado, including 8000 ha of big-leaf mahogany stands. The article documents past levels of illegal big-leaf mahogany logging, which affected 85% of the fruiting populations of the species, and the conservation benefits achieved by supporting sustainable development of indigenous people in the Amazon.

Acknowledgments. This work was done in cooperation with the University of Puerto Rico. I thank the authors who shared their publications with me, and Mildred Alayón for her collaboration during the production of the manuscript.

Species Index

Geographical Index

General Index

Ecological Studies

Volumes published since 1996

Volume 119
Freshwaters of Alaska: Ecological Synthesis
(1997)
A.M. Milner and M.W. Oswood (Eds.)

Volume 120
**Landscape Function and Disturbance
in Arctic Tundra** (1996)
J.F. Reynolds and J.D. Tenhunen (Eds.)

Volume 121
**Biodiversity and Savanna Ecosystem
Processes: A Global Perspective** (1996)
O.T. Solbrig, E. Medina, and J.F. Silva (Eds.)

Volume 122
**Biodiversity and Ecosystem Processes in
Tropical Forests** (1996)
G.H. Orians, R. Dirzo, and J.H. Cushman
(Eds.)

Volume 123
**Marine Benthic Vegetation: Recent Changes
and the Effects of Eutrophication** (1996)
W. Schramm and P.H. Nienhuis (Eds.)

Volume 124
**Global Change and Arctic Terrestrial
Ecosystems** (1996)
W.C. Oechel (Ed.)

Volume 125
**Ecology and Conservation of Great Plains
Vertebrates** (1997)
F.L. Knopf and F.B. Samson (Eds.)

Volume 126
**The Central Amazon Floodplain: Ecology of
a Pulsing System** (1997)
W.J. Junk (Ed.)

Volume 127
**Forest Design and Ozone: A Comparison of
Controlled Chamber and Field Experiments**
(1997)
H. Sanderman, A.R. Wellburn, and
R.L. Heath (Eds.)

Volume 128
**The Productivity and Sustainability of
Southern Forest Ecosystems in a Changing
Environment** (1998)
R.A. Mickler and S. Fox (Eds.)

Volume 129
Pelagic Nutrient Cycles (1997)
T. Andersen

Volume 130
Vertical Food Web Interactions (1997)
K. Dettner, G. Bauer, and W. Völkl (Eds.)

Volume 131
**The Structuring Role of Submerged
Macrophytes in Lakes** (1998)
E. Jeppesen, M. Søndergaard,
M. Søndergaard, and K. Christoffersen (Eds.)

Volume 132
Vegetation of the Tropical Pacific Islands
(1998)
D. Mueller-Dombois and F.R. Fosberg

Volume 133
Aquatic Humic Substances (1998)
D.O. Hessen and L.J. Tranvik

Volume 134
**Oxidant Air Pollution Impacts in the
Montane Forests of Southern California:
A Case Study of the San Bernardino
Mountains** (1999)
P.R. Miller and J.R. McBride (Eds.)

Volume 135
**Predation in Vertebrate Communities:
The Bialowieza Primeval Forest as a
Case Study** (1998)
B. Jedrzejewska and W. Jedrzejewski

Volume 136
**Landscape Disturbance and Biodiversity in
Mediterranean-Type Ecosystems** (1998)
P.W. Rundel, G. Montenegro, and F.M. Jaksic
(Eds.)

Volume 137
**Ecology of Mediterranean Evergreen Oak
Forests** (1999)
F. Roda, J. Retana, C.A. Gracia, and J. Bellot
(Eds.)

Volume 138
**Fire, Climate Change, and Carbon Cycling
in the Boreal Forest** (2000)
E.S. Kasischke and B.J. Stocks (Eds.)

Volume 139
**Responses of Northern U.S. Forests to
Environmental Change** (2000)
R.A. Mickler, R.A. Birdsey, and J. Hom
(Eds.)